Mycotoxin Detection and Management

ABOUT EDITORS

PROF. RISHI KUMAR BEHL currently working as Professor (R&D) and Head of the Crop Improvement Division at MM (DU), Mullana, District Ambala, Haryana. He previously served as Professor of Plant Breeding and Associate Dean at the College of Agriculture, CCS Haryana Agricultural University, Hisar, and as Dean, Faculty of Sciences, Jagan Nath University, Bahadurgarh, India. He obtained his B.Sc. (Agri.) from Rajasthan University, Jaipur, and his M.Sc. (Agri., Plant Breeding) and Ph.D. (Genetics) from CCS Haryana Agriculture University, Hisar, India, with a distinguished academic and research career. He specializes in Genetics & Plant Breeding by training and Sustainable Agriculture by practice. His work includes the development of wheat, triticale, barley, and sugar beet genotypes with specific traits for stress tolerance, produce quality, and favorable plant-microbe interactions. With 45 years of teaching and research experience, he has received numerous accolades, including the Bharat Ratna C. Subhramaniyam Award (2002) for Outstanding Teacher from ICAR, New Delhi. He was an Honorary DAAD Advisor at the German Academic Exchange Service (2006-09). He has been Editor-in-Chief of *Annals of Biology* for 36 years and has held editorial roles in journals such as *Annals of Agri Bio Research*, *Archives of Agronomy and Soil Science* (Germany), *International Journal of Crop Breeding and Genetics* (Turkey), *Tropics* (Japan), *Cereal Research Communication* (Hungary), *South Pacific Journal of Natural Science* (Fiji/Australia), and *Journal of Basic Microbiology* (Germany). He has published around 250 research papers in international and national journals and edited 40 books and manuals. He is the Founder Secretary of the *Society for Sustainable Agriculture and Resource Management* (SSARM), India, and is associated with the *International Council for Sustainable Agriculture* (ICSA) and the *International Foundation for Sustainable Development in Africa and Asia* (IFSDAA), Germany. He has participated in and organized numerous international conferences in India and abroad. Currently, he is engaged in research and development activities focusing on the application of technology for sustainable development. **Email:** rkbehlprof@gmail.com

DR. ABHISHEK KUMAR Ph.D. in Plant Pathology from Chaudhary Charan Singh Haryana Agricultural University, Hisar, Haryana, India in 2023. He obtained his B.Sc. (Agri.) Hons from Sardar Vallabhbhai Patel University of Agriculture and Technology, Modipuram, Meerut, and his M.Sc. (Plant Pathology) from Acharya Narendra Deva University of Agriculture and Technology, Ayodhya, India, with a distinguished academic and research career. After completing his Ph.D. he joined Department of Agriculture, Maharishi Markandeshwar (Deemed to be University), Mullana, Ambala, India in 2023 as an Assistant Professor. His contributions to academia include teaching courses on plant pathology, plant-microbe interactions, and disease management, ensuring a balance between theoretical knowledge and practical applications. Dr. Abhishek Kumar has published good number of papers in high impact international journals of repute and book chapters in books with ISBN numbers. He is also actively working as reviewer in several international journals including Heliyon, Frontiers in Plant Science, Frontiers in Microbiology and others since 2020. Currently he is working on plant diseases of various vegetable and field crops and their control measures. His special interest lied in soil microbiome such as impact of organic carbon (biochar) and biostimulants on rhizospheric microflora and cross talks among microbes. His work aligns with global agricultural sustainability goals by promoting eco-friendly, effective plant disease management strategies. **Email:** abhishekpathology9511@gmail.com

MR. ANKIT KUMAR is a dedicated researcher and scholar in the field of Food Technology. Currently pursuing his Ph.D. at CCS Haryana Agricultural University, Hisar, he holds a Master's degree from Punjab Agricultural University and a Bachelor's degree in Agriculture from CCS Haryana Agricultural University, Hisar, Haryana. He has excelled in competitive examinations, securing top ranks in both the Junior Research Fellowship (JRF) and Senior Research Fellowship (SRF) conducted by ICAR. His research focuses on food processing and preservation techniques, and he has contributed to multiple research papers published in high-impact journals, including *Trends in Food Science and Technology* (Impact Factor: 15.1) and *Food Chemistry* (Impact Factor: 8.5), and others. Additionally, he has authored book chapters published by renowned international publishers. With extensive experience in academic writing, research analysis, and technical documentation, he ensures that content is well-structured, precise, and impactful. His disciplined approach and passion for knowledge enable him to carefully review and refine written work, making complex ideas more accessible and comprehensible. **Email:** ankitnagpal1995@gmail.com

Mycotoxin Detection and Management

– Editors –

Rishi Kumar Behl

Abhishek Kumar

Ankit Kumar

2026

Daya Publishing House®
A Division of

Astral International Pvt. Ltd.
New Delhi – 110 002

ISBN: 978-93-5919-467-7

Published by : **Daya Publishing House®**
A Division of
Astral International Pvt. Ltd.
– ISO 9001:2015 Certified Company –
4736/23, Ground Floor, Ansari Road,
Darya Ganj, New Delhi-110 002
Ph. 011-43549197, 8130496929
e-mail: info@astralint.com
Website: www.astralint.com

Printed at : **Replika Press Pvt Ltd**

Prof. (Dr.) B.R. Kamboj
Vice chancellor
CCS Haryana Agricultural University, Hisar, Haryana

Foreword

Mycotoxins are secondary metabolites toxic compounds produced by a variety of fungal pathogens such as *Aspergillus*, *Fusarium*, and *Penicillium* that contaminate agricultural products, posing serious threats to human and livestock health and food safety. The occurrence and intensity of mycotoxin contamination are influenced by environmental and management conditions, including temperature, humidity, moisture, and inadequate storage facilities. To combat such hazards, accurate, rapid and reliable detection and quantification of mycotoxins to effectively manage and control them is crucial. Techniques like thin-layer chromatography (TLC), gas chromatography mass spectrometry (GC-MS,) high-performance liquid chromatography (HPLC), enzyme-linked immunosorbent assay (ELISA) have been widely adopted to identify and measure the presence of mycotoxins in agricultural products. Moreover, new approaches such as biosensors, lab-on-chip technology and hyperspectral imaging in conjugation with artificial intelligence are also proven as a benchmark for the detection and diagnosis of fungal pathogens and mycotoxins.

The impact of mycotoxin contamination extends far beyond health concerns, significantly affecting the economy and trade. Contaminated crops lead to losses in yield and quality, causing economic setbacks for farmers, food processors, and exporters. Mitigating mycotoxin production requires a multi-pronged approach involving chemical, biological, and genetic management strategies. Additionally, regulatory systems set by national and international agencies play a vital role in enforcing safety limits and establishing guidelines for maximum allowable mycotoxin levels in food and agricultural products.

This book, *Mycotoxins: Detection and Management*, offers a comprehensive examination of the critical aspects of mycotoxin research covering fungal pathogens, types of mycotoxins, environmental factors causing mycotoxin production, their detection techniques, and management. It is a valuable resource for a diverse range of stakeholders, including researchers, students, policy planners, regulatory authorities, and industry professionals for systematic reading as it provides the latest scientific knowledge and practical insights. This book aims to support rationale decision-making and enhance the safety of our food supply chain.

I extend my sincere congratulations to the contributors and editors for their outstanding efforts in creating this valuable compendium. Their dedication and expertise have culminated in a resource that will serve as a cornerstone in advancing the understanding and management of mycotoxins globally.

Prof. (Dr.) B.R. Kamboj

Preface

The presence of mycotoxins in agricultural products is a global concern, with serious implications for food safety, public health, and economic stability. Mycotoxins are produced by fungal pathogens such as *Aspergillus, Fusarium, Penicillium* and molds both under field and storage conditions. Wide variety of various mycotoxins are produced in varying quantities in food grains and foodstuffs, The need for a comprehensive understanding of these hazardous compounds and the means to effectively detect, control, and manage them has never been more critical as a huge quantity of food grains and food stuffs are wasted globally due to mycotoxins. This inspired us to create the book "*Mycotoxins: Detection and Management*".

The book begins by providing a foundational understanding of mycotoxins, what they are, the fungal species responsible for their production, and the environmental factors that influence their occurrence. Given the potential health risks posed by mycotoxins, accurate detection and quantification are paramount. The book covers an array of modern techniques employed for this purpose. These methods are vital for ensuring food and feed safety, as well as for compliance with regulatory standards.

Managing and mitigating the risks associated with mycotoxins requires a diverse set of strategies. The book explores various control methods, ranging from chemical, biological, and genetic approaches that focus on breeding resistant crop varieties. Furthermore, it addresses the role of regulatory frameworks in setting permissible limits for mycotoxins, highlighting the importance of national and international guidelines to protect consumer health and facilitate trade.

In compiling this work, we aimed to provide a resource that is both comprehensive and accessible. The scope of the book spans the entire spectrum of mycotoxin research and management, making it a valuable reference for researchers, students, industry professionals, policy makers, and regulatory authorities. It serves as a guide to understanding current challenges, the latest scientific advancements, and the practical steps needed to ensure food safety and economic resilience.

We would like to express our deepest gratitude to the contributing authors who have shared their knowledge and expertise. Their collaborative efforts have made this book a reliable and thorough resource for researchers engaged in various aspects of mycotoxin production and their effect on food safety and trade. It is our hope that this book will serve useful purpose to researchers, research scholars, students, and policy planners for enhanced food safety, better management practices, and a stronger regulatory framework for mycotoxin control.

Editors
Rishi K. Behl, Abhishek Kumar and Ankit Kumar

Contents

Chapter 1

Post-Harvest Fungal Pathogens and Mycotoxins; Their Significance in Agriculture

Pooja[1]*, Prashant Chauhan[1], Rajbala Yadav[2], Poonam[3], Ankit Kumar[4] and Abhishek Kumar[5]

[1]Department of Plant Pathology, [2]Department of Microbiology, [3]Department of Entomology, [4]Centre of Food Science and Technology, CCS Haryana agricultural University, Hisar-125004

[5]Department of Agriculture, Maharishi Markandeshwar (Deemed to be University), Mullana, Ambala, HR, India

Corresponding author: poojadusad2018@gmail.com

Abstract

Various filamentous fungal genera like *Claviceps, Drechslera, Penicillium, Fusarium, Nigrospora, Aspergillus, Monascus, Alternaria, Cephalosporium*, and *Trichoderma* produce mycotoxins which are toxic secondary metabolites. Among them, *Fusarium and Aspergillus* genera are saprophytic phytopathogens known to cause infections and produce mycotoxins in various crops during pre- and post-harvest stages. More than 400 mycotoxins have been identified to date, while trichothecenes, aflatoxin, ochratoxin A, fumonisins, ergot alkaloids, zearalenone, citrinin, and patulin being the most dominant. These compounds are linked to various health disorders in both humans and animals. In this chapter, we aim to provide consolidated information on the mycotoxins present in crops and their significance on human health.

Keywords: Mycotoxins, secondary metabolites, saprophyte, contamination, DON

Introduction

Cereal crops (maize, wheat, and rice), are critical staple foods that provide significant dietary energy and nutrients globally (Singh, 2015; Jimenez-Garcia *et al.*, 2018; Deepa *et al.*, 2021). The cultivation and resilience of these crops are essential for global food security (Adeyeye, 2016; Lenzi *et al.*, 2021). However, climate change and plant pathogens threaten sustainable cereal production and food safety (Deepa and Sreenivasa, 2019a; Nagaraja *et al.*, 2022). Cereal grains are particularly vulnerable to pathogenic contamination by various fungal and bacterial species (Wu *et al.*, 2014; Adeyeye, 2016; Kluczkovski, 2019), with fungi being the primary contaminants which degrade the nutritional qualities of foods (Juraschek *et al.*, 2022).

Fungi like *Alternaria*, *Fusarium*, *Aspergillus*, *Penicillium*, and *Claviceps* infect food crops and produce mycotoxins that compromise food safety (Lenzi *et al.*, 2021). These microorganisms thrive under different environmental conditions during and after harvest. Post-harvest handling, combined with climatic factors, affects the quality of grains and influences their germinability. Even in safe storage conditions, microbial contamination persists, slowly impacting the grains over time.

Poor post-harvest management accelerates nutrient deterioration in seeds due to microbial activity, leading to discoloration, heating, dry matter loss, lipid and protein degradation, and reduced germination. This also affects grains used for animal feed or as seeds. Particularly, filamentous fungi, produce mycotoxins which may be carcinogenic to both animals and humans, and certain fungal spores may pose respiratory risks to workers (Baidhe *et al.*, 2024).

"Mycotoxin" term is derived from two Greek words namely "mykes" and "toxicum" meaning fungus or mold, and poison respectively (Ingle *et al.*, 2020). Mycotoxins are secondary metabolites which have low molecular weight produced by fungi in response to climatic factors such as moisture, temperature, and food composition (Dieye *et al.*, 2024). More than 400 mycotoxins have been identified, with contamination occurring both directly in food and indirectly through animal products (Mahboubifar *et al.*, 2024). These toxins can severely impact human health, leading to kidney diseases, reproductive issues, intestinal damage, and cancer by altering gene expression (Rodríguez-Melcón *et al.*, 2024). Mycotoxins remain stable even during food processing and may correlate with fungal oxidative stress responses.

Mycotoxin production may start during the preharvest phase and continues to accumulate during harvesting, transportation, and storage of a crop (Agriopoulou *et al.*, 2023). Globally, approximately 25% of grain production is accumulated with mycotoxins, contributing significantly to food loss and global food insecurity (Yu *et al.*, 2021). The mycotoxins can infiltrate crops like cereals, nuts, spices, and dairy products at all stages of the food cycle, impacting both profitability and human health (Rausch *et al.*, 2021). Notable mycotoxins include aflatoxins, zearalenone, deoxynivalenol, and ochratoxin A (Agriopoulou *et al.*, 2023).

The presence of mycotoxins in food and feed is a serious global issue, affecting international trade and food security (Liu *et al.*, 2024). Approximately, 5 billion people are exposed to mycotoxins regularly through contaminated foods, leading to health issues such as cancer and immune suppression (Wu *et al.*, 2014; Khodaei *et al.*, 2021). Fungal contamination during crop growth, storage, and transport can lead to severe losses in agricultural productivity, with aflatoxins being particularly concerning due to their toxicity and prevalence (Deepa and Sreenivasa, 2019b).

Field fungi, such as *Fusarium* and *Alternaria*, infect crops pre-harvest, while storage fungi, including *Aspergillus* and *Penicillium*, thrive post-harvest in drier conditions (George *et al.*, 2022). Mycotoxin containing food grains and other products has wide-reaching consequences for both health and agriculture, contributing to cancer, gastrointestinal problems, and kidney disorders (Table 1) (Adeyeye, 2016; Jimenez-Garcia *et al.*, 2018). Mycotoxins, accumulated by saprophytic fungi, poses a multidisciplinary issue involving specialists from various fields, including environmental science, medicine, veterinary science, agriculture, chemistry, and plant pathology (Spevakova *et al.*, 2020). Mycotoxicosis can occur in both industrialized and developing nations, exacerbated by social, environmental, and economic factors combined with meteorological conditions (high temperatures and humidity), favor fungal growth.

Table 1: List of important mycotoxins

Mycotoxins	*Related Moulds*	*Most Prone Food Products to Be Contaminated*	*Symptoms/ Toxicology*	*References*
Aflatoxins	*Aspergillus parasiticus, A. nomius, A. flavus*	Grain, cherries, strawberries, groundnut, raspberries, maize, peanuts, etc.	Depressed immune response, liver tumors, hepatotoxicity, mutagenicity, teratogenicity, vomiting, and pulmonary convulsions	Liu et al. (2006)
Cyclopiazonic acid	*A. flavus, A. oryzae, A. versicolor, A. tamarii, P. patulum, P. verrucosum, P. cyclopium, P. griseofulvum, P. puberulum*	Peanuts, maize, cheese, etc.	Neurotoxicity, cytotoxicity, weight loss, immunotoxicity, diarrhea, muscle necrosis, and convulsions	Gonçalez et al. (2008)
Deoxynivalenol, Vomitoxin, Zearalenone	*Fusarium graminearum, F. culmorum, F. subglutinans*	Wheat, maize, oats, maize rice, sorghum, and barley	Diarrhea, weight loss, hepatotoxicity, genotoxicity, and immunotoxic effects	Nakagawa et al. (2011)

Fumonisin B1 and Fumonisin B2	*F. moniliforme, F. verticillioides*	Maize, rice, and wheat	Porcine pulmonary edema, equine leukoencephalomalacia, hepatotoxicity, nephrotoxicity, cytotoxicity, and oesophageal cancer	Topi et al. (2021)
Trichothecenes	*F. culmorum, Trichoderma, F. graminearum, F. poae, Cephalosporium, Trichothecium*	Wheat, oats, and maize	Food toxicity, necrosis, oral lesions, vomiting, diarrhea, growth retardation, fever, dizziness, and neurotoxicity	Jimenez and Mateo (1997)
Ochratoxin	*A. ochraceus, P. verrucosum, A. carbonarius*	Wheat, spices, grapes, and coffee	Nephropathy, genotoxicity, immunotoxicity, teratogenicity, neurotoxicity, and inhibition of protein, RNA, and DNA synthesis	Iqbal et al. (2018)
Patulin and Citrinin	*P. expansum*	Apple, orange, grapes, and related products	Kidney damage, nephrotoxicity, immunotoxicity, teratogenicity, hepatotoxicity, and fetotoxicity	Oteiza et al. (2017)
Sterigmatocystin	*A. parasiticus, A. versicolor, A. flavus, A. nidulans, A. rugulosus, A. ruber, A. chevalieri, P. griseofulvum, P. commune*	Maize, rice, wheat, and hay	Carcinogenic, mutagenic, immunotoxicity, cytotoxicity, diarrhea, nausea, and weight loss	Iqbal et al. (2018)
Alternaria toxins	*Alternaria species*	Grains, oilseeds, spices, fruits, and vegetables	Cytotoxicity, genotoxicity, teratogenicity, mutagenicity, fetotoxicity, and dermal toxicity	Dong et al. (2019)

Climate change further complicates the One Health agenda, as it has profound effects on food security and increases the possibility of mycotoxin accumulation in food crops. Warmer temperatures and higher humidity levels create optimal conditions for the growth of toxigenic fungi, particularly those that produce aflatoxin B1 (AFB1) (Molnár *et al.*, 2023). Climate change may not directly inhibit the acclimatization of mycotoxigenic fungi or reduce their production of toxins, thereby sustaining contamination in food chains and causing health concerns in both animals and humans (Medina *et al.*, 2017). This chapter provides an overview of the presence and impact of mycotoxins in food crops, emphasizing the threats they pose to humankind.

Edaphic factors and their impact on mycotoxin contamination

Agricultural commodities provide optimal conditions for growth of fungus and mycotoxin accumulation (Gurikar *et al.*, 2023). Edaphic factors, including physical and chemical properties of soil, play a critical role in influencing fungal infections and mycotoxin contamination. Soil pH, nutrient availability, and environmental conditions contribute to contamination, alongside pest infestations and improper harvesting or storage practices (Juraschek *et al.*, 2022).

Mycotoxin contamination in food grains can originate from contaminated soils during harvest (Wu *et al.*, 2014; Adeyeye, 2016). Risk of fungal growth and toxin production might be due to high temperatures and humidity during storage. Soil pH is a critical factor, with levels between 4.0 and 4.5 promoting mycotoxin synthesis, while pH levels above 6.5 may inhibit it (Kang *et al.*, 2014). Additionally, insect infestations and physical damage to crops facilitate the colonization of toxigenic fungi (*Aspergillus*), thereby increasing contamination risks (Adeyeye, 2016).

Contamination of *Aspergillus* and its derived mycotoxins

Worldwide, the *Aspergillus* genus includes approximately 200 species, making it the most prevalent fungal genus in the environment (Jimenez-Garcia *et al.*, 2018). Notable *Aspergillus* species, *i.e.*, *A. flavus*, *A. repens*, *A. parasiticus*, *A. niger*, *A. fumigatus*, and *A. ochraceus*, are among the most prominent phytopathogens affecting a range of food crops (Chakranarayan and Pati, 2013; Adeyeye, 2016; Gurikar *et al.*, 2023). These species exhibit diverse morphological characteristics, including green-colored colonies, round vesicles, greenish-yellow conidiophores, and spherical, light green conidia (Sankeerthana *et al.*, 2013; Wu *et al.*, 2014; Cinar and Onbaşı, 2019).

Aflatoxins

Aflatoxins, produced by various *Aspergillus* species, are potent mycotoxins commonly found in staple crops (corn, peanuts, sorghum, and wheat), both during pre- and post-harvest stages (Gurikar *et al.*, 2023). The first major outbreak linked to aflatoxins was found in England in Turkey X disease during the 1960s, resulting in over a million turkey deaths due to peanut meal contaminated with *Aspergillus flavus* (Adeyeye, 2016; Grozeva et al., 2017; Jimenez-Garcia et al., 2018).

Aflatoxins primarily affect crops such as cereals and tree nuts, are highly vulnerable to contamination by *Aspergillus* species. (Adeyeye, 2016; Cinar and

Onbaşı, 2019). These fungi exist as soil saprophytes, infecting and colonizing plant tissues, which lead to contamination (Jimenez-Garcia *et al.*, 2018; Lenzi *et al.*, 2021). In semi-arid tropical regions, aflatoxin contamination is particularly severe during drought conditions, although elevated humidity can also increase contamination risks. Fungal infections are exacerbated when moisture levels in food commodities exceed 14% during harvest (Deepa *et al.*, 2021).

There are 18 different aflatoxins, with aflatoxin B1, B2, G1, and G2 being the most prominent, while M1 and M2 being less common (Jimenez-Garcia *et al.*, 2018; Gurikar *et al.*, 2023). B1 aflatoxin is classified as a secondary metabolite, particularly toxic and carcinogenic and is found in various agricultural commodities. Aflatoxins are categorized into two groups based on their chemical structure: B types (difurocoumarocyclopentenone) and G types (difurocoumarolactone). M types aflatoxins are hydroxylated derivatives of aflatoxin B1 and are found in milk and dairy products (Jimenez-Garcia *et al.*, 2018).

Aflatoxins are soluble in various solvents *i.e.*, chloroform, acetone, and methanol and exhibit distinct fluorescence under UV light, B1 and B2 fluoresce emits blue, while G1 and G2 emit green fluorescence (Wu *et al.*, 2014). These toxins are carcinogenic, mutagenic, and immunosuppressive (Sankeerthana *et al.*, 2013; Jimenez-Garcia *et al.*, 2018). They cause immunosuppression by reducing the capacity of antigen-presenting cells and increasing pro-inflammatory cytokines (IFN-γ and TNF-α), thereby, accelerating inflammation (Mahaki *et al.*, 2024).

Both humans and animals are affected by aflatoxicosis, which lead to immunosuppression, liver damage, and an increased risk of cancer. Chronic exposure, especially to aflatoxin B1, can induce hepatocellular carcinoma. In animals, it can also reduce feed efficiency, reproductive performance, and overall health (Weinstock *et al.*, 2024). Aflatoxins are excreted in milk, with dairy animals exposed to contaminated feed producing milk containing aflatoxin M1. Although acute aflatoxicosis is rare, it can lead to sudden death due to liver failure (Weinstock *et al.*, 2024).

Aflatoxins, particularly AFB1, cause liver necrosis, hepatomegaly, and immune suppression in broiler chickens, resulting in decreased feed intake, altered behavior, and increased mortality (Balan *et al.*, 2024). AFB1 residues accumulate in liver and kidney tissues, impairing detoxification. In cattle, aflatoxins can lead to liver cirrhosis, teratogenic effects, and decreased milk production. Nursing animals are more susceptible due to the excretion of AFB1 as AFM1 in milk (Mutlu and Yenice, 2024). High doses reduce feed efficiency and induce liver dysfunction. While sheep are relatively resistant, aflatoxin contamination can still lead to hepatotoxicity and nephritis, altering mineral metabolism and plasma concentrations (Sadariya *et al.*, 2023). Additionally, in horses, aflatoxicosis manifests as hepatomegaly, jaundice, and kidney damage, with clinical sign including depression and lameness (Alnuimy, 2024). Dogs are particularly vulnerable to aflatoxicosis, with exposure linked to anorexia, depression, and acute liver failure (Yang *et al.*, 2023).

Ochratoxins

Ochratoxins are a group of mycotoxins primarily produced by species of *Aspergillus* (*A. ochraceus*) and *Penicillium* (*P. verrucosum*) in various agricultural crops, including cereal grains, peanuts, dried beans, and coffee beans (Cinar and Onbaşı, 2019). These mycotoxins are closely related to isocoumarin derivatives and are linked to the amino acid L-β-phenylalanine through an amide bond. The ochratoxin family includes more than 20 different metabolites, with ochratoxin A (OTA) being the most abundant and toxic (Adeyeye, 2016; Becker-Algeri et al., 2016; Cinar and Onbaşı, 2019).

Food crops, particularly those grown in northern temperate regions, are the main source of OTA. This mycotoxin is suspected to be a leading cause of human Balkan Endemic Nephropathy (BEN) and is also linked to tumors in the human urinary tract (Adeyeye, 2016). Striking similarities exist between OTA-induced porcine nephropathy and BEN in humans. The International Agency for Research on Cancer (IARC) classifies OTA as a possible human carcinogen (Wu *et al.*, 2014; Deepa and Sreenivasa, 2019a; Pickova *et al.*, 2020). From a public health perspective, OTA is the second most significant mycotoxin due to its nephrotoxic, hepatotoxic, genotoxic, teratogenic, and immunosuppressive effects (Becker-Algeri *et al.*, 2016; Pickova *et al.*, 2020).

Ochratoxin A is produced by *Aspergillus* and *Penicillium* species, including *A. ochraceus*, *A. sclerotiorum*, *A. melleus*, *A. sulphureus*, and *P. verrucosum* (Nazareth *et al.*, 2024). Various food items such as cereals, nuts, cocoa beans, cassava flour, peanuts, dried fruits, poultry eggs, and milk are naturally contaminated by OTA (Marrez and Ayesh, 2022). Contamination levels can be significant, with cocoa beans showing a 50% contamination rate and cocoa powder at 22% (Abballe *et al.*, 2021). OTA is nephrotoxic in all animal species and is likely toxic to humans as well. It is also considered hepatotoxic, neurotoxic, immunosuppressive, teratogenic, and carcinogenic (Więckowska *et al.*, 2024). OTA induces oxidative stress in human kidney cells by regulating the translocation of transcription factors like the aryl hydrocarbon receptor (AhR) and pregnane X receptor (PXR), leading to renal damage, immunosuppression, and cancer (Gul *et al.*, 2023).

The first outbreaks of ochratoxicosis were reported in laying hens, broiler chickens, and turkeys. In turkeys, clinical signs included decreased feed intake, poor feed conversion ratio (FCR), and increased mortality. Pathological lesions included nephritis and mild hepatitis. In broilers, ochratoxicosis caused reduced growth, poor FCR, nephrotoxicity, and air-sacculitis. In laying hens, it resulted in decreased egg production and nephritis (Imran *et al.*, 2020). Necropsy findings typically revealed inflamed, pale-yellow kidneys and friable livers, while microscopic lesions included vacuolar degeneration, megaleukocytosis of liver cells, and kidney epithelial hypertrophy. OTA is more nephrotoxic than hepatotoxic in broilers and also impacts other organs, including the spleen, thymus, and bursa, often leading to lymphocytic depletion (Tahir *et al.*, 2022). The immunosuppressive effects include reduced weight of immune organs and decreased numbers of antibody-bearing immune cells. Hematological effects include anemia, lowered red and white blood cell counts, and decreased serum proteins. Additionally, OTA increases susceptibility to other diseases, such as *Eimeria tenella*, fowl typhoid, *Salmonella gallinarum*, and *E. coli* infections (Imran *et al.*, 2020).

Fusarium contamination and derived mycotoxins

Fusarium is one of the most diverse and widespread genera of seed-borne fungi associated with food grains (Gurikar *et al.*, 2023). This genus is responsible for significant plant diseases, including head blight, scab, stalk rot, and ear rot, which have considerable economic impacts on grain crops (Deepa and Sreenivasa, 2019b). While all *Fusarium* species produce secondary metabolites, certain species generate a wide range of physiologically active mycotoxins, such as trichothecenes, zearalenone, fusaric acid, moniliformin, and fumonisins, each with unique chemical structures (Jimenez-Garcia *et al.*, 2018; Nagaraja *et al.*, 2022).

These mycotoxins are classified into two main categories: trichothecenes and non-trichothecenes. Contaminated food and feed can cause severe health issues in humans and animals, including cancer, mutagenicity, and disorders affecting estrogen levels, gastrointestinal health, and kidney function (Moretti *et al.*, 2017) (Figure 1). Globally, *Fusarium* and other mycotoxigenic fungi damage approximately 25% of crops annually (Assunção *et al.*, 2016).

Several *Fusarium* species, such as *F. verticillioides*, *F. graminearum*, *F. equiseti*, *F. proliferatum*, *F. culmorum*, and *F. sporotrichioides*, are major

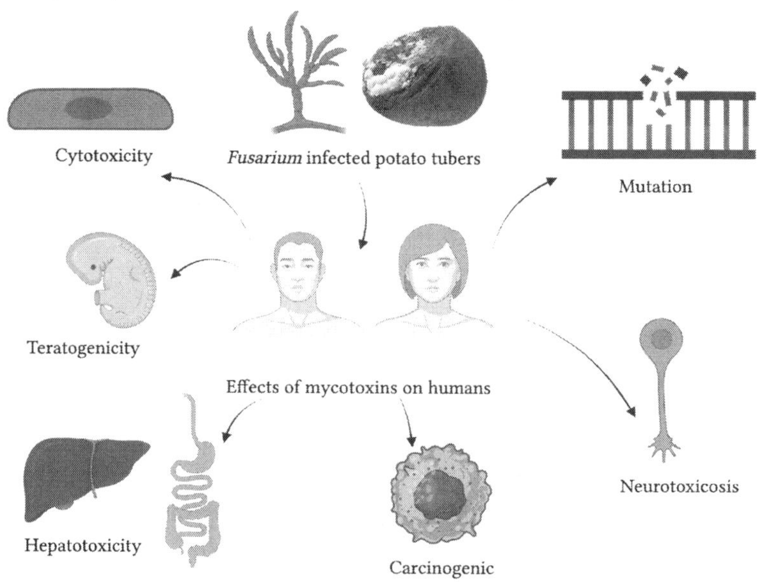

Figure 1. Effects of mycotoxins on humans

phytopathogens that pose serious threats to cereal crops worldwide. Among *Fusarium* mycotoxins, trichothecenes, zearalenone (ZEN), and fumonisins are the most prevalent and are often found together in cereal grains and animal feed at dangerously high levels (Adeyeye, 2016; Nagaraja *et al.*, 2016; Jimenez-Garcia *et al.*, 2018).

Fumonisins

Fumonisins, a group of secondary metabolites primarily produced by *F. verticillioides*, *F. proliferatum*, and a few other species of *Fusarium*, predominantly affecting maize (Adeyeye, 2016; Deepa and Sreenivasa, 2019a). These mycotoxins are commonly found in maize and maize-based products worldwide (Kluczkovski, 2019). In Africa, fumonisins were initially identified in the *F. moniliforme* strain (MRC 826), followed by the discovery of similar metabolites from species such as *F. verticillioides*, *F. proliferatum*, and *F. nygamai* (Wu *et al.*, 2014; Becker-Algeri *et al.*, 2016; Pickova *et al.*, 2020). To date, over 15 structurally related fumonisin analogs have been identified, with FB1, FB2, and FB3 being the most prevalent (Jimenez-Garcia *et al.*, 2018).

Fumonisins are classified into four groups based on their functional groups: A, B, C, and P, with group B being the most common contaminant in maize and cereal crops globally. FB1 accounts for 70%-80% of the total fumonisin content, while FB2 and FB3 contribute 15%-25% and 3%-8%, respectively (D'Arco *et al.*, 2009; Deepa *et al.*, 2021). Elevated levels of FB1 in maize have been associated with an increased incidence of esophageal cancer in humans and neural tube defects in fetuses due to chronic exposure during pregnancy (Summerell and Leslie, 2011). Consequently, the International Agency for Research on Cancer (IARC) has classified FB1 as a possible human carcinogen. Fumonisins have also been linked to several animal diseases, including hepatocarcinogenesis in rats, pulmonary edema in pigs, and leukoencephalomalacia in horses (Müller *et al.*, 2012; Wu *et al.*, 2014; Adeyeye, 2016; Cinar and Onbaşı, 2019). In response to these health risks, the European Union has implemented strict regulations, limiting fumonisin levels in maize-based foods to 2 mg/kg, while the Joint Expert Committee on Food Additives (JECFA) has established a provisional maximum tolerable daily intake of 2 µg/kg body weight (European Union Commission, 2006).

Fumonisins, particularly FB1 and FB2, are highly toxic due to their carcinogenic properties, with FB1 exhibiting greater toxicity than FB2. FB1 has been implicated in neoplastic activity in rats and is a causative agent of equine leukoencephalomalacia (Sertova and Ignatova, 2023). The toxicity mechanism of fumonisins involves the inhibition of sphinganine-N-acetyl transferase, an enzyme crucial for sphingolipid metabolism, resulting in the accumulation of sphinganine and sphingosine and a concomitant reduction in complex sphingolipids (Sertova and Ignatova, 2023). Fumonisin toxicity affects multiple organs, including the gut, brain, liver, and lungs, across various species such as poultry, cattle, pigs, and horses. In chicken embryos, fumonisins lead to high mortality rates, pathological lesions in the liver and other organs, brain edema, and subcutaneous hemorrhages (Awuchi et al., 2021). In broilers, fumonisin exposure results in reduced weight gain and increased mortality, while in laying hens, it causes diarrhea (Awuchi *et al.*, 2021). In horses, fumonisins induce equine leukoencephalomalacia, characterized by anorexia, incoordination, brain and lung edema, liquefactive necrosis of white matter, and eventual death (Gupta and Gupta, 2019). In sheep, fumonisin doses of 11.1–45.5 mg/kg body weight cause liver and kidney inflammation (Farhadi et al., 2019). Furthermore, in cattle, *Fusarium* mycotoxins have been shown to impair reproductive functions by disrupting granulosa cell proliferation and steroid production (Chiminelli *et al.*, 2022).

Trichothecenes (TCTs)

Trichothecenes (TCTs) are a group of toxic sesquiterpenoids produced by various fungal species, including *Fusarium*, *Myrothecium*, *Stachybotrys*, and *Trichoderma*. Over 200 types of TCTs have been identified, primarily contaminating cereal crops such as maize, wheat, rye, barley, and oats worldwide (Yadav *et al.*, 2023). The term "trichothecenes" is derived from "Trichothecin," which was first identified in 1949 by Freeman and Morrison from an isolate of *Trichothecium roseum* (Zhu *et al.*, 2023). TCTs are classified into four groups (types A-D) based on their functional groups. HT-2 toxin, T-2 toxin, and T-2 triol are Type A, while Type B comprises deoxynivalenol (DON), 3-acetyl-DON, 15-acetyl-DON, and nivalenol. *Fusarium poae* and *F. sporotrichioides* produce Type A TCTs, while *F. graminearum* and *F. culmorum* are responsible for production of Type B (Leannec-Rialland, 2022).

Trichothecenes disrupt mitochondrial and cell membrane functions, as well as protein, DNA, and RNA biosynthesis, and may also exhibit immunosuppressive effects (Bojanowski *et al.*, 2013). Twenty-four *Fusarium* species, majorly *F. graminearum*, *F. sporotrichioides*, and *F. culmorum*, are known producers of TCTs (Foulongne-Oriol *et al.*, 2024). TCT contamination has been documented globally, affecting food crops in America, Asia, Europe, and Africa. These mycotoxins often co-occur, leading to exposure to multiple TCTs through contaminated food and feed. Symptoms of acute oral exposure include nausea, vomiting, headaches, and alimentary hemorrhage, while chronic exposure may cause immunosuppression, anemia, anorexia, and neuroendocrine alterations. The first incident reported mycotoxicosis due to TCTs occurred in the USSR in 1932, with mortality exceeding 60%, linked to *F. sporotrichioides* contamination (Bojanowski *et al.*, 2013).

Deoxynivalenol (DON)

DON, a Type B trichothecene, is one of the most significantly distributed *Fusarium* mycotoxins (Birr *et al.*, 2021). It predominantly contaminates cereal grains, especially in temperate regions. Also known as "vomitoxin," DON induces gastroenteritis and vomiting in humans and animals. Chronic exposure leads to reduced feed intake, poor feed efficiency, reproductive issues, and even death. While T-2 toxin is more toxic to mammals, DON remains a serious concern at higher concentrations. The use of DON-contaminated grains led to food poisoning in both humans and livestock (Hou *et al.*, 2023). In broilers, DON-contaminated feed increases the relative weight of the heart, gizzard, and lymphoid organs, while nivalenol (NIV) contamination leads to reduced feed intake and feed conversion ratios (Yu and Pedroso, 2023). T-2 and DAS toxins cause weight loss, reduced feed intake, and ulcers in buccal cavity of chicks (Zhang *et al.*, 2022). In cattle, DAS causes gastrointestinal lesions, decreased feed intake, and decreased milk production, while T-2 toxin suppresses immune responses, reduces lymphoproliferative responses, and affects antibody concentrations. This has also been linked to infertility and immune organ necrosis in cattle (Janik *et al.*, 2021). Moreover, in dogs, DON levels in feed must remain below 0.5 mg/kg to avoid toxic effects. Intravenous injection of T-2 toxin in cats causes hypovolemic shock and leukopenia, often resulting in death (Theresa *et al.*, 2021).

Zearalenone (ZEN)

Zearalenone (ZEN) is primarily produced by various *Fusarium* species, including *F. graminearum*, *F. moniliforme*, *F. culmorum*, *F. cerealis*, *F. equiseti*, and *F. verticillioides*. It commonly contaminates cereal grains *i.e.*, maize, wheat, barley, oats, and sorghum, majorly in temperate regions (Nagaraja *et al.*, 2022). ZEN levels tend to significantly increase during storage if the moisture content exceeds 30%. Despite its usually low levels in the field, ZEN contamination remains a significant food safety concern.

ZEN is classified as a nonsteroidal estrogenic compound due to its phenolic resorcylic acid lactone structure, exhibiting strong estrogenic activity by binding to estrogen receptors. It frequently co-occurs with other *Fusarium* mycotoxins (deoxynivalenol (DON) and trichothecenes (TCTs) (Adeyeye, 2016; Pickova *et al.*, 2020). Upon ingestion, ZEN is metabolized into β-zearalenol and α-zearalenol, with α-zearalenol being the most estrogenic variant (Becker-Algeri *et al.*, 2016).

In animals, ZEN is absorbed through the gastrointestinal tract, with high absorption rate of 80%-85% in swine. The estrogenic effects of ZEN differ between adult males and females. In males, ZEN causes immunosuppression, reduced testosterone levels, decreased testis weight, impaired feminization, spermatogenesis, and lowered libido (Deepa and Sreenivasa, 2017; Jimenez-Garcia *et al.*, 2018). In females, particularly poultry, ZEN leads to immune suppression, decreased weight, vulvar dilation, vulvovaginitis, rectal prolapse, and reductions in embryo size and quantity. Additionally, ZEN has been linked to cancer, genotoxicity, and endocrine disruption (Awad *et al.*, 2014). ZEN has structural similarity to estradiol which causes reproductive problems across various species, making it a significant risk factor in agriculture and animal husbandry (Liu and Applegate, 2020). ZEN is particularly toxic to pigs. In boars, it leads to feminization, decreased libido, reduced serum testosterone levels, decreased testis weight, and impaired spermatogenesis. In sows, ZEN causes vulvovaginitis, embryonic mortality, anestrus, delayed post-weaning estrus, and degeneration of germinal epithelium

(Roach, 2023). Piglets exposed to ZEN may experience reduced conception rates, smaller litter sizes, and enlargement of vulva, uterus, and ovaries (Zhou *et al.*, 2022). In cattle, ZEN reduces milk production, causes infertility, and increases serum estrogen concentrations, negatively impacting conception rates (D'mello *et al.*, 1998).

Penicillium contamination and derived mycotoxins

Penicillium is a huge genus comprising approximately 200 known species, with at least 50 of these being majorly found in nature (George *et al.*, 2022). All common species of *Penicillium* impart green colonies and grow slowly. Cereal food crops cultivated in temperate regions (northern and central Europe and Canada), are particularly vulnerable to *Penicillium* infections (George *et al.*, 2022). These fungi pose significant challenges in postharvest situations, especially when storage conditions are inadequate. Grain infections are primarily attributed to contamination during post-harvest handling and storage equipment (Adeyeye, 2016; Cinar and Onbaşı, 2019; George *et al.*, 2022).

Citrinin (CTN)

Citrinin, a secondary metabolite produced by several toxigenic strains of fungi belonging to the genera *Penicillium* (e.g., *P. expansum*, *P. verrucosum*), *Aspergillus* (e.g., *A. terreus*), and *Monascus* (e.g., *M. ruber*). It contaminates various plant-based products, especially cereal grains, and frequently co-occurs with another nephrotoxic mycotoxin, ochratoxin A (OTA) (George *et al.*, 2022). While citrinin has been demonstrated to be nephrotoxic in animal studies, its acute toxicity varies between species. In addition to nephrotoxicity, citrinin can cause embryocidal and fetotoxic effects in species such as mice, swine, and poultry, with the kidney identified as the principal target organ. Moreover, citrinin has teratogenic, carcinogenic, and depressant effects on the human central nervous system. Excessive ingestion of citrinin has been associated with altered liver metabolism, increased water intake, severe diarrhea, and changes in the morphology of intestinal villi (Adeyeye, 2016; Pickova *et al.*, 2020). The pH of grains plays a critical role in the biosynthesis of citrinin by toxigenic fungi. Nitrogen sources can affect the pH range of substrates, while nitrogen-containing *Monascus* pigments (such as monascine, rubropunctatine, and monascorubrine) regulate the final pH and promote citrinin production. Historically, citrinin was believed to have antibacterial properties and was employed to inhibit the growth of various bacteria, fungi, protozoa, and yeasts (Alcaide-Molina *et al.*, 2009; Adeyeye, 2016; Pickova *et al.*, 2020).

Patulin (PAT)

Patulin produced by various fungal species, is a secondary metabolite, primarily from the genus *Penicillium* (such as *P. crustosum*, *P. expansum*, and *P. roqueforti*). More recently, patulin has been detected in cereal-based foods, with concentrations ranging from 2.33 to 4.5 μg/kg (Assunção *et al.*, 2016). The co-occurrence of patulin with other fungal metabolites, such as citrinin, cyclopiazonic acid, ochratoxin A, and roquefortine C, has also been noticed (Adeyeye, 2016). Patulin is a water- and alcohol-soluble cyclic chemical compound, though it loses its toxicity under alkaline conditions (Adeyeye, 2016). *P. expansum* is responsible for causing various plant diseases, particularly in apples, pears, and cherries, making it one of the most common plant pathogens. Patulin exhibits nephrotoxic and immunotoxic properties, resulting in duodenal lesions, intestinal hemorrhages, gastric ulcers, and alterations in intestinal function (Assunção *et al.*, 2016; George *et al.*, 2022).

***Claviceps* Contamination and Derived Mycotoxins**

Genus *Claviceps* is a parasitic fungus which includes more than fifty species. This fungus predominantly infects cereals, with rye being the most common host. Ergot is linked with a number of poisonous alkaloids such as ergotamine, ergocrystinine, ergometrimine, ergonovin, and ergocrystine (Jimenez-Garcia *et al.*, 2018). Infestations result in decreased quality and quantity of grain yields, and consumption of ergot-contaminated crops poses significant health risks to animals (Adeyeye, 2016).

The metabolites produced by this fungus, referred to as ergot alkaloids, are classified as indole alkaloids and are characterized by a tetracyclic ergoline ring structure (Jimenez-Garcia *et al.*, 2018). Ergotism is a human disease caused by the ingestion of cereal products contaminated with ergot sclerotia. It manifests in two forms: gangrenous and convulsive. Gangrenous ergotism disrupts blood flow, while convulsive ergotism affects the central nervous system.

Cattle, sheep, pigs, and poultry are particularly susceptible to ergotism, which is a significant concern in veterinary medicine (Gurikar *et al.*, 2023). Clinical signs of ergotism in animals include abortion, gangrene, convulsions, hypersensitivity, suppression of lactation, and ataxia (Jimenez-Garcia *et al.*, 2018; Nagaraja *et al.*, 2022).

Conclusion

Mycotoxins are predominantly found in cereals and other processed food items, particularly mycotoxins produced by genus *Fusarium*, prevalent in heavily consumed grains (wheat, maize, barley, and oats). Various mycotoxins in food and feed may occur both in their natural and altered forms, however, their presence and co-occurrence remain barely understood. This lack of understanding complicates efforts to realistically demonstrate co-exposure patterns in animals and humans. To address this issue, it is essential to conduct more research on the combined risks of multiple mycotoxin exposures, as well as to clarify dose-response relationships and toxicology. Additionally, maintaining and, where possible, enhancing surveillance and biomonitoring programs is crucial for understanding the alteration of pattern of mycotoxic fungi and mycotoxin contamination, as well as for monitoring the exposure levels in animals and humans. The occurrence and concentration of mycotoxins is affected by several other factors, such as geographic and climatic conditions. Therefore, a more comprehensive study of the co-occurrence and reactivity of mycotoxins is necessary.

References:

Abballe, C., Gomes, F. M. L., Lopes, B. D., de Oliveira, A. P. F., Berto, M. I., Efraim, P., & Tfouni, S. A. V. (2021). Cocoa beans and derived products: Effect of processing on polycyclic aromatic hydrocarbons levels. *Lwt, 135*, 110019.

Adeyeye, S. A. (2016). Fungal mycotoxins in foods: A review. *Cogent Food & Agriculture, 2*(1), 1213127.

Agriopoulou, S., Tarapoulouzi, M., Varzakas, T., & Jafari, S. M. (2023). Application of encapsulation strategies for probiotics: from individual loading to co-encapsulation. *Microorganisms, 11*(12), 2896.

Alcaide-Molina, M., Ruiz-Jiménez, J., Mata-Granados, J. M., & de Castro, M. L. (2009). High through-put aflatoxin determination in plant material by automated solid-phase extraction on-line coupled to laser-induced fluorescence screening and determination by liquid chromatography–triple quadrupole mass spectrometry. *Journal of chromatography A, 1216*(7), 1115-1125.

Alnuimy, A. (2024). Toxopathological And Histopathological Effects Of Aflatoxins. *Magazine of Al-Kufa University for Biology, 16*(1).

Assunção, R., Martins, C., Dupont, D., & Alvito, P. (2016). Patulin and ochratoxin A co-occurrence and their bioaccessibility in processed cereal-based foods: A contribution for Portuguese children risk assessment. *Food and Chemical toxicology, 96*, 205-214.

Awad, W. A., Ghareeb, K., Dadak, A., Hess, M., & Böhm, J. (2014). Single and combined effects of deoxynivalenol mycotoxin and a microbial feed additive on lymphocyte DNA damage and oxidative stress in broiler chickens. *PloS one, 9*(1), e88028.

Awuchi, C. G., Ondari, E. N., Ogbonna, C. U., Upadhyay, A. K., Baran, K., Okpala, C. O. R., ... & Guiné, R. P. (2021). Mycotoxins affecting animals, foods, humans, and plants: Types, occurrence, toxicities, action mechanisms, prevention, and detoxification strategies—A revisit. *Foods, 10*(6), 1279.

Baidhe, E., Clementson, C. L., Senyah, J., & Hammed, A. (2024). Appraisal of Post-Harvest Drying and Storage Operations in Africa: Perspectives on Enhancing Grain Quality. *AgriEngineering, 6*(3), 3030-3057.

Balan, B., Dhaulaniya, A. S., Kumar, M., Kumar, M., & Kumar, P. (2024). Aflatoxins in food: Prevalence, health effects, and emerging trends in its mitigation—An updated review. *Food Safety and Health, 2*(1), 39-71.

Becker-Algeri, T. A., Castagnaro, D., de Bortoli, K., de Souza, C., Drunkler, D. A., & Badiale-Furlong, E. (2016). Mycotoxins in bovine milk and dairy products: A review. *Journal of food science, 81*(3), R544-R552.

Birr, T., Jensen, T., Preußke, N., Sönnichsen, F. D., De Boevre, M., De Saeger, S., ... & Klink, H. (2021). Occurrence of Fusarium mycotoxins and their modified forms in forage maize cultivars. *Toxins, 13*(2), 110.

Bojanowski, A., Avis, T. J., Pelletier, S., & Tweddell, R. J. (2013). Management of potato dry rot. *Postharvest biology and technology, 84*, 99-109.

Chakranarayan, M., & Pati, A. (2013). Comparison of microscopic, macromorphological and aflatoxin producing capabilities of Aspergillus species associated with rhizosphere of groundnut (A. hypogaea L.). *Journal of Chemical, Biological and Physical Sciences (JCBPS), 3*(2), 1327.

Chiminelli, I., Spicer, L. J., Maylem, E. R., & Caloni, F. (2022). Emerging mycotoxins and reproductive effects in animals: A short review. *Journal of Applied Toxicology, 42*(12), 1901-1909.

Cinar, A., & Onbaşı, E. (2019). Mycotoxins: The hidden danger in foods. *Mycotoxins and food safety*, 1-21.

D'Arco, G., Fernández-Franzón, M., Font, G., Damiani, P., & Mañes, J. (2009). Survey of fumonisins B1, B2 and B3 in conventional and organic retail corn products in Spain and Italy and estimated dietary exposure. *Food Additives and Contaminants, 2*(2), 146-153.

Deepa, N., & Sreenivasa, M. Y. (2017). Fumonisins: A review on its global occurrence, epidemiology, toxicity and detection. *Journal of Veterinary Medicine and Research, 4*(6).

Deepa, N., & Sreenivasa, M. Y. (2019a). Molecular methods and key genes targeted for the detection of fumonisin producing Fusarium verticillioides–An updated review. *Food Bioscience, 32*, 100473.

Deepa, N., & Sreenivasa, M. Y. (2019b). Sustainable approaches for biological control of mycotoxigenic fungi and mycotoxins in cereals. In *New and Future Developments in Microbial Biotechnology and Bioengineering* (pp. 149-161). Elsevier.

Deepa, N., Adkar-Purushothama, C. R., & Sreenivasa, M. Y. (2021). Molecular technologies for the early detection of fungal phytopathogens associated with cereal crops. In *Microbes in Land Use Change Management* (pp. 137-154). Elsevier.

Dieye, C. A. T., Durand, N., Schorr-Galindo, S., Strub, C., & Fontana, A. (2024). Impacts of abiotic factors on the growth of three commercial biological control agents, on the growth and mycotoxinogenesis of Fusarium graminearum and on their interaction. *Journal of the Science of Food and Agriculture, 104*(2), 932-941.

Dong, H., Xian, Y., Xiao, K., Wu, Y., Zhu, L., and He, J. (2019). Development and comparison of single-step solid phase extraction and QuEChERS clean-up for the analysis of 7 mycotoxins in fruits and vegetables during storage by UHPLC-MS/MS. Food Chem. 274, 471–479.

European Union Commission. (2006). Commission Regulation (EC) No. 1881/2006 of 19th December 2006. *Setting maximum levels for certain contaminants in foodstuffs. Off J Eur Union L, 364*, 15-24.

Farhadi, A., Nowrozi, H., & Kachuei, R. (2019). Metabolism, toxicity, detoxification, occurrence, intake and legislations of fumonisins-a review. *Journal of Pharmaceutical Research International, 29*(6), 35.

Foulongne-Oriol, M., Garcia, M. A., Mahmoud, R., Bancal, M. O., Bancal, P., Bernillon, S., ... & Richard-Forget, F. (2024). Ecophysiological Behaviour Of Major Fusarium Species In Response To Combinations Of Temperature And Water Activity Constraints. *bioRxiv*, 2024-09.

George, M. M., Nisha, K., Lekhana, S. M., & Gurikar, C. (2022). Patulin: a potentially harmful food contaminant. *Int. J. Chem. Stud, 10*, 11-18.

Gonçalez, E., Nogueira, J. H. C., Fonseca, H., Felicio, J. D., Pino, F. A., and Correa, B. (2008). Mycobiota and mycotoxins in Brazilian peanut kernels from sowing to harvest. Int. J. Food Microbiol. 123, 184–190.

Grozeva, N., Valchev, I., Binev, R., Lazarov, L., Hristov, T., & Kanakov, D. (2017). Pathomorphological changes in the spleen of turkey broilers challenged with Aflatoxin B1 alone or co-administered with Mycotox NG. *Int. J. Vet. Sci. Technol, 1*, 1-6.

Gul, A., Jahangeer, M., Murtaza, B., Ahmed, J., Nadeem, A. A., Yongping, X., ... & Mehmood, M. (2023). Different molds and their toxins: as contaminants in one health perspective. *One Health Triad, Unique Scientific Publishers, Faisalabad, Pakistan, 2*, 231-241.

Gupta, P. K., & Gupta, P. K. (2019). Mycotoxicoses. *Concepts and Applications in Veterinary Toxicology: An Interactive Guide*, 203-225.

Gurikar, C., Shivaprasad, D. P., Sabillón, L., Gowda, N. N., & Siliveru, K. (2023). Impact of mycotoxins and their metabolites associated with food grains. *Grain & Oil Science and Technology, 6*(1), 1-9.

Hou, S., Ma, J., Cheng, Y., Wang, H., Sun, J., & Yan, Y. (2023). The toxicity mechanisms of DON to humans and animals and potential biological treatment strategies. *Critical Reviews in Food Science and Nutrition, 63*(6), 790-812.

Imran, M., Cao, S., Wan, S. F., Chen, Z., Saleemi, M. K., Wang, N., ... & Munawar, J. (2020). Mycotoxins–a global one health concern: A review. *Agrobiological records, 2*, 1-16.

Ingle, A. P., Gupta, I., Jogee, P., & Rai, M. (2020). Role of nanotechnology in the detection of mycotoxins: A smart approach. In *Nanomycotoxicology* (pp. 11-33). Academic Press.

Iqbal, S. Z., Malik, S., Asi, M. R., Selamat, J., and Malik, N. (2018). Natural occurrence of patulin in different fruits, juices and smoothies and evaluation of dietary intake in Punjab, Pakistan. Food Control 84, 370–374.

Janik, E., Niemcewicz, M., Podogrocki, M., Ceremuga, M., Stela, M., & Bijak, M. (2021). T-2 toxin—The most toxic trichothecene mycotoxin: Metabolism, toxicity, and decontamination strategies. *Molecules, 26*(22), 6868.

Jimenez, M., and Mateo, R. (1997). Determination of mycotoxins produced by Fusarium isolates from banana fruits by capillary gas chromatography and high performance liquid chromatography. J. Chromatogr. A 778, 363–372.

Jimenez-Garcia, S. N., Garcia-Mier, L., Garcia-Trejo, J. F., Ramirez-Gomez, X. S., Guevara-Gonzalez, R. G., & Feregrino-Perez, A. A. (2018). Fusarium mycotoxins and metabolites that modulate their production. *Fusarium—Plant diseases, pathogen diversity, genetic diversity, resistance and molecular markers.* Intech Open

Juraschek, L. M., Kappenberg, A., & Amelung, W. (2022). Mycotoxins in soil and environment. *Science of the Total Environment, 814*, 152425.

Kang, B., Zhang, X., Wu, Z., Wang, Z., & Park, S. (2014). Production of citrinin-free Monascus pigments by submerged culture at low pH. *Enzyme and microbial technology*, *55*, 50-57.

Khodaei, D., Javanmardi, F., & Khaneghah, A. M. (2021). The global overview of the occurrence of mycotoxins in cereals: A three-year survey. *Current Opinion in Food Science*, *39*, 36-42.

Kluczkovski, A. M. (2019). Fungal and mycotoxin problems in the nut industry. *Current Opinion in Food Science*, *29*, 56-63.

Leannec-Rialland, V. (2022). *Efficacy and mechanisms of action of tick defensins against the phytopathogenic and toxinogenic fungus Fusarium graminearum* (Doctoral dissertation, Université de Bordeaux, France).

Lenzi, A., Marvasi, M., & Baldi, A. (2021). Agronomic practices to limit pre- and post-harvest contamination and proliferation of human pathogenic Enterobacteriaceae in vegetable produce. *Food Control*, *119*, 107486.

Liu, J., & Applegate, T. (2020). Zearalenone (ZEN) in livestock and poultry: dose, toxicokinetics, toxicity and estrogenicity. *Toxins*, *12*(6), 377.

Liu, M., Zhang, X., Luan, H., Zhang, Y., Xu, W., Feng, W., & Song, P. (2024). Bioenzymatic detoxification of mycotoxins. *Frontiers in Microbiology*, *15*, 1434987.

Liu, Z., Gao, J., and Yu, J. (2006). Aflatoxins in stored maize and rice grains in Liaoning province, China. J. Stored Prod. Res. 42, 468–479.

Mahaki, H., Ravari, H., Kazemzadeh, G., Lotfian, E., Daddost, R. A., Avan, A., ... & Tanzadehpanah, H. (2024). Pro-inflammatory responses after peptide-based cancer immunotherapy. *Heliyon*, *10*(11).

Mahboubifar, M., Karami-Osboo, R., & Jassbi, A. R. (2024). Essential Oils and Their Eco-friendly Encapsulated Products for Mycotoxins Reduction: Current Situation and Future Prospects. *Food and Bioprocess Technology*, 1-33.

Marrez, D. A., & Ayesh, A. M. (2022). Mycotoxins: The threat to food safety. *Egyptian Journal of Chemistry*, *65*(1), 353-372.

Medina, A., Akbar, A., Baazeem, A., Rodriguez, A., & Magan, N. (2017). Climate change, food security and mycotoxins: Do we know enough?. *Fungal biology reviews, 31*(3), 143-154.

Molnár, K., Rácz, C., Dövényi-Nagy, T., Bakó, K., Pusztahelyi, T., Kovács, S., ... & Dobos, A. (2023). The effect of environmental factors on mould counts and AFB1 toxin production by Aspergillus flavus in maize. *Toxins, 15*(3), 227.

Moretti, A., Logrieco, A. F., & Susca, A. (2017). Mycotoxins: An underhand food problem. *Mycotoxigenic Fungi: Methods and Protocols*, 3-12.

Müller, S., Dekant, W., & Mally, A. (2012). Fumonisin B1 and the kidney: Modes of action for renal tumor formation by fumonisin B1 in rodents. *Food and chemical toxicology*, *50*(10), 3833-3846.

Mutlu, N., & Yenice, G. (2024). The relationship between aflatoxin M1 and immunoglobulin levels in cows' colostrum. *Mycotoxin Research*, 1-9.

Nagaraja, H., Chennappa, G., Deepa, N., Naik, M. K., Ajithkumar, K., Amaresh, Y. S. and Sreenivasa, M. Y. (2022). Antifungal potential of Azotobacter salinestris strain Azt 31 against phytopathogenic Fusarium spp. associated with cereals. *Journal of Fungi*, *8*(5), 473.

Nagaraja, H., Chennappa, G., Rakesh, S., Naik, M. K., Amaresh, Y. S., & Sreenivasa, M. Y. (2016). Antifungal activity of Azotobacter nigricans against trichothecene-producing Fusarium species associated with cereals. *Food science and biotechnology*, *25*, 1197-1204.

Nakagawa, H., Ohmichi, K., Sakamoto, S., Sago, Y., Kushiro, M., Nagashima, H., et al. (2011). Detection of a new Fusarium masked mycotoxin in wheat grain by high resolution LC–Orbitrap™ MS. Food Addit. Contam: Part A 28, 1447–1456.

Nazareth, T. D. M., Soriano Pérez, E., Luz, C., Meca, G., & Quiles, J. M. (2024). Comprehensive Review of Aflatoxin and Ochratoxin A Dynamics: Emergence, Toxicological Impact, and Advanced Control Strategies. *Foods*, *13*(12), 1920.

Oteiza, J. M., Khaneghah, A. M., Campagnollo, F. B., Granato, D., Mahmoudi, M. R., Sant'Ana, A. S., et al. (2017). Influence of production on the presence of patulin and ochratoxin A in fruit juices and wines of Argentina. LWT Food Sci. Technol. 80, 200–207.

Pickova, D., Ostry, V., Malir, J., Toman, J., & Malir, F. (2020). A review on mycotoxins and microfungi in spices in the light of the last five years. *Toxins*, *12*(12), 789.

Roach, C. M. (2023). *Characterizing the ovarian response to xenobiotic and environmentally induced metabolic challenges* (Doctoral dissertation, Iowa State University, USA).

Rodríguez-Melcón, C., Alonso-Calleja, C., & Capita, R. (2024). The One Health approach in food safety: Challenges and opportunities. *Food Frontiers.*

Sadariya, K. A., Bhavsar, S. K., & Solanki, T. H. (2023). Poisoning in Goats. *Principles of Goat Disease and Prevention*, 165-181.

Sankeerthana, C., Pinjar, S., Jambagi, R., Bhavimani, S., Anupama, S., Sarovar, B., & Inamdar, S. R. (2013). Production and partial characterization of protease from Aspergillus flavus using rice mill waste as a substrate and its comparision with Aspergillus niger protease. *International Journal of Current Engineering and Technology, 1*, 143-147.

Sertova, N., & Ignatova, M. (2023). Fumonisins: origin and impact on animal and human health. *Journal of Mountain Agriculture on the Balkans*, 26(1), 97-118.

Singh, G. (2015). Agriculture diversification for food, nutrition, livelihood and environmental security: Challenges and opportunities. *Indian Journal of Agronomy*, *60*(2), 172-184.

Spevakova, I., Fernandez-Cruz, M. L., Tokarova, K., Greifova, H., & Capcarova, M. (2020). The protective effect of stilbenes resveratrol and pterostilbene individually and combined with mycotoxin citrinin in human adenocarcinoma HT-29 cell line in vitro. *Journal of Environmental Science and Health, Part A*, *56*(1), 75-88.

Summerell, B. A., & Leslie, J. F. (2011). Fifty years of Fusarium: how could nine species have ever been enough?. *Fungal Diversity, 50*, 135-144.

Tahir, M. A., Abbas, A., Muneeb, M., Bilal, R. M., Hussain, K., Abdel-Moneim, A. M. E., ... & Alagawany, M. (2022). Ochratoxicosis in poultry: Occurrence, environmental factors, pathological alterations and amelioration strategies. *World's Poultry Science Journal,* 78(3), 727-749.

Theresa, M., Unni, A. S., Geevarghese, A., Sebastian, S. K., Pareek, S., & Krishnankutty, R. E. (2021). Foodborne Pathogens and Food-Related Microorganisms. In *Sequencing Technologies in Microbial Food Safety and Quality* (pp. 25-108). CRC Press.

Topi, D., Babic, J., Pavsic-Vrtac, K., Tavcar-Kalcher, G., and Jakovac-Strajn, B. (2021). Incidence of Fusarium mycotoxins in wheat and maize from Albania. Molecules 26:172.

Tripathi, A., & Alam, A. (2020). Mycotoxins, Mycotoxicosis and Managing Mycotoxin Contamination: A Review. *Bio-management of postharvest diseases and mycotoxigenic fungi*, 161-180.

Weinstock, L. B., Campbell, A. W., Curtis, L., & Gutovich, J. (2024). Mycotoxin-Triggered Attacks of Nausea, Vomiting, and Abdominal Pain and Episodes of Pseudo-Obstruction. *Universal Library of Medical and Health Sciences, 2*(1).

Więckowska, M., Cichon, N., Szelenberger, R., Gorniak, L., & Bijak, M. (2024). Ochratoxin A and Its Role in Cancer Development: A Comprehensive Review. *Cancers, 16*(20), 3473.

Wu, F., Groopman, J. D., & Pestka, J. J. (2014). Public health impacts of foodborne mycotoxins. *Annual review of food science and technology, 5*(1), 351-372.

Yadav, P., Debnath, N., Arora, S., & Yadav, A. K. (2023). Occurrence, Detection, and Management of T-2 Toxin and HT-2 Toxin in Food and Feed. In *Mycotoxins in Food and Feed* (pp. 157-187). CRC Press.

Yang, L., Yang, L., Cai, Y., Luo, Y., Wang, H., Wang, L., & Liu, N. (2023). Natural mycotoxin contamination in dog food: a review on toxicity and detoxification methods. *Ecotoxicology and Environmental Safety*, *257*, 114948.

Yu, J., & Pedroso, I. R. (2023). Mycotoxins in cereal-based products and their impacts on the health of humans, livestock animals and pets. *Toxins*, *15*(8), 480.

Zhang, J., Liu, X., Su, Y., & Li, T. (2022). An update on T2-toxins: metabolism, immunotoxicity mechanism and human assessment exposure of intestinal microbiota. *Heliyon*, *8*(8).

Zhou, J., Zhao, L., Huang, S., Liu, Q., Ao, X., Lei, Y., ... & Ma, Q. (2022). Zearalenone toxicosis on reproduction as estrogen receptor selective modulator and alleviation of zearalenone biodegradative agent in pregnant sows. *Journal of Animal Science and Biotechnology*, *13*(1), 36.

Zhu, M., Duan, X., Cai, P., Qiu, Z., & Li, Z. (2023). Genome sequence resource of Trichothecium roseum (ZM-Tr2021), the causal agent of postharvest pink rot. *Plant Disease*, *107*(1), 205-209.

Chapter 2

Mycotoxin Biosynthesis and Factors Affecting its Production in Post-Harvest Stages

M. Chaithra[1*], *Shreedevasena S.*[2], *Akshay Kumar H. M.*[3] *and Gangaraj R.*[4]

[1]*ICAR-Central Plantation Crop Research Institute, Research Center, Kahikuchi, Guwahati, Assam, India-781017*
[2]*ICAR-Directorate of Medical and Aromatic Plant Research, Anand, Gujarat, India- 387310*
[3]*ICAR-Indian Agricultural Research Institute, Assam, India-734301*
[4]*ICAR-Indian Institute of Vegetable Research, Regional Research Station, Sargatia, Kushinagar, Uttar Pradesh, India-274406*

**Corresponding author: M. Chaithra; chaithram06@gmail.com*

Abstract

Mycotoxins are toxic compounds produced by certain fungi and represent a major concern for food safety, particularly during the post-harvest stages of crop production. Understanding the toxicity of four key agriculturally important mycotoxins *i.e.*, trichothecenes, aflatoxins, fumonisins, and sterigmatocystin is crucial, along with their biosynthesis pathways and regulatory mechanisms. Antimicrobial proteins and peptides (AMPs) with antifungal activity present a promising approach, as they can inhibit mycotoxin production at low concentrations by targeting regulatory mechanisms. Similarly, numerous other factors affect mycotoxin production during post-harvest handling, including environmental conditions such as temperature and humidity, as well as storage practices and mechanical damage to crops. Advances in genomics, transcriptomics, and genome-editing technologies have provided powerful tools for fully elucidating the regulatory and biosynthesis pathways of these mycotoxins. This chapter explains the biosynthesis of mycotoxins, focusing on the primary fungal species responsible and the biochemical pathways involved. By providing a detailed understanding of the factors influencing mycotoxin biosynthesis in the post-harvest stage, this chapter aims to inform effective interventions to safeguard food quality and public health.

Keywords: Aflatoxins; Fusarium toxins; Mycotoxin pathways; Ochratoxins, Post-harvest mycotoxins

Introduction

Mycotoxins are toxic secondary metabolites produced by filamentous fungi, primarily from the genera *Aspergillus, Penicillium*, and *Fusarium*. These compounds pose significant health risks to humans and animals by contaminating various agricultural products, particularly during the post-harvest stage. The biosynthesis of mycotoxins is a complex process regulated by various genetic, environmental, and nutritional factors, and it is often triggered under stress conditions that fungi encounter during storage. In the post-harvest environment, factors such as moisture content, temperature, oxygen availability, and substrate composition significantly influence mycotoxin production. For example, high moisture levels and warm temperatures create ideal conditions for fungal growth and mycotoxin biosynthesis. Poor storage practices, such as inadequate drying, improper packaging, and lack of ventilation, exacerbate these conditions, leading to increased mycotoxin contamination. Moreover, interactions among different fungal species and the availability of specific nutrients in stored products further modulate mycotoxin production levels.

Recent studies underscore the importance of controlling post-harvest conditions to minimize mycotoxin contamination. Advances in understanding the molecular pathways involved in mycotoxin biosynthesis have opened new avenues for developing targeted interventions. For instance, research has shown that managing storage environments by controlling temperature and humidity can significantly reduce the risk of mycotoxin accumulation. Additionally, antifungal treatments and bio-control agents are being explored as strategies to inhibit fungal growth and mycotoxin production during storage. For example, Palencia *et al.* (2010) demonstrated that maintaining low moisture levels in stored maize effectively reduces the production of aflatoxins, one of the most potent mycotoxins produced by *Aspergillus* species. Similarly, Milicevic *et al.* (2020) highlighted the critical role of environmental factors in regulating mycotoxin biosynthesis, emphasizing the need for stringent control of storage conditions to prevent fungal contamination and mycotoxin production.

Types Of Mycotoxins And Their Producers

The most significant mycotoxins include aflatoxins, ochratoxins, fusarium toxins, and other notable compounds. Below is an overview of these mycotoxins, their primary fungal producers, and their associated impacts, supported by recent studies.

Aflatoxins: Aflatoxins are among the most dangerous and well-studied mycotoxins, primarily produced by *Aspergillus flavus* and *A. parasiticus*. These fungi frequently contaminate crops such as maize (Mutegi *et al.*, 2024), peanuts, cottonseed, and tree nuts. Aflatoxin B1 is highly carcinogenic and has been classified as a Group 1 carcinogen by the International Agency for Research on Cancer (IARC). Aflatoxins are known to suppress the immune system and are strongly linked to liver cancer in humans and animals (Wild *et al.*, 2007). Recent research emphasizes the importance of factors such as moisture and temperature during post-harvest storage in influencing aflatoxin production (Klich, 2007).

Ochratoxins: Ochratoxins, particularly Ochratoxin A, are produced by *A. ochraceus* and *Penicillium verrucosum*. These toxins are commonly found in cereals, coffee, dried fruits, and wine. Ochratoxin A has been classified as a possible human carcinogen (Group 2B) by the IARC. During food processing, ochratoxin A is particularly difficult to eliminate from contaminated products, presenting a significant challenge for food safety. Advances in controlling ochratoxin contamination, such as improved storage conditions and the use of bio-control agents, have been the focus of recent studies (Magan and Aldred, 2008).

***Fusarium* Toxins:** *Fusarium* toxins are produced by various *Fusarium* species, with the most common being Trichothecenes (e.g., Deoxynivalenol (DON) and Fumonisins) (Miller, 2005). These toxins frequently contaminate cereals like wheat, maize, and barley. Trichothecenes such as DON cause immunosuppression and gastrointestinal disturbances, while zearalenone is an estrogenic mycotoxin that disrupts reproductive functions. Fumonisins, particularly fumonisin B1, have been linked to esophageal cancer and neural tube defects (Marasas *et al.*, 2008). Recent research continues to explore how environmental conditions during post-harvest stages influence the production of *Fusarium* toxins, necessitating effective storage management.

Other Notable Mycotoxins: In addition to the aforementioned mycotoxins, other significant toxins include patulin, citrinin, and ergot alkaloids. Patulin, produced by *Penicillium expansum*, is commonly found in rotting apple products and is known for its neurotoxic and genotoxic effects. Citrinin, produced by *Penicillium* and *Aspergillus* species, primarily contaminates cereals and exhibits nephrotoxic properties. Ergot alkaloids, produced by *Claviceps purpurea*, infect rye and other cereals, leading to ergotism, which is characterized by severe vasoconstriction and gangrene. The diversity of mycotoxins and their widespread occurrence in the food supply underscore the importance of monitoring and effective control measures. **Table 1** lists various types of mycotoxins and their producers.

Table 1: list of various types of mycotoxins and their primary fungal producers.

Mycotoxins type	Primary Producers	References
Aflatoxins	*Aspergillus flavus, Aspergillus parasiticus*	Palencia *et al.*, 2010; Klich, 2007; Mutegi *et al.*, 2024
Ochratoxins	*Aspergillus ochraceus, Penicillium verrucosum*	Magan and Aldred, 2008;
Deoxynivalenol (DON)	*Fusarium graminearum, Fusarium culmorum*	Marasas *et al.*, 2008; Chen *et al.*, 2019
Fumonisins	*Fusarium verticillioides, Fusarium proliferatum*	Marasas *et al.*, 2008; Chen *et al.*, 2019
Zearalenone	*Fusarium graminearum, Fusarium culmorum*	Chen et al., 2019; Palencia *et al.*, 2010
T-2 Toxin	*Fusarium sporotrichioides, Fusarium poae*	Marasas *et al.*, 2008; Chen *et al.*, 2019
Patulin	*Penicillium expansum, Aspergillus clavatus*	Brase *et al.*, 2013; Marasas *et al.*, 2008
Citrinin	*Penicillium citrinum, Aspergillus terreus*	Frisvad *et al.*, 2019; Ahmed *et al.*, 2015
Ergot Alkaloids	*Claviceps purpurea, Claviceps africana*	Frisvad *et al.*, 2019; Marasas *et al.*, 2008
Sterigmatocystin	*Aspergillus versicolor, Aspergillus nidulans*	Frisvad *et al.*, 2019; Ahmed *et al.*, 2015
Alternariol	*Alternaria alternata*	Frisvad *et al.*, 2019; Marasas *et al.*, 2008
Moniliformin	*Fusarium verticillioides, Fusarium proliferatum*	Marasas *et al.*, 2008; Brase *et al.*, 2013
Enniatins	*Fusarium avenaceum, Fusarium tricinctum*	Jansen *et al.*, 2018; Marasas *et al.*, 2008
Cyclopiazonic Acid	*Aspergillus flavus, Penicillium griseofulvum*	Palencia *et al.*, 2010; Lappa *et al.*, 2017
Roquefortine C	*Penicillium roqueforti, Penicillium crustosum*	Lappa *et al.*, 2017; Frisvad *et al.*, 2019
Penitrem A	*Penicillium crustosum*	Palencia *et al.*, 2010; Ahmed *et al.*, 2015
Penicillic Acid	*Penicillium puberulum, Aspergillus ochraceus*	Brase *et al.*, 2013; Frisvad *et al.*, 2019
Brevianamide A	*Penicillium brevicompactum*	Frisvad *et al.*, 2019; Jansen *et al.*, 2018

Biosynthetic Pathways Of Major Mycotoxins

Mycotoxins, which are secondary metabolites produced by filamentous fungi, present significant health risks to both humans and animals. To develop effective control strategies, it is crucial to understand the biosynthetic pathways, enzymatic steps, genetic regulation, and key biosynthetic genes involved in mycotoxin production (Magan *et al.*, 2008).

Aflatoxins, produced by Aspergillus species such as *Aspergillus flavus* and *Aspergillus parasiticus*, are synthesized through a polyketide pathway beginning from acetate units. Key enzymes in this pathway include polyketide synthase (PKS), norsolorinic acid reductase, and versicolorin B synthase (Ayofemi, 2020; Kolawole *et al.*, 2021). The gene cluster responsible for aflatoxin biosynthesis includes over 25 genes, among them aflR, a transcriptional activator that regulates the expression of other genes in the cluster. Genetic regulation of aflatoxin biosynthesis involves the transcriptional activator AflR, encoded by the aflR gene, which binds to the promoter regions of cluster genes to activate their transcription. AflS (formerly known as AflJ) is a co-activator that enhances AflR ability to transcribe biosynthetic genes (Figure1).

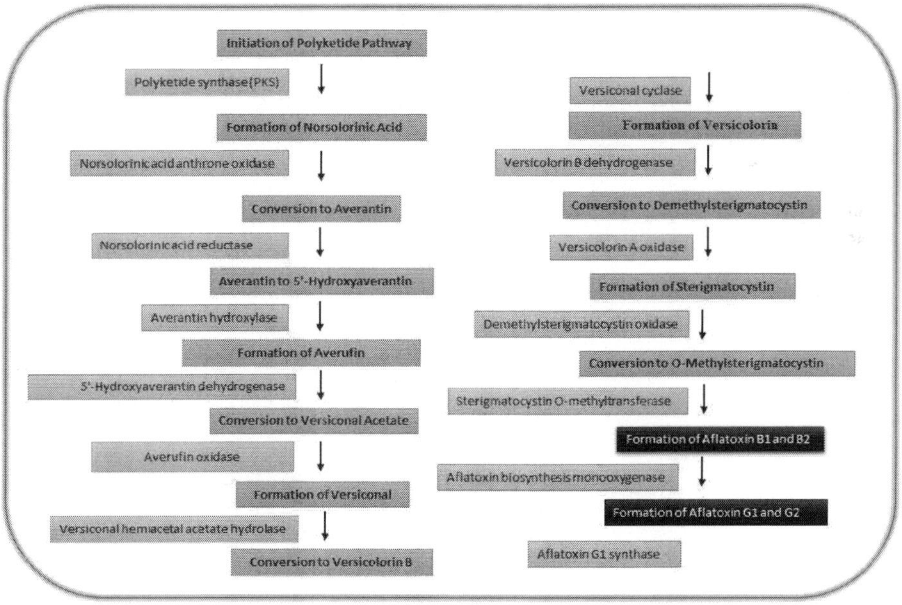

Figure 1: Graphical representation of the biosynthesis pathway of Aflatoxin

Trichothecenes, notably deoxynivalenol (DON), are produced by *Fusarium* species. The trichothecene biosynthetic pathway begins with the cyclization of farnesyl pyrophosphate (FPP). Key enzymes include trichodiene synthase (Tri5), trichothecene C-3 hydroxylase (Tri11), and acetyltransferase (Tri101) (Brown *et al.*, 2021). The trichothecene biosynthetic gene cluster contains TRI genes, with Tri6 and Tri10 being essential for pathway regulation (Alexander *et al.*, 2009) (Figure 2). Tri6 acts as a pathway-specific transcription factor, while Tri10 regulates Tri6 and other genes within the cluster (Brown *et al.*, 2021).

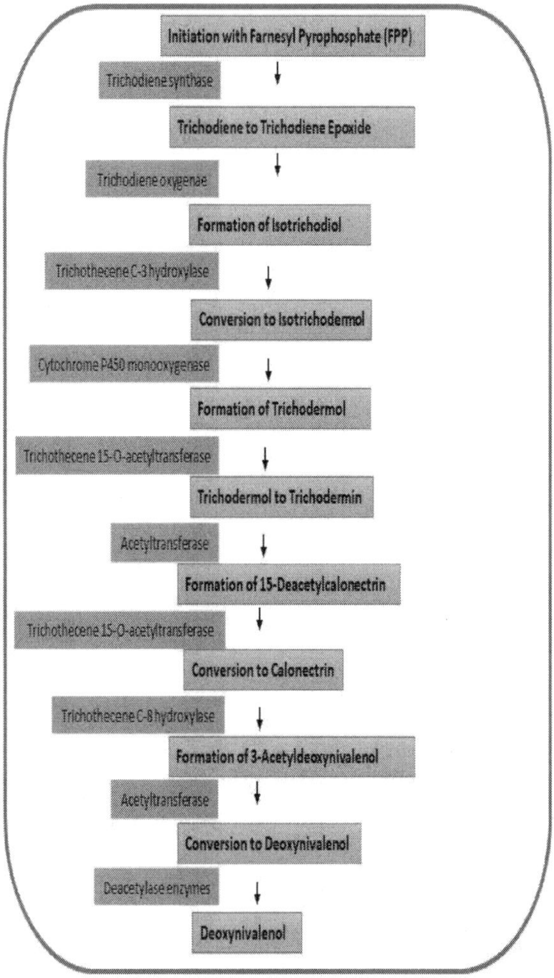

Figure 2: Graphical representation of the biosynthesis pathway of Deoxynivalenol

Fumonisin, produced by *Fusarium verticillioides*, are synthesized through the condensation of acetate units, involving key enzymes such as polyketide synthase (Fum1p) and non-ribosomal peptide synthetase (NRPS). The Fumonisin biosynthetic gene cluster comprises 15 genes, with FUM1, FUM6, and FUM8 playing critical roles in the pathway (Ferrara *et al.*, 2022). Ochratoxins, particularly ochratoxin A (OTA), are produced by *Aspergillus and Penicillium* species. The biosynthesis of OTA involves a polyketide pathway, with key enzymes including polyketide synthase and non-ribosomal peptide synthetase (Wang, *et al.*, 2016).

The OTA biosynthetic gene cluster includes otaA, otaB, and otaC, which are involved in forming the ochratoxin molecule. Zearalenone, produced by *Fusarium* species, is synthesized through the cyclization of a polyketide precursor. Key enzymes include polyketide synthase (ZEB1) and a cytochrome P450 monooxygenase (ZEB2). The zearalenone biosynthetic gene cluster includes several genes, with ZEB1 and ZEB2 being essential for toxin production (Gonzalez *et al.*, 2021).

The regulation of mycotoxin biosynthesis is complex, involving multiple genetic and environmental factor. Transcription factors such as AflR for aflatoxins, Tri6 for trichothecenes, and FUM21 for Fumonisin are critical in regulating gene expression within their respective clusters. Environmental factors, including temperature, pH, and nutrient availability, significantly impact mycotoxin production. Understanding the detailed mechanisms of biosynthetic pathways, enzymatic steps, and genetic regulation of major mycotoxins is essential for developing strategies to mitigate their production and contamination in food and feed. Future research should focus on elucidating the intricate details of genetic regulation and exploring novel approaches for controlling mycotoxin contamination.

Environmental and Biological Factors Influencing Mycotoxin Production

The production of mycotoxins is influenced by environmental conditions, and even optimal growth and storage environments cannot entirely prevent their formation (Qu *et al.*, 2022). Mycotoxin contamination can detrimentally affect the quality and processing performance of agricultural commodities like wheat. It is estimated that up to 25% of the world's food and feed crops may be contaminated with mycotoxins (Cheli *et al.*, 2017). One consequence of this contamination is reduced agricultural yield, leading to food shortages (Sokefun *et al.*, 2018).

The growth of mycotoxin-producing species is closely related to grain moisture levels, which enable biological activity in the grain ecosystem (Fleurat-Lessard, 2017). As a result, mould growth in stored grain can be anticipated through early detection of grain and mould respiration. Preventing the spoilage of stored grain by mycotoxigenic fungi requires a preventive strategy that includes a combination of methods, materials, and practices (Fleurat-Lessard, 2017). The latest research in cereal contamination focuses on using microbial methods to inhibit toxin-producing moulds and detoxify mycotoxins. Microorganisms such as yeasts, lactic acid bacteria, bacilli, non-toxic moulds, and marine microorganisms play a crucial role in preventing and controlling toxic mould and mycotoxin production in cereals (Zhang *et al.*, 2023). Additionally, emerging decontamination technologies are gaining attention as consumers demand high-quality and safe food products (Mir *et al.*, 2021). However, some conventional methods fall short of industry expectations due to limitations such as safety concerns, inefficiency, and the destruction of food quality attributes (Mir *et al.*, 2021).

1. Environmental Factors

(a) Temperature: Temperature plays a crucial role in fungal growth and mycotoxin production. Each species has an optimal temperature range, which varies depending on the specific mycotoxin (Paterson and Lima, 2017; Mannaa and Kim, 2017). Generally, warmer temperatures within the 25-30°C range favour the growth of many mycotoxigenic fungi and increase mycotoxin production. For instance, *A. flavus*, which produces aflatoxins, thrives at temperatures between 25-35°C, while Fusarium graminearum, a producer of deoxynivalenol (DON), prefers temperatures between 15-30°C, with the optimal temperature for DON production averaging approximately 25°C (Hay *et al.*, 2021). However, extreme temperatures can inhibit fungal growth and toxin synthesis.

(b) Moisture Content: Moisture content, often expressed as water activity (aw), is another critical factor influencing fungal growth and mycotoxin production. Most mycotoxigenic fungi thrive in environments with high water activity (aw > 0.8). While drought stress can limit fungal growth, some species, like *Aspergillus flavus* (aflatoxin producer), are more tolerant of low aw conditions (Zhang *et al.*, 2023).

(c) Relative Humidity: The surrounding air's relative humidity (RH) significantly impacts the substrate's moisture content. High RH levels (above 70%) create favourable conditions for fungal growth and mycotoxin production. Conversely, low RH can inhibit fungal development, although some species can tolerate drier conditions (Paterson and Lima, 2017).

(d) pH: The pH of the substrate also influences fungal growth and mycotoxin production. Most mycotoxigenic fungi prefer slightly acidic to neutral pH conditions (pH 4-7). However, some species, like *Penicillium verrucosum* (ochratoxin A producer), can tolerate a wider pH range (Awuchi *et al.*, 2021).

(e) Oxygen Availability: As aerobic organisms, fungi require oxygen for growth and mycotoxin production. Adequate aeration is essential to prevent the accumulation of carbon dioxide, which can inhibit fungal growth. However, some fungi can tolerate low oxygen levels, and certain mycotoxins, like fumonisins, are produced under limited oxygen availability (Milani, 2013).

(f) Nutrient Availability: The availability of nutrients such as carbon, nitrogen, and trace elements significantly influences fungal growth and mycotoxin production. Different fungal species have varying nutrient requirements, and imbalances in nutrient availability can stimulate mycotoxin production in some cases. For example, *A. flavus*, a well-known aflatoxin producer, produces higher levels of these toxins when there is an imbalance in the carbon-to-nitrogen ratio in its growth substrate, similar to how *Alternaria* toxins (e.g., alternariol) are produced in cereals (Daou *et al.*, 2021; Brzonkalik *et al.*, 2012).

(g) Insect and Pest Damage: Insect and pest infestations, such as those caused by corn borers, aphids, or weevils, can damage crops, creating wounds that serve as entry points for fungi. Additionally, insects can act as vectors, carrying fungal spores and facilitating their dispersal within the field or storage facilities, like flies or grain beetles transporting *Aspergillus* spores (Ochsenreither, 2011).

(h) Climate Change: Climate change, associated shifts in temperature, precipitation patterns, and extreme weather events, is expected to significantly impact mycotoxin contamination. Warmer temperatures and increased humidity in some regions may favor certain fungi's growth and toxin production, such as *A. flavus* (aflatoxins) and *Fusarium graminearum* (deoxynivalenol). Additionally, extreme weather events like droughts and floods can stress crops, making them more susceptible to fungal infection, as seen with the increased risk of *Aspergillus* and *Fusarium* infestations in maize and wheat during drought and flood conditions, respectively (Paterson and Lima, 2017; Milani, 2013).

Biological Factors

(a) Fungal Species and Strain Variability: Different fungal species, and even strains within a species, can produce varying types and amounts of mycotoxins. For example, *A. flavus* produces aflatoxins (Klich, 2007), while *F. graminearum* produces deoxynivalenol and zearalenone. Genetic variability within fungal populations contributes to variations in mycotoxin production (Saharan *et al.*, 2021).

(b) Microbial Interactions: Interactions with other microorganisms in the environment can influence mycotoxin production. Some bacteria and fungi can inhibit the growth of mycotoxigenic fungi or degrade mycotoxins, while others may stimulate their production. Furthermore, the production of mycotoxins can be affected by the complex interactions between fungi and other microorganisms present in the same ecological niche (Ismaiel and Papenbrock, 2015). The presence of certain non-toxic *Aspergillus* species can influence the production of aflatoxins by *A. flavus* (Reddy *et al.*, 2009).

(c) Plant Host Factors: The plant species, variety, and overall health status can influence susceptibility to fungal infection and mycotoxin contamination. For example, certain maize and peanut varieties exhibit natural resistance to *A. flavus*, the fungus that produces aflatoxins (Bhatnagar *et al.*, 2015). Additionally, stressed or damaged plants, such as those affected by drought or insect pests, are more vulnerable to fungal infection and mycotoxin contamination.

Mycotoxin production is a complex process influenced by numerous interacting environmental and biological factors. Understanding these factors is crucial for developing effective strategies to prevent mycotoxin contamination and ensure food safety. Integrated approaches that combine good agricultural practices, proper storage conditions, resistant crop varieties, and biological control methods are essential to minimizing the risks associated with mycotoxins in the food and feed chain.

Post-Harvest Conditions And Mycotoxin Contamination

Mycotoxins are toxic compounds generated by specific fungi that can contaminate grains. The main mycotoxins present in grains are aflatoxin, fumonisin, deoxynivalenol (DON), ochratoxin (OT), and zearalenone (ZEN). These toxins can develop at different stages: DON, ZEN, and fumonisins might form in the field or during pre-harvest, while aflatoxin and OT typically produce during storage due to improper postharvest practices. The crops such as maize, peanuts, sorghum, millet, wheat, and rice are commonly affected (Neme and Mohammed, 2017). Mycotoxin-related postharvest losses are a growing global issue, worldwide nearly 25% of food crops are contaminated by mycotoxins, resulting in an annual loss of nearly 1 billion tons of food.

Harvesting practices to reduce mycotoxin contamination

Fungi can invade, colonize, and produce mycotoxins during pre and post-harvest stages. The key to managing mycotoxin contamination effectively lies in prevention, which involves minimizing fungal growth in both the field and during storage (Hamad *et al.* 2023). Generally, mainly *Fusarium* species, impact during pre-harvest stage only but *Aspergillus* and *Penicillium* species infection take place after the harvest. Therefore, adopting sound agricultural and harvesting practices is essential for reducing mycotoxin contamination in grains, fodder, and other crops.

Precautions to Follow During Harvesting: Key factors contributing to mycotoxin contamination include harvest timing, mechanical damage, insect infestations, drying methods, storage conditions, and processing practices. The following sections provide a detailed discussion of these factors.

(a) Optimal harvest timing: To minimize the risk of fungal infection, harvest crops at the optimal stage of maturity. Crops that are either overripe or underdeveloped are more prone to fungal attacks. Ensure that harvesting occurs when crops are dry, with low moisture or water activity (Aw) levels. Delays in harvesting, especially during rainy periods, can lead to higher mycotoxin levels (such as deoxynivalenol, zearalenone, and T-2 toxin) and deteriorate maize quality (Kochiieru *et al.*, 2021). High humidity (>85%), elevated temperatures (25°C), and the presence of insects or rodents can promote the growth of *A. flavus* and *A. parasiticus*, which are responsible for aflatoxin production in maize. These conditions also encourage the proliferation of pests like *Mussidia nigrivenella*, *Sitophilus zeamais*, and *Carpophilus dimidiatus*, which further spread *Aspergillus* spp. Early harvesting and proper drying are essential to reducing mycotoxin producing fungus contamination.

(b) Reduced mechanical damage: Utilize appropriate harvesting equipment and methods to minimize crop damage, as physical injuries can create entry points for fungi. Employing lighter machinery helps to minimize seed damage. Adjusting harvesting settings, such as fan speeds and shutter openings, can reduce the incidence of *Fusarium*-damaged wheat kernels and lower DON levels in the harvested wheat. While removing damaged grains may reduce overall yield, it enhances storage conditions and improves grain safety, thereby mitigating economic losses (Salgado *et al.*, 2011).

(c) Moisture condition: Harvest crops when their moisture content meets or falls below the recommended levels for each type of grain, generally it should be below 14-15%. After harvesting, dry the crops to a safe moisture level. For example, 14% for maize, 13-14% for rice, 14-14.5% for barley, and 7-8% for canola or rapeseed (Adegoke and Letuma, 2013).

(d) Management practices: Effective field management *ie.*, timely irrigation, employing crop rotation, weeding, use of clean equipment and regular inspection for the signs of fungal infection (Spores or resting structures) need to be carried out. Remove the affected area or portion to prevent contamination of the entire batch.

Strategies for the Elimination and Detoxification of mycotoxins after harvest

a) Use of CO_2 sensors for detection: CO_2 sensors are effective for early detection of spoilage during storage. They identify insect infestations and mould-induced spoilage earlier than traditional methods such as visual inspection, odour detection, or temperature monitoring. Early detection allows timely management decisions, potentially reducing the need for costly spoilage control measures like turning, aeration, and fumigation, thus maintaining grain quality and enabling timely marketing to prevent further deterioration (Choudhary and Kumari 2010)

Elimination and inactivation of mycotoxins: Mycotoxin-contaminated grains can be addressed through physical separation, filtration, and milling.

- Physical separation: This method identifies damaged kernels based on size, shape, color, and visible mould growth. Significant amounts of aflatoxins can be removed from shelled peanuts through electronic sorting and hand-picking. Floating and density segregation can also be effective for instance, 95% of aflatoxin in 21 out of 29 peanut samples was found in kernels that floated in water.
- Filtration and separation: Membrane techniques, like ultrafiltration, can separate mycotoxins from contaminated liquids. Liquid-liquid extraction using solvents can also reduce mycotoxin concentrations. For example, single filtration methods have removed up to 90% of aflatoxins from peanuts (Choudhary and Kumari, 2010).

Inactivation of mycotoxins: Various methods can be used to inactivate mycotoxins, ensuring the safety of food and feed products.

- **Chemical methods:** Chemical treatments help to break down mycotoxins into less harmful substances. Ozone, hydrogen peroxide, and ammonia are common chemicals. Ozone has been effective in degrading aflatoxins and ochratoxins (Sipos *et al.*, 2021). Alkaline treatments using lime or sodium hydroxide and acidic conditions can also degrade certain mycotoxins (Abou *et al.*, 2022).
- **Biological methods:** Certain microorganisms, including bacteria and fungi, can degrade mycotoxins. For example, *Aspergillus niger* and *Bacillus*

species can metabolize and detoxify mycotoxins like ochratoxins and zearalenone. Various bacterial species, including lactic acid bacteria, have demonstrated the ability to degrade mycotoxins. For instance, *Tetragenococcus halophilus, Rhodococcus erythropolis,* and *Mycobacterium fluoranthenivorans* have been shown to degrade aflatoxin B1 (AFB1). *Pediococcus parvulus* and *Lactobacillus acidophilus* are effective in controlling ochratoxin A (OTA), AFB1, and aflatoxin M1 (AFM1). *Bifidobacterium animalis* is particularly useful for managing patulin, while *Pseudomonas otitidis* and *Bacillus velezensis* Strain ANSB01E are capable of detoxifying zearalenone (ZEN) (Sajid *et al.*, 2019; Ndiaye *et al.*, 2022). In addition to these enzymes such as laccases, peroxidases, and esterases can also break down mycotoxins like aflatoxins and fumonisins (Lyagin and Efremenko, 2019).

- **Physical methods:** Heat treatment and UV radiation can degrade mycotoxins. High temperatures can reduce aflatoxin levels, although this method must be carefully controlled to avoid nutrient loss. UV light can disrupt the chemical structure of mycotoxins, effectively degrading aflatoxins and ochratoxins on contaminated surfaces (Choudhary and Kumari 2010; Sipos 2021).

Influence of storage conditions to reduce the contamination after harvest

Proper storage conditions are critical for controlling mycotoxin production. Once crops are infected in the field, fungal growth typically continues with increased vigor during post-harvest and storage. Effective storage practices focus on controlling environmental factors that contribute to fungal growth and mycotoxin production:

1. **Moisture control: M**oisture level should be maintained below 14-15% in stored products. Use drying techniques such as natural sun drying, use of dryers (Batch Dryers, Continuous Flow Dryers, and Drying with Desiccants) before storage will reduce moisture. Employ dehumidifiers or moisture-absorbing materials like silica gel or clay desiccants which maintain low humidity, preventing mold growth and mycotoxin production.
2. **Temperature management:**

   The cool storage facility during storage keeps temperatures below 15°C to inhibit fungal growth and slow mycotoxin production. In addition to this, the use of sensors and monitoring systems helps to ensure temperatures remain within the optimal range for preventing contamination.
3. **Ventilation:** Proper ventilation promotes airflow and prevents moisture accumulation. The use of fans or ventilation systems will ensure even temperature and humidity distribution, especially in bulk storage facilities.

4. **Storage container design:**

   The moisture-resistant, airtight sealed containers help to protect against external humidity and contamination. Regularly inspect and repair containers to maintain optimal storage conditions.

5. **Cleanliness and regular monitoring**

   Regularly clean and sanitize storage areas and implements which prevents insect infestations and fungal spreads, which can damage stored products and facilitate infections. Regular inspections need to be done, routine sampling and testing allows early detection of infection samples so that the corrective actions could be taken.

(a) Transportation:

Mycotoxin contamination can occur throughout the entire chain of production for grain and oil crops, including growing, harvesting, storage, transportation, and processing, as long as the conditions are suitable. Some of the strategies to be followed during transportation are discussed below.

1. **Cleaning and regular inspection of transportation vehicles:** Before loading, sanitize the floors and surroundings of transportation vehicles using detoxifying chemicals such as Sodium hypochlorite, Hydrogen peroxide, Quaternary ammonium compounds (Quats), Acetic acid, or Formaldehyde. Regular inspection of vehicles and containers is essential to prevent the spread of mycotoxins (Lineback *et al.*, 2018).
2. **Use of mycotoxin absorbents:** Mycotoxin absorbents are substances designed to bind or neutralize mycotoxins in food or feed products, thereby reducing their harmful effects. These absorbents are effective at removing or inactivating mycotoxins without leaving toxic residues or affecting the product's technological properties, nutritional value, or palatability. Commonly used absorbents include:
   - ☆ **Clay-based absorbents:** Activated clays such as bentonite or montmorillonite, and zeolites, are effective in adsorbing various mycotoxins, including aflatoxins
   - ☆ **Silica gel and other organic absorbents:** Silica gel and activated carbon can reduce fungal growth and mycotoxin contamination. These absorbents can be incorporated into packaging materials or used in sachets within transport containers.
3. **Proper packaging:** Use hermetic (airtight) packaging to limit exposure to oxygen and moisture, which promote fungal growth. Hermetic seals help maintain a stable internal environment, reducing fungal development. They are a safe, cost-effective, and environmentally friendly alternative to chemicals for controlling insects and mycotoxin contamination in stored products (Kumar and Kalita, 2017). Additionally, incorporating antifungal

compounds into films and coatings has proven to be an innovative and effective solution for various products, including fresh fruits and vegetables (Akhila *et al.*, 2022).

Viii. Future Perspectives And Research Directions

1. Advanced Detection and Quantification Method

(a) Rapid and Sensitive Detection: Ongoing research focuses on developing more sensitive, rapid, and cost-effective methods for detecting mycotoxins at trace levels in food and feed products. Innovations such as biosensors, nano-based detection systems, and portable devices are crucial for early detection and prevention of contamination. For instance, recent advances in electrochemical biosensors have shown promise for on-site mycotoxin detection, offering high sensitivity and specificity (Zhang et al., 2023).

(b) Non-Destructive Techniques: There is increasing interest in non-destructive testing methods, such as hyperspectral imaging and Raman spectroscopy, which can assess mycotoxin contamination without compromising the integrity of stored products. These methods are particularly valuable in ensuring food quality and safety during post-harvest stages (Zhang et al., 2023).

2. Understanding the Molecular Mechanisms of Mycotoxin Bio-synthesis

(a) Genomic and Transcriptomic Studies: Further exploration into the genomic and transcriptomic profiles of mycotoxin-producing fungi is vital for understanding the regulation of biosynthetic gene clusters. Recent studies have identified key genes involved in mycotoxin biosynthesis, providing potential targets for disrupting toxin production through genetic or chemical means (Zhang et al., 2023).

(b) Epigenetic Regulation: Investigating the role of epigenetic modifications in regulating mycotoxin biosynthesis genes could open new avenues for control. Recent research has highlighted the potential of epigenetic inhibitors to suppress mycotoxin production, offering a novel approach to mitigate contamination (Zhang et al., 2023).

3. Development of Biocontrol Strategies

(a) Endophytes and Biopesticides: Leveraging plant-associated microorganisms, such as endophytes, to inhibit mycotoxin-producing fungi is a growing area of research. These bio-control agents can outcompete pathogens or produce antifungal compounds, thereby reducing mycotoxin contamination in crops. Recent field trials have demonstrated the efficacy of certain bio-control strains in reducing mycotoxin levels in maize and peanuts (Battilani et al., 2016).

(b) Genetically Modified Organisms (GMOs): The development of genetically engineered crops resistant to fungal infections and mycotoxin contamination is an ongoing research focus. While promising, this approach requires careful consideration of environmental safety and public acceptance (Zhang et al., 2023).

4. Impact of Climate Change on Mycotoxin Production

(a) Modelling and Prediction: Climate change significantly influences mycotoxin production, making predictive models that account for climate variables essential. These models help anticipate and mitigate risks associated with increased mycotoxin production due to rising temperatures and altered precipitation patterns (Battilani et al., 2016).

(b) Adaptive Storage Solutions: Research into climate-resilient storage technologies and management practices is critical to minimize mycotoxin risks. Innovations in storage solutions, such as hermetic storage and controlled atmosphere techniques, have shown effectiveness in limiting fungal growth under varying environmental conditions (Chilaka et al., 2017).

5. Sustainable Post-Harvest Management Practices

(a) Integrated Pest Management (IPM): Refining IPM strategies to include physical, chemical, and biological control measures is essential for preventing mycotoxin contamination. Recent studies highlight the importance of integrating these approaches to manage mycotoxin risks effectively (Pitt ***et al***., 2013).

(b) Improved Drying and Storage Technologies: Innovations in drying methods, such as solar drying and the use of natural preservatives, are crucial for reducing moisture content and inhibiting fungal growth during storage. Recent advances emphasize the need for context-specific solutions tailored to different agricultural systems (Rodriguez ***et al***., 2023).

6. Policy and Regulation Enhancements

(a) Global Standards Harmonization: Harmonizing global mycotoxin regulations and standards is crucial for facilitating international trade and ensuring food safety. Research continues to explore ways to align these standards across regions, considering differences in dietary exposure and agricultural practices (Egmond and Jonker, 2007).

(b) Consumer Awareness and Education: Educating farmers, food processors, and consumers about mycotoxins, their risks, and prevention strategies is an ongoing priority. Recent initiatives emphasize the need for comprehensive education programs that are accessible and culturally relevant (Milicevic ***et al***., 2020).

7. Data-Driven Approaches

(a) Big Data and AI: The use of big data and artificial intelligence (AI) to analyze trends in mycotoxin contamination, predict outbreaks, and optimize post-harvest management practices is gaining traction. AI-driven tools can provide real-time insights and decision support to stakeholders across the food supply chain (Dos Santos et al., 2021).

(b) Blockchain for Traceability: Implementing blockchain technology to track and trace agricultural products through the supply chain is another emerging area of interest. Blockchain ensures transparency, reduces the risk of contamination, and enhances consumer trust (Dos Santos et al., 2021).

8. Collaboration and Interdisciplinary Research

(a) Global Research Networks: Encouraging collaboration between researchers, industry stakeholders, and policymakers at an international level is vital to address the global challenge of mycotoxins. Interdisciplinary research that integrates microbiology, agriculture, and climatology is particularly valuable for developing comprehensive solutions (Milicevic ***et al.***, 2020).

These future research directions aim to mitigate the risks associated with mycotoxin contamination during post-harvest stages, ensuring food safety and enhancing global food security.

Conclusion

Mycotoxin biosynthesis and the factors influencing its production during post-harvest stages represent a complex and multifaceted challenge that demands an integrated approach. Advances in detection technologies, understanding of molecular mechanisms, and the development of bio-control strategies offer promising avenues for mitigating mycotoxin contamination. The impact of climate change further complicates this issue, necessitating adaptive storage solutions and predictive modelling to anticipate and manage risks. Sustainable post-harvest management practices, combined with harmonized global regulations and increased stakeholder collaboration, are essential to ensuring food safety and security. Continued interdisciplinary research and the application of innovative technologies, such as AI and block chain, will be critical in addressing the evolving challenges posed by mycotoxins in the global food supply chain.

References

Abou, A. R., Liu, H., & Nie, D. (2022). Chemical degradation of aflatoxins: A comprehensive review. *Food Chemistry, 393*, 133–150.

Adegoke, G. O., & Letuma, P. (2013). Strategies for the prevention of mycotoxin contamination in maize during post-harvest handling. *Food and Agricultural Safety Journal, 25*(1), 25–36.

Ahmed, A. F., Sohail, A., & Jabeen, F. (2015). Biological control of mycotoxin-producing fungi. *International Journal of Mycology, 5*(2), 99–115.

Akhila, B., Prasanna, P., & Kumar, M. (2022). Recent trends in antifungal food packaging materials. *Food Packaging and Shelf Life, 33*, 100837.

Alexander, N. J., McCormick, S. P., & Hohn, T. M. (2009). Tri6 and Tri10: Key regulatory genes in trichothecene biosynthesis in Fusarium species. *Journal of Agricultural Food Chemistry, 55*(15), 5233–5244.

Ayofemi, O. J. (2020). Advances in aflatoxin biosynthesis: A review of genetic and biochemical regulation. *Journal of Agricultural Biotechnology, 12*(3), 115–125.

Awuchi, C. G., Igwe, V. S., & Ewa, S. (2021). Impacts of pH on fungal growth and mycotoxin biosynthesis. *Journal of Food Safety, 41*(3), e12992.

Battilani, P., Toscano, P., Van der Fels-Klerx, H. J., Moretti, A., Camardo Leggieri, M., Brera, C., & Robinson, T. (2016). Aflatoxin contamination in maize in Europe increases due to climate change. *Scientific Reports, 6*, 24328.

Bhatnagar, D., Ehrlich, K. C., & Cleveland, T. E. (2015). Genetic regulation of aflatoxin biosynthesis. *Mycopathologia, 162*(2), 105–112.

Brown, D. W., Lee, S., Kim, H.-S., & Busman, M. (2021). Trichothecene biosynthesis and regulation in Fusarium. *Mycotoxin Research, 37*(2), 93–107.

Brase, S., Encinas, A., Keck, J., & Nising, C. F. (2013). Chemistry and biology of mycotoxins and related fungal metabolites. *Chemical Reviews, 109*(9), 3903–3990.

Brzonkalik, K., Herrling, T., Syldatk, C., & Neumann, A. (2012). Influence of carbon-to-nitrogen ratios on Alternaria toxin production. *Journal of Agricultural and Food Chemistry, 60*(23), 6326–6331.

Chen, S., Liu, X., Zhang, Z., & Zheng, Z. (2019). Molecular regulation of fumonisin biosynthesis in Fusarium species. *Fungal Genetics and Biology, 132*, 103267.

Cheli, F., Pinotti, L., Rossi, L., & Dell'Orto, V. (2017). Effect of mycotoxins on food quality and agricultural performance. *Journal of Agricultural Science, 16*(2), 125–145.

Chilaka, C. A., De Boevre, M., Atanda, O. O., & De Saeger, S. (2017). The status of Fusarium mycotoxins in sub-Saharan Africa. *Food Control, 69*, 28–38.

Choudhary, A. K., and Kumari, P. (2010). Management of mycotoxin contamination in preharvest and post-harvest crops: present status and future prospects. *Journal of Phytology*, 2(7).

Daou, R., Joubrane, K., Maroun, R., Khabbaz, L. R., Ismail, A., & El Khoury, A. (2021). Mycotoxins: Factors influencing production and control strategies. *Applied Microbiology, 10*(5), 293–310.

Dos Santos, D. A., Pozza, E. A., & Rodrigues, F. A. (2021). Big data and AI for predicting fungal toxin outbreaks in food systems. *Agricultural Systems, 186*, 102959.

Egmond, H. P. V., & Jonker, M. A. (2007). Worldwide regulations on mycotoxins. *World Mycotoxin Journal, 2*(2), 159–169.

Ferrara, M., Gallo, A., Perrone, G., & Epifani, F. (2022). Fumonisin biosynthesis: Genetic regulation and environmental factors. *Toxins, 14*(1), 67.

Fleurat-Lessard, F. (2017). Role of grain moisture in fungal contamination during storage. *Food and Feed Safety Journal, 12*(3), 1–17.

Frisvad, J. C., Larsen, T. O., & Nielsen, K. F. (2019). Advanced tools for mycotoxin identification. *Mycotoxin Research, 35*(4), 57–68.

Gonzalez, M., Torres, R., & Ramirez, A. (2021). Zearalenone production and its control mechanisms. *Food Mycology Journal, 9*(4), 227–238.

Hamad, G. M., Mehany, T., Simal-Gandara, J., Abou-Alella, S., Esua, O. J., Abdel-Wahhab, M. A., and Hafez, E. E. (2023). A review of recent innovative strategies for controlling mycotoxins in foods. *Food Control*, 144, 109350.

Hay, E. D., Liu, Z., & Babcock, E. J. (2021). Trichothecene biosynthesis and its relationship to fungal development. *Plant Pathology Journal, 19*(3), 162–175.

Ismaiel, A. A., & Papenbrock, J. (2015). Mycotoxins: producing fungi and mechanisms of phytotoxicity. *Agriculture, 5*(3), 492-537.

Janssen, E. M., Liu, C., & Van der Fels-Klerx, H. J. (2018). Fusarium infection and trichothecenes in barley and its comparison with wheat. *World Mycotoxin Journal, 11*(1), 33-46.

Klich, M. A. (2007). Aspergillus flavus: the major producer of aflatoxin. *Molecular plant pathology, 8*(6), 713-722.

Kochiieru, Y., & Zhuravlev, P. (2021). Post-harvest factors affecting mycotoxin accumulation in cereals. *Grain Quality Research Journal, 14*(2), 38–48.

Kolawole, A. O., Bandyopadhyay, R., & Hell, K. (2021). Genetic regulation of aflatoxin biosynthesis. *Toxins, 12*(8), 221.

Kumar, M., & Kalita, P. (2017). Hermetic storage systems for mycotoxin control in grains. *Food Science and Nutrition Journal, 16*(3), 12–18.

Lappa, I. K., Papoutsis, K., & Gekas, V. (2017). Roquefortine C: Biosynthesis and risk assessment. *Toxins, 9*(4), 187.

Lineback, C. B., Nkemngong, C. A., Wu, S. T., Li, X., Teska, P. J., and Oliver, H. F. (2018). Hydrogen peroxide and sodium hypochlorite disinfectants are more effective against *Staphylococcus aureus* and *Pseudomonas aeruginosa* biofilms than quaternary ammonium compounds. *Antimicrobial Resistance & Infection Control, 7*, 1-7.

Lindblad, M., Gidlund, M., & Olsen, M. (2004). Control of ochratoxin contamination in cereals. *Food Safety Journal, 21*(4), 123–129.

Lyagin, I. V., & Efremenko, E. N. (2019). Enzymatic methods for mycotoxin detoxification. *Biotechnology Journal, 35*(2), 171–184.

Magan, N., & Aldred, D. (2008). Strategies for controlling ochratoxin contamination. *Food Science Journal, 22*(1), 98–103.

Marasas, W. F. O., Gelderblom, W. C. A., Shephard, G. S., & Vismer, H. F. (2008). Fumonisins: Biology, chemistry, and significance. *Annual Review of Phytopathology, 46*, 293–316.

Mutegi, C. K., Hell, K., & Ochieng, E. (2024). Pre- and post-harvest strategies for managing aflatoxins in maize. *Journal of Food Protection, 89*(1), 39–47.

Proctor, R. H., Hohn, T. M., & McCormick, S. P. (2003). Fumonisin biosynthesis in Fusarium verticillioides. *Applied and Environmental Microbiology, 69*(1), 671–676.

Mannaa, M. and Kim, K. D. (2017). Influence of temperature and water activity on deleterious fungi and mycotoxin production during grain storage. *Mycobiology, 45*(4), 240-254.

Mir, S. A., Shah, M. A., & Mir, M. M. (2021). Emerging technologies for mycotoxin decontamination. *Food Bioscience, 42*, 101190.

Milani, J. M. (2013). Ecological conditions affecting mycotoxin production in cereals. *World Mycotoxin Journal, 6*(2), 167–192.

Milicevic, D., Udovicki, B., Petrovic, Z., Jankovic, S., Radulovic, S., Gurinovic, M. and Rajkovic, A. (2020). Current status of mycotoxin contamination of food and feeds and associated public health risk in Serbia. *Meat Technology*, 61(1).

Ndiaye, A., Diop, Y., & Ba, F. (2022). Biological detoxification of mycotoxins in food using microbial methods. *African Journal of Biotechnology, 21*(5), 121–130.

Neme, K., & Mohammed, A. (2017). Mycotoxin occurrence in grains: A review. *Food Control, 80*, 300–310.

Ochsenreither, K. (2011). *Process development for the production of Alternaria toxins in a bioreactor* (Doctoral dissertation, Karlsruhe, Karlsruher Institut für Technologie (KIT), Diss., 2011).

Palencia, E., Hinton, D. M., & Bacon, C. W. (2010). The role of fungal ecology in aflatoxin contamination. *Frontiers in Bioscience, 2*, 372–382.

Paterson, R. R. M. and Lima, N. (2017). Thermophilic fungi to dominate aflatoxigenic/ mycotoxigenic fungi on food under global warming. *International journal of environmental research and public health*, 14(2), 199.

Pitt, J. I., & Hocking, A. D. (2013). Mycotoxins in Australia: Biocontrol strategies. *Food Security Journal, 7*(4), 523–536.

Qu, B., Li, H., Zhang, X., & Liu, Z. (2022). Advances in predictive modeling of mycotoxin contamination under environmental conditions. *Environmental Monitoring and Assessment, 194*(3), 143.

Reddy, K. R. N., Salleh, B., & Saad, B. (2009). Biocontrol strategies for aflatoxin mitigation. *Food Safety Journal, 30*(6), 868–875.

Rodriguez, M., Crespo, J. M., & Peña, A. (2023). Post-harvest storage techniques to reduce mycotoxins in grains. *Agricultural Systems and Food Security Journal, 18*(1), 98–108.

Saharan, B. S., & Nehra, V. (2021). Fungal strain variability in mycotoxin production: Implications for food safety. *Journal of Agricultural Mycology, 15*(2), 67–76.

Sajid, M., Khan, M. N., & Khan, T. (2019). Role of bacteria in degrading aflatoxins. *Journal of Microbial Biotechnology, 9*(5), 231–242.

Salgado, J. D., Wallhead, M., Madden, L. V., and Paul, P. A. (2011). Grain harvesting strategies to minimize grain quality losses due to *Fusarium* head blight in wheat. *Plant disease, 95*(11), 1448-1457.

Sipos, P., Peles, F., & Gyori, Z. (2021). Application of ozone for mycotoxin detoxification in cereals. *Toxins, 13*(2), 114.

Sokefun, T., Ogunniyi, A., & Arinze, E. (2018). Economic impact of mycotoxin contamination in African agriculture. *Journal of Agribusiness Research, 45*(3), 143–157.

Wang, Y., Nie, Y., Ji, J., & Fang, Z. (2016). Polyketide biosynthesis pathways in ochratoxin-producing fungi. *Fungal Genetics and Biology, 91*, 42–48.

Zhang, X., Wu, J., & Zhou, Q. (2023). Advances in microbial detoxification of mycotoxins. *Applied Microbiology and Biotechnology, 107*(3), 711–723.

Chapter 3

Emerging Fungal Pathogens and Mycotoxins: Epidemiology and Extraction

Saptarshi Mukherjee[1], Arpita Das[2], Soumyadipta Samanta[2], Sourin Das[2] and Sumit Nath[3]*

[1] *M.Tech Bioinformatics, Maulana Abul Kalam Azad University of Technology*

[2] *MSc Biotechnology, Maulana Abul Kalam Azad University of Technology*

[3] *M.Tech Biotechnology, Maulana Abul Kalam Azad University of Technology*

**Corresponding author: saptarshi5k17@gmail.com*

Abstract

Food safety, agriculture, and global health are seriously threatened by emerging fungal diseases and mycotoxins, which call for immediate attention and creative solutions. Secondary metabolites of filamentous fungus, known as mycotoxins, can potentially harm both human and animal health. The majority of filamentous fungi responsible for producing these mycotoxins are classified into four genera: *Alternaria, Penicillium, Fusarium,* and *Aspergillus*. In addition to their ability to contaminate crops and cause food to spoil, mycotoxigenic fungi pose a persistent and serious threat to agriculture in several ways, including the potential for malnutrition when nutritional quality is compromised. It can be extremely difficult to detect and identify these pathogens and toxins due to their complexity; often, sophisticated molecular and analytical techniques are required.

Keywords: Fungal pathogens, Mycotoxigenic fungi, Mycotoxin determination, Chromatography, Mass spectrometry, Biosensors

Introduction

Invasive mycosis due to emerging opportunistic fungal pathogens and mycotoxins have increased by several folds in recent past (Pfaller & Diekema, 2004). Mycotoxins are low molecular weight (MW≈700 Da) toxic secondary metabolites, produced by a wide variety of filamentous fungi of the genera *Aspergillus, Penicillium, Fusarium, Alternaria, Claviceps, Mucor* and *Rhizopus* (Mateus et al., 2021) under specific conditions like high humidity, poor agricultural practices, poor crop storage conditions, improper transportation facilities during international trade, etc. (Al-Jaal et al., 2019). Mycotoxins are ubiquitous and contaminate a variety of foodstuffs and agricultural crops at any stage of the food production process, leading to direct exposure of consumers to mycotoxin-contaminated food products or an indirect exposure by feeds. Mycotoxins intake through foodstuffs causes severe intoxication on both humans and animals known as mycotoxicosis. Mycotoxin being mutagenic, teratogenic, cytotoxic, neurotoxic, and carcinogenic and also capable of inhibiting DNA and RNA synthesis by interacting with nucleic acids at the cellular level (Smith et al., 2007) can cause several health consequences like central nervous system disorder, hepatotoxicity, cardiotoxicity, nephrotoxicity, gastrointestinal tract damage, cancer and even death in the worst-case scenario (Zhang et al., 2020). This increase in diseases and disorders has also led to an increase in morbidity and mortality rate especially for individuals undergoing or having previously undergone solid-organ transplantation, blood and bone marrow transplantation, major surgery, suffering from AIDS, and premature babies (Pfaller & Diekema, 2004), which is a major concern in public health and welfare worldwide. Aflatoxin, ochratoxin, ergot alkaloids, fumonisins, deoxynivalenol, zearalenone, enniatins, patulin, trichothecenes, *Alternaria* toxins, and T-2 and H-2 toxins are the most common and major mycotoxins (Barac, 2018). About 10,000 fungi have been identified, more than 500 species of fungi are reported which produce mycotoxins, and around 1000 more species are yet to be discovered (Zhang et al., 2020).

Various mycotoxins are very stable against chemical, physical, and biological treatments, and also resistant to heat processes, which make these mycotoxins very difficult to eliminate from the foodstuffs during food processing. Several fungal pathogens and mycotoxins are extremely difficult to detect and identify due to their complexity; often, sophisticated molecular and analytical techniques are required. This chapter contains several integrated techniques and approaches to address these challenges, based on two pillars: modern pre-harvest strategies to prevent contamination and post-harvest detection and elimination of pathogens. Innovative methods include molecular assays like enzyme-linked immunosorbent assay (ELISA), chromatographic methods like high-performance liquid chromatography

(HPLC), gas chromatography, high-resolution mass spectrometry, biosensors, portable detection devices, and immunochemical-based tests. The investigation of natural antifungals and biocontrol agents simultaneously offers hope for safer and more long-lasting therapies.

Epidemiology and Economic Effect

According to the Food and Agriculture Organisation (FAO), each year 25% of global food products and feed is mycotoxin contaminated after pre-harvest or post-harvest strategies, causing an estimated loss of 1.3 billion metric tonnes of global food and food products, which in turn wastes 1.47-1.96 Gha of arable land, 0.75-1.25 trillion cubic litres of water, and 1-1.5% of total global energy (Ishangulyyev et al., 2019). Approximately 25-50% of global cereal products suffer mycotoxin contamination, out of which 5-10% are irreversible.

Several studies revealed that more than 5 billion people get exposed to mycotoxins daily by some unknown sources and even consume mycotoxin-contaminated food on a daily basis (Mateus et al., 2021). Mycotoxin contamination of food became a global concern in 1960, when 1,00,000 turkeys in the United Kingdom died on feeding foods having molds and contaminated with aflatoxin (Pitt & Miller, 2016). In India, Malaysia, China, and Kenya, severe aflatoxicosis outbreaks due to poisoning from aflatoxin contaminated maize have seen hundreds of affected people with around several fatalities besides hepatocellular carcinoma or liver cancer and also cause vomiting, edema, and abdominal pain (Liew & Mohd-Redzwan, 2018; Haque et al., 2020). Mycotoxin- contaminated corn possessed chronic health and economic concerns in the United States. Mycotoxigenic fungi and production of mycotoxins not only provide potential threat to human health and welfare but also result in several economic losses by- A) Crop output reduction by increasing plant diseases and removal or discard of affected plants and its products. B) Drop in quality and nutritional value and thus reducing the commercial value of the product. C) Increase in animal health problems related to mycotoxin-contaminated feed thus reducing their productivity. D) Increase investment in use of selected resistant cultivars and use of fungicide to restrain mycotoxigenic fungi contamination, to maintain optimum sterile storage conditions and transportation facilities, to set up proper surveillance networks to ensure animal and human health and welfare. All these add to the extra production cost resulting in an increase in price of final goods. Thus, besides being a potential threat to human health, it is also an alarming threat to the global economy and needs to be resolved through utmost urgency using precision detection and purification techniques.

Table 1: Overview of Mycotoxins: Types, Producers, Effects, and Sources
The following table provides a comprehensive overview of the major types of mycotoxins, their primary producers, associated toxic effects on humans and animals, and the common food sources they contaminate, along with relevant references.

Type of Mycotoxin	*Description*	*Main Sources*	*Effects*	*References*
Aflatoxins (AFTs)	The most significant mycotoxins worldwide due to occurrence, toxicity, and economic impact. Primary toxin: Aflatoxin B1 (AfB1). Metabolized to less toxic AfM1 and AfM2.	Peanuts, nuts, cereals, spices, dried fruits, edible oils, animal products (milk, meat).	Acute/chronic liver diseases, teratogenic, mutagenic, hepatocarcinogenic effects.	Varga et al. (2011), Luo et al. (2018), Agriopoulou et al. (2020)
Ochratoxin A (OTA)	Produced by *Aspergillus* and *Penicillium* species. Classified into A, B, C, with OTA being the most toxic and prevalent.	Cereals, dried vine fruits, wine, coffee, meat, cocoa, spices.	Nephrotoxicity, neurotoxicity, carcinogenesis, genotoxicity, immunotoxicity.	Wang et al. (2016), Alshannaq & Yu (2017), Agriopoulou et al. (2020), Eskola et al. (2019)
Trichothecenes (TCTs)	Produced by *Fusarium* and other species. Over 200 types, categorized into A–D.	Cereals (rice, wheat, barley, oats), cereal-derived products, sunflower seeds, bananas, soybeans.	Bleeding, diarrhea, vomiting, necrotic lesions, myelotoxicity, hematotoxicity.	Ferrigo et al. (2016), Haque et al. (2020), Luo et al. (2018)
Zearalenone (ZEN)	A macrocyclic β-resorcylic acid lactone produced by *Fusarium* species.	Barley, corn, rye, sorghum, wheat, maize.	Oestrogenic effects, early puberty, oxidative stress, ROS production.	Bhatnagar et al. (2002), Luo et al. (2018), Alshannaq & Yu (2017)

Ergot Alkaloids (EAs)	Produced by *Claviceps* species. Includes ergopeptides, lysergic acids, and clavines.	Rye, wheat, triticale, oats, barley, millet.	Convulsions, hallucinations, gangrene, hypersensitivity, abortion.	Topi et al. (2017), Grusie et al. (2018), Haque et al. (2020)
Patulin (PAT)	A polyketide mycotoxin primarily produced by *Aspergillus* and *Penicillium* species.	Fruits (apples, oranges, pears), fruit juices.	Gastrointestinal disturbances, mutagenicity, neurotoxicity, carcinogenesis.	Walravens et al. (2014), Marin et al. (2013), Vidal et al. (2019)
Citrinin (CIT)	Secondary benzopyran metabolite produced by *Penicillium* and *Aspergillus* species.	Stored grains, dairy products, rice, beans, nuts, spices.	Nephropathy, RNA synthesis inhibition, cancer.	Rocha et al. (2014), Tanaka et al. (2007), Haque et al. (2020)
Fumonisins (FUMs)	Produced by Fusarium species, mainly *F. proliferatum* and *F. verticillioides.*	Corn grains, soybeans, black tea, grapes, asparagus, barley, figs.	Esophageal cancer, neural tube defects, pulmonary edema syndrome.	Pitt et al. (2012), Alberts et al. (2016), Luo et al. (2018)
Emerging Mycotoxins	Includes enniatins, beauvericin, fusaproliferin, and moniliformin. Produced by *Fusarium* spp.	Cereals, grains, feed ingredients.	Cytotoxicity, phytotoxicity, zootoxicity.	Logrieco et al. (1996, 1998), Jestoi (2008), Streit et al. (2013)

Extraction Methodology and Solution

First, samples must be prepared by extracting and purifying mycotoxins from food samples using the proper solvents. In order to free the mycotoxins from the matrix and get rid of the things that might impede the identification of mycotoxins, extraction is necessary. Finding ecologically acceptable solvents, optimising the procedure, and getting fast results are the main goals of sample preparation. Before the analysis of mycotoxins, extraction and cleanup methods are the most crucial. By employing the proper solvents, the contaminated food and feed samples are extracted with the goal of eliminating mycotoxins from the material. The right solvents and proper choice of extraction are the most crucial for significantly increasing the success of the extraction process. A suitable extraction solvent can efficiently remove only the mycotoxins with the highest possible efficiency. It lessens the matrix impact, is affordable, and safe for the user. The two most utilised extraction solvents in mycotoxin analysis are mixtures of methanol and water, as well as acetonitrile and water in varying ratios. Several extraction solvents, such as a combination of ethyl acetate and formic acid, must be used due to the presence of pigments, essential oils, and fatty acids in the samples, which make extraction challenging. Mycotoxins have also been extracted using different extraction solvents, such as 1-octanol and toluene, dichloromethane, acetone, and chloroform. While polar mycotoxins, such as FBs, are soluble in water, hydrophobic mycotoxins are readily dissolved in all of the previously mentioned solvents. He et al. (2019) have reported the effective extraction of aflatoxins from rice samples using deep eutectic solvent (DES), a novel green solvent. Two inexpensive, safe, and biodegradable ingredients—malonic acid and tetramethylammonium chloride—were combined to create this extractant. Additionally, using this extract has reduced the need for solvents and derivatization chemicals, in line with green chemistry principles.

QuEChERS (Quick, Easy, Cheap, Effective, Rugged, and Safe) method was earlier introduced for the analysis of pesticides. To identify mycotoxins in various matrices, QuEChERS is the technique employed. Using acetonitrile and water, the toxin is initially extracted, and mycotoxins are subsequently transferred to the organic phase by adding inorganic salts to induce liquid-liquid partitioning (Gonzalez-Jartin et al., 2019). It generates high quality results of mycotoxin analysis and is pretty fast-paced. Juan et al. (2017) reported that QuEChERS has been utilized in jam and juice made from berries to analyze mycotoxins like OTA and AFs.

SLE stands for **liquid-solid extraction**. It is a technique that involves weighing a homogenized sample in an extraction solvent and shaking it. Xie et al. (2016) and Rubert et al. (2012) have both employed this to extract mycotoxins linked to grains. SLE mostly employs ultrasonic extraction, homogenization, and shaking for mycotoxin analysis in agricultural goods with solid structures.

The **liquid-liquid extraction (LLE)** approach relies on the fact that toxins are soluble in distinct aqueous and organic phases. Moreover, LLE has been used to analyze OTA and AFs (Andrade et al., 2013; Turner et al., 2015). Mycotoxins have varying solubilities in the two immiscible phases that are employed in LLE from separate solvents. The nontarget materials are eliminated in one step, and the mycotoxins are eliminated in the subsequent phase.

Pressurized Liquid Extraction (PLE) or **Accelerated Solvent Extraction (ASE)** is a technique that enhances the extraction of mycotoxins by operating at higher temperatures (100–180°C) and pressures (1,500–2,000 psi) in a vessel. Using this technique, mycotoxins in tomato samples were found (Rico-Yuste et al., 2018; Miklos et al., 2020). Utilizing conventional solvents and temperatures above the boiling point, ASE is a quick, easy, and environmentally friendly extraction method.

The extraction of non-polar chemical compounds, like ZEA found in maize flour, is accomplished by the use of **supercritical fluid extraction (SFE)** (Zougagh and Ríos, 2008). The benefit of the SFE approach is that supercritical CO2 reduces and eliminates the need for organic solvents (Xie et al., 2016).

MAE is considered an environmentally friendly method that needs less time for extraction and few organic solvents. This method involves quickly heating organic solvents using microwave radiation in order to segregate mycotoxins inside the solution. Water, an extraction solvent, and a dispersive solvent are the three main solvents that the **VALDS-ME** typically employs. Two nontoxic solvents—1-octanol and toluene—were used as the extractant in a recent work by Somsubsin et al. (2018); AFs were extracted from rice samples relatively quickly by using a vortex and adding Na_2SO_4 as a demulsifier.

Table 2. Types of extraction methods and its advantages and disadvantages

Methods of Extraction	*Extraction Solvents*	*Advantages*	*Disadvantages*	*Reference*
QuEChERS	Organic solvents and mixtures (CH3CN, MeOH, CH3CN/ MeOH)	Low ppb level detection that is affordable, quick, easy to use, and has improved repeatability and accuracy	Alterations to the first process and the requirement for a further enrichment phase	Gonzalez-Jartin et al., 2019
SLE	Combination of diluted acids or water and organic solvents	Reduced solvent volumes	Matrix effects	Rubert et al. (2012)
LLE	Combination of diluted acids or water with organic solvents (cyclohexane and hexane)	Efficient when carrying out small-scale preparations	Time-consuming, based on the matrix and identified substances, the sample may be absorbed by the glass apparatus.	Turner et al., 2015
PLE or ASE	Organic solvent mixtures (MeOH/ CH_3CN, CH3CN/ water)	Higher extraction efficiency, less solvent use, fully automated, and quicker extraction than traditional methods	Expensive equipment and heavily coextracted matrix elements	Rico-Yuste et al., 2018

SFE	Supercritical fluids (CO2), acetone, MeOH, ethanol	Quick method, tiny solvent quantities, preconcentration impact, and temperature-sensitive analyte extraction	Requires costly and specialized instruments, unsuitable for routine analysis	Xie et al., 2016
MAE	Aqueous solution	Shorter extraction times and requiring less solvent	Useful only for thermally stable substances and requires expensive equipment	Somsubsin et al. (2018)
VALDS-ME	Combination of water, dispersive solvent, and organic solvents	Utilizing low density solvents is easy, quick, and efficient.	Optimization following extensive parameter control	Somsubsin et al. (2018)

Cleanup/ Purification Procedures Of Mycotoxins

Purification of the extract is the next stage, which improves the quantification's specificity, sensitivity, and accuracy. Solid Phase extraction (SPE) and immunoaffinity columns (IAC) are the two most often utilized techniques for the purification of mycotoxins (Alshannaq and Yu, 2017). The fast and effective SPE approach uses cartridge-packed absorbents (octadecylsilyl (C18), amino-propyl (NH_2), silica gel, and multi-walled carbon nanotubes (MWCNTs) to absorb mycotoxins (Wang M. et al., 2016; Huertas-Perez et al., 2017; Jiang et al., 2018). Type A TCTs in rice, maize, and wheat (Dong et al., 2015) and ZEA in maize (Han et al., 2017) have both been purified using SPE.

In IAC, certain monoclonal antibodies are utilized to detect selected mycotoxins, and the target mycotoxin is bound by particular antibodies on the column. Pure acetonitrile or methanol is used to elute the mycotoxins from the column (Liu et al., 2018). In maize (Lattanzio et al., 2007), cereals (Lattanzio et al., 2014), and wheat bran (Irakli et al., 2017), IAC was utilized in the study of OTA, ZEA, and AFs.

IACs are typically used for cleanup in order to obtain higher quantification, precision, and accuracy limitations. When the test sample passes through the column, the material that IACs have packed with particular antibodies for various mycotoxins interacts with the test sample's antigens. Concurrently, the sample's interfering substances are eliminated using column washing, and the mycotoxin is extracted using a solvent that is typically miscible (Al-Jaal et al., 2019). A popular technique for single- and multi-mycotoxin analyses is the IAC cleanup. Multi-mycotoxin IACs, which evaluate Fusarium toxins, OTA, ZEA, and AFs all at once, have been created for multi-mycotoxin analysis (Irakli et al., 2017; Lattanzio et al., 2014). While using IACs to analyze mycotoxin yields reliable findings, the process is time-consuming and requires the use of several hazardous organic chemicals (Golge et al., 2016). These columns could be unable to fully absorb mycotoxins because of the many matrix components that obstruct the antibodies' capacity to attach to them or because the antibodies are inhibited against mycotoxin structures. IACs are very costly and have a limited lifespan (Castegnaro et al., 2006).

Mycotoxins are also cleaned up using the BSA-based sample clean-up columns. Serum albumins are widely available and reasonably priced, particularly those derived from cows. They can be utilised in the cleanup phase since they have a great affinity for certain mycotoxins. Leal et al. (2019) developed and verified a new BSA-based sample clean-up column that can be used to measure OTA in wine. One significant drawback of using BSA as a blocking agent is that it can cause OTA interference because of its binding to the protein, which can alter or improve the assay results.

Synthetic systems such as peptides, MIPs, and aptamers have been developed to serve as substitutes for the antibodies found in immunoaffinity complexes (IACs) (Huertas-Pérez et al., 2017; Pichon and Combès, 2016). Compared to antibodies, MIPs are synthetic polymers that are highly specialised in recognising certain analytes at a lower cost (Zhang et al., 2018). In vitro selection may be used to generate aptamers (Huang et al., 2015; J. Wang et al., 2016), which are then employed for quick on-site detection of food pollutants and are recommended as a substitute for mycotoxin and toxigenic fungal detection (Xu et al., 2016; Chen et al., 2014).

Conclusion

Global food safety, agriculture, and public health are seriously threatened by emerging fungal pathogens and mycotoxins. These mycotoxins, which are produced by different filamentous fungi, are difficult to detect and manage because of their chemical complexity and resistance to environmental changes. They are also persistent in contaminating foods. The most common mycotoxins are highlighted in this chapter, including fumonisins, ochratoxins, aflatoxins, trichothecenes, and zearalenone. These toxins can have a wide range of harmful effects on both humans and animals, including immunosuppressive, neurotoxic, mutagenic, and carcinogenic effects.

In this chapter, we have covered a number of cutting-edge detection strategies that can help with these problems. These strategies include biosensors, chromatographic techniques, and sophisticated molecular assays that provide accurate and timely detection. Moreover, pre- and post-harvest interventions must be combined in integrated strategies to minimise the effects of these pathogens and mycotoxins. Utilising different extraction and purification techniques to lower contamination levels is one of these tactics, along with improving storage conditions and using biological control agents.

The persistent and widespread nature of these pathogens requires continuous research and development of more efficient, sustainable, and financially viable methods to ensure food safety and public health, even with the advancements in detection and control measures. Developing comprehensive strategies to manage and mitigate the risks associated with fungal pathogens and mycotoxins requires a multidisciplinary approach involving molecular biology, chemistry, agriculture, and biotechnology.

References

Ali, A.M.A., Fahmy, M.F., Metwally, M.M., Hassanin, O., Azazy, H.A. and Mowafy, R.E., 2021. Ameliorative Effects of Cholestyramine and Oxihumate on Aflatoxicosis in Broiler Chickens. *Pakistan Veterinary Journal, 41*(1).

Alshannaq, A. and Yu, J.H., 2017. Occurrence, toxicity, and analysis of major mycotoxins in food. *International journal of environmental research and public health, 14*(6), p.632.

Altomare, C., Logrieco, A.F. and Gallo, A., 2021. Mycotoxins and mycotoxigenic fungi: risk and management. A challenge for future global food safety and security.

Barac, A., 2019. Mycotoxins and human disease. *Clinically relevant mycoses: a practical approach*, pp.213-225.

Cheng, Z. and Jiang, H., 2019. Supported liquid extraction (SLE) in LC-MS bioanalysis. *Sample Preparation in LC⍰MS Bioanalysis*, pp.76-84.

Diao EnJie, D.E., Hou HanXue, H.H., Hu WeiCheng, H.W., Dong HaiZhou, D.H. and Li XiangYang, L.X., 2018. Removing and detoxifying methods of patulin: a review.

El-Sayed, R.A., Jebur, A.B., Kang, W. and El-Demerdash, F.M., 2022. An overview on the major mycotoxins in food products: Characteristics, toxicity, and analysis. *Journal of Future Foods, 2*(2), pp.91-102.

Enyiukwu, D.N., Awurum, A.N. and Nwaneri, J.A., 2014. Mycotoxins in stored agricultural products: Implications to food safety and health and prospects of plant-derived pesticides as novel approach to their management. *Greener Journal of Microbiology and Antimicrobials, 2*(3), pp.32-48.

Gao, J., Wang, J., Wu, C., Hou, F., Chang, S., Wang, Z., Pu, Q., Guo, D. and Fu, H., 2019. Fast screening of aflatoxins in dairy cattle feeds with CE-LIF method combined with preconcentration technique of vortex assisted low density solvent–microextraction. *Electrophoresis, 40*(4), pp.499-507.

Ishangulyyev, R., Kim, S. and Lee, S.H., 2019. Understanding food loss and waste—why are we losing and wasting food?. *Foods, 8*(8), p.297.

Ji, F., He, D., Olaniran, A.O., Mokoena, M.P., Xu, J. and Shi, J., 2019. Occurrence, toxicity, production and detection of Fusarium mycotoxin: A review. *Food Production, Processing and Nutrition, 1*(1), pp.1-14.

Khan, R., Anwar, F. and Ghazali, F.M., 2024. A comprehensive review of mycotoxins: Toxicology, detection, and effective mitigation approaches. *Heliyon*.

Luo, S., Du, H., Kebede, H., Liu, Y. and Xing, F., 2021. Contamination status of major mycotoxins in agricultural product and food stuff in Europe. *Food Control, 127*, p.108120.

McCormick, S.P., Stanley, A.M., Stover, N.A. and Alexander, N.J., 2011. Trichothecenes: from simple to complex mycotoxins. *Toxins, 3*(7), pp.802-814.

Moss, M.O., 1991. Economic importance of mycotoxins-recent incidence in the United States. *Animal Science, 27*, pp.3941-3949.

Mukhtar, K., Nabi, B.G., Ansar, S., Bhat, Z.F., Aadil, R.M. and Khaneghah, A.M., 2023. Mycotoxins and consumers' awareness: Recent progress and future challenges. *Toxicon*, p.107227.

Omotayo, O.P., Omotayo, A.O., Mwanza, M. and Babalola, O.O., 2019. Prevalence of mycotoxins and their consequences on human health. *Toxicological research, 35*, pp.1-7.

Panaccione, D.G., 2005. Origins and significance of ergot alkaloid diversity in fungi. *FEMS Microbiology letters, 251*(1), pp.9-17.

Perestrelo, R., Silva, P., Porto-Figueira, P., Pereira, J.A., Silva, C., Medina, S. and Câmara, J.S., 2019. QuEChERS-Fundamentals, relevant improvements, applications and future trends. *Analytica chimica acta, 1070*, pp.1-28.

Pfaller, M.A. and Diekema, D.J., 2004. Rare and emerging opportunistic fungal pathogens: concern for resistance beyond Candida albicans and Aspergillus fumigatus. *Journal of clinical microbiology, 42*(10), pp.4419-4431.

Pitt, J.I. and Miller, J.D., 2017. A concise history of mycotoxin research. *Journal of agricultural and food chemistry, 65*(33), pp.7021-7033.

Sairam, P., Ghosh, S., Jena, S., Rao, K.N.V. and Banji, D., 2012. Supercritical fluid extraction (SFE)-an overview. *Asian Journal of Research in Pharmaceutical Science, 2*(3), pp.112-120.

Scott, P.M., 2012. Recent research on fumonisins: a review. *Food additives & contaminants: part A, 29*(2), pp.242-248.

Silvestre, C.I., Santos, J.L., Lima, J.L. and Zagatto, E.A., 2009. Liquid–liquid extraction in flow analysis: A critical review. *Analytica chimica acta, 652*(1-2), pp.54-65.

Smaoui, S., Agriopoulou, S., D'Amore, T., Tavares, L. and Mousavi Khaneghah, A., 2023. The control of Fusarium growth and decontamination of produced mycotoxins by lactic acid bacteria. *Critical Reviews in Food Science and Nutrition, 63*(32), pp.11125-11152.

Sowmya, K.L. and Ramalingappa, B., A review Mycotoxins: Extraction, characterization and analysis.

Tsai, J.F., Wu, T.S., Huang, Y.T., Lin, W.J., Yu, F.Y. and Liu, B.H., 2023. Exposure to Mycotoxin Citrinin Promotes Carcinogenic Potential of Human Renal Cells. *Journal of Agricultural and Food Chemistry, 71*(48), pp.19054-19065.

Zhang, X., Li, G., Wu, D., Liu, J. and Wu, Y., 2020. Recent advances on emerging nanomaterials for controlling the mycotoxin contamination: From detection to elimination. *Food Frontiers, 1*(4), pp.360-381.

Chapter 4

Epidemiology and Spread of Fungal Pathogens in Storage and Food Processing

Arushi Aren[1], Deepika Bansal[2] and Mukesh Kumar[1]*

[1]Department of Botany and Microbiology, Gurukula Kangri (Deemed to be University), Haridwar-249404, Uttarakhand
[2]Department of Floriculture and Landscaping, Maharana Pratap Horticultural University, Karnal-132001, Haryana
**Correspondence Email: mukesh.kumar@gkv.ac.in*

Abstract

From farm to fork, food contamination can occur at any stage of the supply chain. It is critical to determine the main sources of contamination. Food contamination may occur rapidly and readily at any-time; thus, it must be prevented and controlled with careful handling and observation. Contamination is often divided into three stages: pre-harvest, harvest, and post-harvest, however it can happen at any time. Fungal pathogens are major culprits behind post-harvest losses, contaminating stored agricultural products and processed foods. The primary goal of food safety is to prevent foodborne illnesses and diseases. Neglecting food safety or mishandling food can lead to outbreaks of foodborne diseases and fatalities in both developing and developed nations. Consuming contaminated food can cause various hazards—physical, chemical, biological, and allergenic—making its prevention and control essential. Food contamination often results from poor hygiene and sanitation practices by workers. Every stage of food preparation is susceptible to contamination, necessitating careful attention at all times. The integration of advanced epidemiological methods and molecular tools is essential for effective detection and tracking of fungal contamination in food storage and processing. These approaches provide the precision and speed needed to manage fungal risks, ensuring the safety and quality of food products. Ongoing advancements in technology and methodology will continue to enhance our ability to combat fungal contamination and safeguard public health. This chapter explores the epidemiology and spread of fungal pathogens in storage and food processing environments, underscoring their significant impact on food safety and economic stability. Moreover, it consolidates current research to highlight the significant influence fungi have on the contemporary food industry and emphasizes the importance of continued scientific investigation in leveraging their potential for sustainable and safe food production.

Keywords: Contamination, Epidemiology, Food-Borne Diseases, Fungal Pathogens, Post-Harvest Losses.

Introduction

Food spoilage, both before and after harvest, involves a variety of fungi that differ widely in their ecological behaviour. These range from plant pathogens and weak pathogens to superficial saprophytes that damage food before harvest, and molds that cause losses in quality and dry matter in storage. Some of these fungi produce toxic metabolites that are of greater concern to consumers than the visual appearance of the food itself. The colonization of plant surfaces by microorganisms begins soon after leaves or flowers are exposed to the air, with bacteria usually arriving first, followed by yeasts and eventually pathogenic and saprophytic filamentous fungi. These filamentous fungi continue to grow throughout the plant's life, especially as it ages and its seeds mature. The harvest process disrupts the ecosystem, transitioning plants from the dynamic environment of the field to the more stable conditions of storage, bringing about a significant shift in the microbial composition. Fungi found on plants before harvest are typically referred to as "field fungi" and include species like *Cladosporium*, *Alternaria*, *Epicoccum*, *Verticillium*, *Fusarium*, and phytopathogens. The distinction between pathogens and saprophytes is often unclear, with some fungi, like Alternaria, displaying a range from being leaf surface saprophytes to weak pathogens.

Field fungi can persist on fresh produce post-harvest, causing visible lesions on fruits and vegetables, but in stored products such as cereal grains, they only remain if the grain is dry enough to inhibit the growth of storage fungi. These storage fungi, primarily species with *Aspergillus* and *Penicillium* anamorphs, are usually present in small numbers before harvest. As moisture levels rise, fungal activity becomes more vigorous, resulting in spontaneous heating and the proliferation of thermophilic fungi, including species from the genera *Absidia*, *Rhizomucor* and *Humicola*, frequently occurring alongside thermophilic actinomycetes. However, the distinction between field and storage fungi is not always clear-cut. Some species, like *Aspergillus flavus*, which is typically associated with storage, also appear in the field in humid tropical regions. Therefore, it is more accurate to classify organisms based on their ecological requirements rather than their location in the field or storage. Both pre-harvest and during storage, the growth of filamentous fungi is influenced by environmental factors such as water availability, temperature, and gas composition; interactions with other microorganisms and arthropods; and control measures applied. This chapter reviews the role of these factors and illustrated them with examples of fungi that cause spoilage in food and stored products.

Epidemiology of Fungal Pathogens

The epidemiology of fungal pathogens in plants is a crucial aspect of plant pathology that explores how these pathogens spread, the conditions that favour their development, and their interactions with host plants. Fungal diseases often arise when a susceptible plant host, a virulent pathogen, and conducive environmental conditions intersect—commonly referred to as the disease triangle. These pathogens can be transmitted through various means, such as airborne spores, soil, water, and even infected seeds, leading to widespread outbreaks across agricultural regions. Environmental factors like temperature, humidity and wind play significant roles in determining the severity and spread of fungal infections. High humidity and moderate temperatures, for example, are typically conducive to spore germination and infection. Additionally, the susceptibility of plants can vary based on genetic resistance, stress levels, and age, with younger and older plants often being more vulnerable. Understanding the epidemiology of these pathogens is vital for developing effective control strategies, such as crop rotation, fungicide application, and the use of resistant plant varieties, all aimed at mitigating the impact of fungal diseases on crops and ensuring food security (Pfaller, 1994).

Food is rarely viewed as an ecosystem, possibly because it isn't seen as a "natural" system. However, food does function as an ecosystem, and a significant one at that, as food plants and the fungi that colonize their reproductive parts (like seeds and fruits) have been co-evolving for thousands of years. The seed and nut caches stored by rodents have created a habitat for the development of storage fungi, while fallen fruits, as they undergo decay and drying, have served as a substrate for various pathogenic and spoilage fungi. Humans have further encouraged the growth of food spoilage fungi by creating large and diverse food storage environments. It can be argued that fast-evolving organisms, such as haploid asexual fungi, are taking advantage of the niches humans have created by cultivating certain plants for food. This chapter discusses the various factors, including water activity, temperature, and pH, that influence fungal growth in food (Oraon *et al.,* 2024).

Common Fungal Pathogens in Food and Storage Environments

Common fungal pathogens in food and storage environments pose significant risks to food safety and quality, leading to spoilage and potential health hazards. Among these, *Aspergillus* and *Penicillium* species are particularly prevalent, often found in grains, nuts, and dried fruits. *Aspergillus* species, such as *Aspergillus flavus* and *Aspergillus niger*, are notorious for producing mycotoxins like aflatoxins, which are highly toxic and carcinogenic. *Penicillium* species, including *Penicillium expansum* and *Penicillium verrucosum*, are also common in stored foods, contributing to spoilage and producing mycotoxins like ochratoxin A. Additionally, *Fusarium* species are frequent contaminants of cereals and legumes, causing diseases like Fusarium head blight and producing harmful mycotoxins such as fumonisins and trichothecenes. These fungi thrive in environments with high humidity and improper storage conditions, where they can grow rapidly and contaminate large quantities of food. Controlling these pathogens requires stringent storage practices, regular monitoring, and, when necessary, the use of antifungal agents to minimize their impact on food safety and quality (Jurick *et al.,* 2024).

Aspergillus spp.

Aspergillus molds are extremely common, playing a significant role in food spoilage, though some species are also beneficial in the production of fermented foods. This genus thrives in environments with high sugar and salt concentrations, making it well-suited to growing in many foods with low moisture content. The *Aspergillus* genus consists of 344 species (Samson *et al.,* 2014), and there is considerable chemical diversity among these species.

Penicillium spp.

Penicillium is a well-known and highly prevalent fungus, found across the globe and having a significant economic influence on human life. Its primary role in nature is breaking down organic matter, with some species causing severe rot as pre- and post-harvest pathogens on food crops (Pitt & Hocking, 2009), and producing various mycotoxins (Frisvad & Samson, 2004). However, certain species have beneficial uses, particularly in the food industry, where they are used in the production of specialty fermented sausages (Ludemann *et al.,* 2010) and cheeses like Camembert and Roquefort (Giraud *et al.,* 2010).

Fusarium spp.

Fusarium species are widely distributed molds, commonly found as soil saprophytes. They are significant plant pathogens, responsible for diseases such as crown rot, head blight, and scab in cereal crops. Occasionally, *Fusarium* can infect animals, causing both invasive and mycotoxin-related illnesses. In humans, *Fusarium* species can lead to a variety of infections, ranging from superficial infections like keratitis and onychomycosis to locally invasive and even disseminated infections, the latter primarily affecting severely immunocompromised individuals. *Fusarium* can also trigger allergic conditions like sinusitis in healthy individuals, and mycotoxicosis through ingestion of food contaminated with its toxins. These fungi are prevalent in various environments, including soil, plant material, organic substrates, water systems (as biofilms), and even the air. Their ability to thrive on diverse substrates and disperse efficiently contributes to their widespread occurrence. Fusariosis can occur year-round but is more common during rainy seasons, as wind and rain play key roles in spreading Fusarium spores. While over 50 species of *Fusarium* are known, only a few are associated with human infections, with *F. solani* being the most common (accounting for around 50% of cases), followed by *F. oxysporum* (20%), and *F. verticillioides* and *F. moniliforme* (10% each). Other species linked to human infection include *F. dimerum*, *F. proliferatum*, *F. chlamydosporum*, and others. *F. solani* is the primary cause of fusarial keratitis and, along with *F. oxysporum*, is responsible for most Fusarium-related cases of onychomycosis (Garnica and Nucci, 2013).

Sources and Routes of Contamination

Sources and routes of contamination by fungal pathogens in food and storage environments are diverse and multifaceted, often beginning in the field and continuing throughout the food supply chain. Field contamination occurs when crops are exposed to fungal spores in the soil, water, or air, particularly during growth, harvesting, and post-harvest handling. Fungal spores can be introduced through contaminated seeds, irrigation water, or by insects that serve as vectors. Once harvested, food products are vulnerable to storage contamination, where improper conditions such as high humidity, inadequate ventilation, and fluctuating temperatures create ideal environments for fungal growth. Storage facilities themselves can become sources of contamination if they are not properly cleaned and maintained, allowing fungal spores to persist and spread. Additionally, cross-contamination during processing, transportation, and packaging is a significant route, where previously uncontaminated products come in contact with contaminated surfaces, equipment, or packaging materials. Insect infestations in storage areas can also contribute to the spread of fungal pathogens by transferring spores from one product to another. Understanding these sources and routes of contamination is crucial for implementing effective prevention and control measures to protect food quality and safety (Alegbeleye *et al.,* 2018).

Factors Influencing Fungal Growth

Fungal growth in food storage and processing environments is influenced by several environmental and biological factors. Understanding these factors is essential for implementing strategies to control fungal contamination, prolonging shelf life, and maintaining food safety. The following subsections detail the major factors influencing fungal proliferation.

Temperature

Temperature is a critical factor for fungal growth. Most fungi grow optimally within a moderate temperature range of 20 to 30°C. Temperatures above or below this range can slow down fungal metabolism or even kill spores. However, certain fungi can grow at lower temperatures (psychrophilic fungi) or higher temperatures (thermophilic fungi). Effective control of storage temperatures is essential to inhibit fungal activity, as low temperatures can suppress growth, while high temperatures may destroy fungal spores. Cold storage and freezing are common methods to limit fungal contamination in food products (Seib *et al.,* 2023).

Humidity

Fungi require moisture to grow, and high humidity provides the ideal environment for fungal spores to germinate and multiply. Humidity levels above

65% significantly enhance fungal growth, especially in stored crops and perishable food products. Controlling humidity in storage areas is critical for reducing moisture content, which can prevent mold formation. Techniques such as proper ventilation, dehumidification, and maintaining dry storage conditions are employed to minimize fungal contamination in food storage (Davies *et al.,* 2021).

Substrate Composition

Fungi feed on organic matter, making the composition of food and storage substrates a significant factor for fungal growth. Foods rich in carbohydrates, proteins, and lipids provide ample nutrients for fungi. Certain food types, such as cereals, fruits, nuts, and dairy products, are more prone to fungal contamination due to their nutrient content. The substrate's pH also influences fungal proliferation, with slightly acidic environments favouring fungal growth. Understanding the composition of stored food products and altering conditions, such as reducing nutrient availability, can help prevent fungal growth (Pouris *et al.,* 2024).

Storage Conditions

Storage conditions, including light, ventilation, and oxygen levels, play a vital role in fungal development. Fungi thrive in dark, poorly ventilated areas where oxygen is available. Reducing oxygen exposure through vacuum or modified atmosphere packaging can inhibit fungal growth. Additionally, exposure to UV light can help reduce fungal contamination on surfaces and equipment. Proper storage practices, such as using clean containers, regulating airflow, and maintaining sanitation, are essential to keep fungal growth in check (Tang *et al.,* 2015).

Impact of Fungal Pathogens on Food Quality and Safety

Mycotoxin Production

Mycotoxins are secondary metabolites produced by specific filamentous fungi, often forming in foods due to fungal contamination. When ingested by higher vertebrates and other animals, these compounds trigger a toxic response known as mycotoxicosis. In humans, mycotoxin exposure mainly occurs through plant-based foods or through residues and metabolites found in animal-derived products, potentially leading to liver and kidney damage. Some mycotoxins are neurotoxic, while others interfere with protein synthesis, causing effects ranging from skin sensitivity or tissue necrosis to severe immunodeficiency. The primary mycotoxigenic fungi in the human food supply belong to the *Aspergillus, Fusarium,* and *Penicillium* genera. *Fusarium* species are destructive plant pathogens that produce mycotoxins before or soon after harvest, whereas *Penicillium* and *Aspergillus* species are more commonly found contaminating food and commodities during the drying and storage phases (Singh, K. and Kumari, 2022).

Fungi are abundant in nature, with an estimated 19 Phyla, 83 classes, 1,220 families, 10,685 genera and ca 140,000 species (Hyde *et al.,* 2024). They inhabit diverse ecosystems and can produce a variety of extracellular compounds known as secondary metabolites. Some of these metabolites are beneficial, especially in pharmaceuticals, while others are harmful and known as mycotoxins (Magan and Aldred, 2007). Mycotoxins, produced by filamentous fungi (molds), contaminate agricultural products before or after harvest, posing significant food safety risks, public health concerns, and economic losses by lowering crop value and limiting grain marketability. Despite the identification of hundreds of mycotoxins, only a few significantly impact global agriculture. The most important mycotoxins with agro-economic relevance include aflatoxins, ochratoxins, trichothecenes, zearalenone, fumonisins, tremorgenic toxins, and ergot alkaloids (Atanda *et al.,* 2011). The acute and chronic health effects of these mycotoxins on humans and animals have been well documented (Bennett and Klich, 2003; Richard, 2007). Among the vast array of fungi, some mycotoxins are highly carcinogenic and toxic to vertebrates. The most important mycotoxigenic molds in food contamination belong to the *Aspergillus, Penicillium,* and *Fusarium* genera (Moss, 2000).

Spoilage and Sensory Changes

Fungal contamination in food products leads to spoilage, causing significant alterations in both the physical and sensory qualities of the affected items. Spoilage occurs as fungi metabolize nutrients within the food, resulting in visible and chemical changes that affect its edibility, appearance, texture, flavour, and aroma. One of the most noticeable effects of fungal spoilage is the change in appearance, with mold growth often visible as white, green, black, or gray patches on the surface of food products. Over time, these molds penetrate deeper into the food, causing discoloration and textural degradation. Fruits and vegetables may develop soft spots or become mushy and slimy, while grains and baked goods can turn brittle, crumbly, or soggy. Fungal growth also results in the breakdown of cellular structures within food, leading to significant textural changes that render the food inedible (Pouris *et al.,* 2024).

In addition to the visible signs of spoilage, fungi can drastically alter the flavour of food products. As they break down carbohydrates, proteins, and fats, they release metabolic byproducts that impart sour, musty, or unpleasant flavours. These undesirable taste changes are caused by the release of organic acids, alcohols, and other fermentation byproducts as fungi degrade the food matrix. Odor is another key sensory change associated with fungal contamination. Fungal activity produces distinct odours, ranging from musty, earthy smells to pungent and off-putting aromas. These odours are due to the production of volatile compounds, such as alcohols, ketones, and sulphur-containing chemicals, during fungal metabolism.

Fungal spoilage not only affects the sensory attributes of food but also diminishes its nutritional value. As fungi consume essential nutrients within the food, including carbohydrates, proteins, and vitamins, the overall nutritional quality is reduced. This depletion of nutrients, along with the negative changes in appearance, texture, flavour, and odour, makes the spoiled food unfit for consumption. The combination of these sensory and nutritional changes highlights the importance of preventing fungal contamination in food storage and processing environments to maintain product quality, safety, and nutritional integrity (Pouris *et al.,* 2024).

Spread of Fungal Pathogens in Storage

The spread of fungal pathogens in storage environments is a significant challenge in the preservation of food products. Fungi, including molds and yeasts, can easily contaminate food during storage, leading to spoilage, reduced shelf life, and in some cases, the production of harmful mycotoxins that pose health risks to consumers. Several factors contribute to the spread of fungal pathogens in storage facilities, including environmental conditions, the physical state of the food, and the presence of spores or other sources of contamination (Martín *et al.,* 2022).

Airborne fungal spores are a primary means of spread in storage environments. These microscopic spores are carried by air currents and can settle on food products, packaging, and surfaces within the storage area. Once settled, they can begin to grow and reproduce if the conditions such as temperature and humidity are favourable. Poor ventilation and inadequate air filtration systems in storage facilities can exacerbate this problem, allowing spores to spread rapidly and contaminate large quantities of food (Pakdel *et al.,* 2023).

Cross-contamination from infested products is another common way fungal pathogens spread in storage. When one batch of food is already contaminated, the fungi can easily move to adjacent products, especially in bulk storage or tightly packed environments. This can occur through direct contact between contaminated and uncontaminated products or via equipment, tools, and surfaces that have not been properly sanitized. For example, grains, nuts, and dried fruits are particularly vulnerable to fungal cross-contamination if not stored in a clean, dry environment (Akinsemolu and Onyeaka, 2024).

Moisture plays a crucial role in the proliferation of fungal pathogens. Fungi thrive in humid environments, and even a small increase in moisture levels can lead to rapid fungal growth. Condensation in storage facilities, leaks, or improper packaging can create moist conditions that encourage the spread of fungi. Once fungi begin to grow on one product, they can quickly spread to others in close proximity, particularly in warm, humid storage facilities (Awuchi *et al.,* 2021).

Inadequate storage conditions such as poor temperature control, improper packaging, and insufficient cleaning and maintenance of storage areas also contribute to the spread of fungal pathogens. Warm temperatures accelerate fungal growth, while packaging that does not provide a proper barrier to moisture and air can facilitate fungal contamination. Storage facilities that are not regularly cleaned or maintained may harbour fungal spores on walls, floors, and equipment, creating a continuous source of contamination.

To prevent the spread of fungal pathogens in storage, it is essential to implement effective control measures such as maintaining optimal temperature and humidity levels, using appropriate packaging materials, regularly cleaning storage areas, and ensuring that storage equipment and tools are sanitized. Additionally, monitoring for fungal contamination and taking immediate action when contamination is detected can help to minimize the spread of fungal pathogens and protect food quality and safety (Ahmed et al., 2024).

Modes of Transmission in Storage Facilities

Fungal pathogens can spread through various modes of transmission in storage facilities, leading to contamination, spoilage, and potential health risks. These modes of transmission include direct contact, airborne spread, vector-based contamination, and environmental factors. Understanding these transmission mechanisms is crucial to implementing effective preventive measures and maintaining food quality in storage environments (Sagar and Pareek, 2020).

Air-borne Transmission

Airborne fungal spores are one of the most common ways fungi spread in storage facilities. These microscopic spores are released into the air by contaminated food products, packaging, or surfaces and can be carried by air currents throughout the facility. Poor ventilation systems, lack of air filtration, and stagnant air increase the likelihood of spores settling on food items and initiating fungal growth. For instance, molds such as *Aspergillus* and *Penicillium* are known to spread via airborne spores, leading to widespread contamination in bulk storage areas (Pakdel *et al.,* 2023).

Cross-Contamination Through Direct Contact

Fungal pathogens can spread when contaminated and uncontaminated food products come into direct contact with each other. This occurs frequently in bulk storage systems, where large quantities of food are packed tightly together. Once one item becomes contaminated, the fungi can easily spread to adjacent products. Inadequate handling, such as using the same containers, tools, or equipment for both contaminated and uncontaminated products, exacerbates this transmission mode. Fruits, grains, and vegetables are particularly vulnerable to direct-contact contamination in storage (Akinsemolu and Onyeaka, 2024).

Transmission *via* Packaging Materials

Improper or damaged packaging can facilitate the transmission of fungal pathogens. Packaging that does not provide an adequate barrier against moisture, air, and external contaminants allows fungal spores to enter and infect the food. In some cases, fungi can grow on the packaging material itself, especially if it retains moisture or organic residue. Plastic, paper, and cardboard packaging that is not sealed tightly or becomes compromised during storage can provide a pathway for fungal growth and spread (Vermelho *et al.,* 2024).

Moisture-Driven Transmission

High humidity and moisture levels in storage facilities promote the growth of fungi, which can then spread across stored products. Condensation, leaks, or improper drying of products before storage can create moist conditions that encourage fungal colonization. Fungi thrive in moist environments, and once established on one product, they can quickly spread to others. Moisture-driven transmission is particularly problematic in products with high moisture content, such as fresh produce and grains, as well as in storage areas with poor moisture control (Awuchi *et al.,* 2021).

Contamination from Surfaces and Equipment

Storage surfaces, such as floors, walls, and shelves, as well as handling equipment, can serve as reservoirs for fungal pathogens. If not properly cleaned and sanitized, these surfaces can harbour fungal spores that contaminate food products upon contact. Forklifts, conveyor belts, and storage bins are common culprits in spreading fungi if not regularly disinfected. Contamination from these surfaces can occur either directly, through contact with food, or indirectly, by contaminating packaging materials or air in the facility (Morris and Dynasty, 2024).

Transmission by Pests (Vectors)

Rodents, insects, and other pests that infest storage facilities can act as vectors for fungal transmission. These pests can carry fungal spores on their bodies and transfer them to food products or surfaces they come into contact with. Pests may also cause physical damage to packaging or food items, creating entry points for fungal pathogens. Rodent droppings and insect activity can further contribute to the spread of fungi, particularly in poorly sealed or infested storage environments (Chava *et al.,* 2024).

Role of Packaging in Preventing or Facilitating Spread

Packaging plays a crucial role in both preventing and facilitating the spread of fungal pathogens in storage environments. Properly designed and maintained packaging can serve as a physical barrier that protects food products from contamination, while inadequate or compromised packaging can create ideal conditions for fungal growth and transmission.

Prevention Through Packaging

a. Moisture Control

One of the primary functions of packaging is to regulate the moisture content of stored food products. Fungal pathogens thrive in humid conditions, and packaging materials that prevent moisture buildup are essential in preventing fungal growth. Vacuum-sealed or moisture-resistant packaging can reduce the risk of condensation, which often creates a breeding ground for fungi. Proper packaging can also help maintain the dryness of low-moisture food products like grains and cereals, minimizing their susceptibility to fungal contamination (Davies *et al.*, 2021).

b. Oxygen Barrier

Many fungi require oxygen to grow and spread. Packaging that is designed to limit oxygen exposure—such as modified atmosphere packaging (MAP)—can significantly reduce the growth of aerobic fungal species. MAP works by replacing the oxygen in the packaging with gases like carbon dioxide or nitrogen, creating an inhospitable environment for fungal growth. This technique is especially effective for fresh produce, which is vulnerable to fungal spoilage in high-oxygen environments (Kandeepan *et al.*, 2022).

c. Physical Barrier

Packaging serves as a physical barrier that protects food products from external contaminants, including airborne fungal spores. Tightly sealed, robust packaging can prevent fungal spores from coming into direct contact with food products, reducing the risk of contamination. Properly sealed packaging also limits direct contact between food items and contaminated surfaces or handling equipment in storage facilities, further preventing the spread of fungi (Jildeh *et al.*, 2021).

d. Temperature Regulation

Some advanced packaging solutions incorporate insulation or temperature-regulating materials that help maintain a stable internal temperature. Since fungal pathogens often thrive in warm, humid environments, packaging that keeps food products cool can inhibit fungal growth. For instance, cold-chain packaging systems designed for perishable goods such as fruits, vegetables, and dairy products help maintain optimal storage temperatures and reduce fungal contamination (Seib *et al.*, 2023).

Facilitation of Fungal Spread Through Poor Packaging

a. Compromised Seals and Damaged Packaging

Damaged or poorly sealed packaging is one of the leading factors facilitating the spread of fungal pathogens in storage. Small punctures, tears, or unsealed edges allow airborne spores to infiltrate the packaging and contaminate the food products inside. Additionally, moisture and oxygen can enter through damaged areas, creating ideal conditions for fungal growth. Once inside, the fungus can quickly spread within the packaging, affecting the entire batch (Okori *et al.*, 2022).

b. Inadequate Material Selection

Certain packaging materials, such as low-quality plastics or thin cardboard, may not provide sufficient protection against environmental factors like moisture and oxygen. These materials can also degrade over time, especially in humid or fluctuating temperature conditions, allowing fungal spores to enter. Some packaging materials may also absorb moisture, creating an environment that promotes fungal growth on the packaging surface, which can then spread to the food products (Kohli et *al.,* 2024).

c. Reusable Packaging and Cross-Contamination

In some storage facilities, reusable packaging materials like crates, bins, and pallets are commonly used. If these materials are not properly cleaned and sanitized between uses, they can harbour fungal spores and act as vectors for cross-contamination between batches of food products. This is especially true for porous materials like wood, which can retain moisture and fungal spores, making them more likely to spread fungi to other products (Accorsi *et al.,* 2022).

d. Lack of Proper Ventilation in Packaging

While some food products require sealed packaging, others especially fresh produce need ventilation to maintain quality and reduce moisture buildup. Packaging that lacks adequate ventilation can trap humidity inside, creating a damp environment that encourages fungal growth. Poorly designed packaging without proper airflow can accelerate the spoilage process, making food products more vulnerable to fungal pathogens (Bandyopadhaya *et al.,* 2024).

Cross-contamination Risks During Storage

Cross-contamination during storage is a critical concern in maintaining food safety and quality, particularly due to the susceptibility of storage facilities to fungal pathogens. Fungi can spread easily through airborne spores, physical contact, or shared equipment. Improper cleaning of reusable containers, pallets, and surfaces increases the risk of cross-contamination between batches of stored products. Additionally, fungal spores may be introduced via infested raw materials or packaging materials, which can serve as vectors for fungal spread throughout storage facilities. Packaging integrity plays a key role in either preventing or facilitating fungal contamination. Damaged or improperly sealed packaging can allow spores to enter and infect the stored product, while protective, moisture-resistant packaging can limit fungal growth (Akinsemolu and Onyeaka, 2024).

Storage conditions, including temperature, humidity, and ventilation, further impact the risk of cross-contamination. If environmental conditions are not adequately controlled, high humidity or fluctuating temperatures can create favourable conditions for fungal growth. Moreover, improper stacking and handling can lead to the transmission of fungi between products, especially in facilities where segregating contaminated from uncontaminated goods is not practiced. In cases where infected products are not immediately identified and quarantined, the risk of fungal spread increases significantly, compromising the entire storage inventory. Therefore, strict sanitation protocols, regular monitoring, and well-maintained storage environments are crucial for minimizing cross-contamination risks and preventing the spread of fungal pathogens in storage facilities (Possas and Pérez-Rodríguez, 2023).

Spread of Fungal Pathogens During Food Processing

Fungal pathogens can spread at various stages of food processing, from raw material handling to post-processing storage. Each phase introduces critical control points where contamination risks must be managed to prevent the proliferation and spread of fungi. The role of fungal contamination in food safety is significant, as it can lead to spoilage, reduced shelf life, and health risks due to the production of mycotoxins by certain fungi. By identifying critical control points and implementing strict hygiene and safety measures, the spread of fungal pathogens can be minimized (Davies *et al.,* 2021).

Critical Control Points in Food Processing

Critical control points (CCPs) in food processing are stages where the risk of contamination is particularly high, and where preventive or corrective measures can be applied to eliminate or reduce fungal presence. These points may include raw material reception, processing equipment, cooking stages, and packaging. Monitoring temperature, humidity, and cleanliness at these stages is essential to ensure that fungal spores are not introduced or allowed to proliferate. Preventive actions such as proper sanitation of surfaces, air filtration systems, and timely equipment maintenance are necessary to reduce the risk of contamination (Awuchi *et al.,* 2023).

Contamination Risks in Different Stages of Processing

Raw Material Handling

Contamination risks begin at the raw material stage, where crops, grains, fruits, or other ingredients can arrive already contaminated with fungal spores. Inadequate cleaning, sorting, and handling can spread these spores to other materials or processing equipment. Poor storage before processing, such as exposure to moisture or fluctuating temperatures, can also encourage fungal growth. Ensuring the cleanliness of raw materials and maintaining proper storage conditions before processing are crucial to limiting fungal contamination (Bari *et al.,* 2024).

Processing and Cooking

During the processing and cooking phases, contamination risks arise from contact between raw materials, surfaces, and equipment. Fungal spores can survive if cooking temperatures are not consistently high enough to kill them, especially in cases where foods are not uniformly cooked. Moreover, cross-contamination can occur when equipment is not adequately cleaned between batches. Critical processes, such as blanching, pasteurization, and sterilization, should be carefully monitored to ensure that they effectively kill fungal spores and minimize contamination risks (Pinto *et al.,* 2020).

Post-processing Contamination

After food has been processed, it remains vulnerable to fungal contamination during packaging, storage, and transport. Improperly sealed packaging can allow fungal spores to enter and contaminate the food. Additionally, exposure to high humidity or temperature fluctuations can accelerate fungal growth on finished products. Hygienic packaging, proper sealing, and storage in controlled environments are necessary to prevent fungal contamination post-processing. Furthermore, handling by workers, improper transportation, or contact with contaminated surfaces can introduce fungi after processing, highlighting the importance of strict post-processing hygiene measures.

Influence of Processing Techniques on Fungal Survival and Spread

The effectiveness of various processing techniques in reducing fungal contamination depends on the specific method used. Heat-based methods, such as pasteurization, are effective in killing most fungal spores, while cold storage inhibits their growth but may not eliminate them. Techniques like fermentation and drying alter the water activity levels in food, making it less conducive for fungal growth. However, improper application of these methods can result in incomplete fungal inactivation. Properly designed and maintained processing equipment, along with continuous monitoring of environmental conditions, plays a vital role in controlling fungal survival and preventing their spread throughout the food processing chain (Pinto *et al.,* 2020).

Impact of Climate Change on Fungal Pathogens

Climate change has a significant impact on the prevalence, behaviour, and management of fungal pathogens in food storage and processing environments. Changes in temperature, humidity, and precipitation patterns directly influence fungal growth, survival, and the ability to infect food products. As global temperatures rise, fungal pathogens may thrive in regions previously unsuitable for their growth, leading to the expansion of their geographic range and increasing the risk of contamination during both pre-harvest and post-harvest stages.

Temperature fluctuations associated with climate change create conditions conducive to fungal proliferation in storage environments. Many fungal species grow rapidly at higher temperatures, which accelerates spoilage of food products. This is particularly problematic in regions that already experience warm climates, as inadequate cooling or ventilation systems may fail to prevent fungal colonization. Temperature shifts also disrupt the natural seasonality of certain fungal pathogens, causing earlier or prolonged periods of activity, thus increasing the likelihood of contamination. Increased humidity due to rising global temperatures and altered precipitation patterns also contributes to fungal growth. High moisture levels in storage facilities provide ideal conditions for molds and fungi to thrive, particularly in grains, fruits, and vegetables. When humidity levels are not properly controlled, fungal spores can spread quickly, leading to significant spoilage and economic losses. Moreover, in areas prone to extreme weather events, such as flooding, food processing facilities may experience an influx of moisture, further exacerbating fungal contamination risks (Nnadi and Carter, 2021).

Changes in atmospheric carbon dioxide (CO_2) levels are another factor that affects the growth of fungal pathogens. Elevated CO_2 levels can stimulate the growth of some fungal species, making them more resilient and potentially increasing their resistance to fungicides. Additionally, climate change can alter the composition of crops, potentially weakening their natural defense against fungal infection, which further increases the vulnerability of food products in storage. The interactions between climate change and fungal pathogens pose challenges for food safety and security. As fungi become more adaptable and aggressive under changing environmental conditions, the efficacy of traditional preservation and processing methods may decline. This necessitates the development of improved storage technologies and better management practices to mitigate the effects of climate change on fungal contamination (Hunjan and Lore, 2020).

Conclusion

Fungal contamination presents significant challenges to both food storage and processing environments, with risks that begin at the pre-harvest stage and continue throughout the entire supply chain. The epidemiology of fungal pathogens in storage and food processing highlights the critical factors influencing fungal growth, such as temperature, humidity, substrate composition, and storage conditions. Understanding the modes of transmission within storage facilities, as well as the role of packaging, plays a crucial role in preventing or facilitating the spread of fungal pathogens. In food processing, fungal contamination can occur at multiple stages, from raw material handling to post-processing storage. Each stage presents unique risks, with critical control points that must be carefully monitored to mitigate contamination. Techniques such as proper handling, sanitation, temperature control, and airtight packaging are essential in curbing fungal proliferation. Additionally, the influence of processing techniques on fungal survival underscores the need for efficient heat treatments, drying, and other preventive measures that effectively inhibit fungal growth. The impact of climate change further complicates the spread of fungal pathogens in storage and food processing, as fluctuating temperatures and humidity levels create more favourable environments for fungal growth. Addressing these challenges requires a comprehensive approach, integrating proper site selection, genetic optimization of crops, and effective post-harvest and processing strategies. Only through strict adherence to hygiene protocols, environmental controls, and technological interventions can the risks associated with fungal pathogens be significantly minimized, ensuring food safety, quality, and longevity.

References

Accorsi, R., Battarra, I., Guidani, B., Manzini, R., Ronzoni, M. and Volpe, L. (2022). Augmented spatial LCA for comparing reusable and recyclable food packaging containers networks. *Journal of Cleaner Production*, 375, 134027.

Ahmed, M.A.E.G.E.S., Abbas, H.S. and Kotakonda, M. (2024). Fungal diseases caused by serious contamination of pharmaceuticals and medical devices, and rapid fungal detection using nano-diagnostic tools: a critical review. *Current Microbiology*, 81(1), 10.

Akinsemolu, A.A. and Onyeaka, H.N. (2024). Microorganisms Associated with Food Spoilage and Foodborne Diseases. In *Food Safety and Quality in the Global South* (pp. 489-531). Singapore: Springer Nature Singapore.

Alegbeleye, O.O., Singleton, I. and Sant'Ana, A.S. (2018). Sources and contamination routes of microbial pathogens to fresh produce during field cultivation: A review. *Food Microbiology*, 73, 177-208.

Awuchi, C.G. (2023). HACCP, quality, and food safety management in food and agricultural systems. *Cogent Food & Agriculture*, 9(1), 2176280.

Awuchi, C.G., Ondari, E.N., Eseoghene, I.J., Twinomuhwezi, H., Amagwula, I.O. and Morya, S. (2021). Fungal growth and mycotoxins production: Types, toxicities, control strategies, and detoxification. *Fungal reproduction and growth*, 100207.

Bandyopadhaya, S., Sarkar, U. and Bera, D. (2024). Food Stability Simulation: Accelerated Shelf-Life Mechanism. *Food Coatings and Preservation Technologies*, 259-300.

Bari, M.L. and Uddin, M.N. (2024). 8 Food and Vegetables Contaminants in Fruits. *Food Safety: Contaminants and Risk Assessment*, 118.

Chava, N.R., Lakshmi, K.S.I., & Kumar, V.S. (2024). THEIR MANAGEMENT.

Davies, C.R., Wohlgemuth, F., Young, T., Violet, J., Dickinson, M., Sanders, J.W., Vallieres, C. and Avery, S.V. (2021). Evolving challenges and strategies for fungal control in the food supply chain. *Fungal Biology Reviews*, 36, 15-26.

Frisvad, J.C. and Samson, R.A. (2004). Polyphasic taxonomy of Penicillium subgenus Penicillium. A guide to identification of food and air-borne terverticillate Penicillia and their mycotoxins. *Studies in Mycology*, 49(1), 1-174.

Garnica, M. and Nucci, M. (2013). Epidemiology of fusariosis. *Current Fungal Infection Reports*, 7, 301-305.

Giraud, F., Giraud, T., Aguileta, G., Fournier, E., Samson, R., Cruaud, C., Lacoste, S., Ropars, J., Tellier A. and Dupont, J. (2010). Microsatellite loci to recognize species for the cheese starter and contaminating strains associated with cheese manufacturing. *International Journal of Food Microbiology*, 137(2-3), 204-213.

Hunjan, M.S. and Lore, J.S. (2020). Climate change: Impact on plant pathogens, diseases, and their management. *Crop Protection Under Changing Climate*, 85-100.

Hyde, K.D., Noordeloos, M.E. and Savchenko, K.G. (2024). The 2024 Outline of Fungi and fungus-like taxa. *Mycosphere*, 15(1), 5146–6239.

Jildeh, Z.B., Wagner, P.H. and Schöning, M.J. (2021). Sterilization of objects, products, and packaging surfaces and their characterization in different fields of industry: The status in 2020. *Physica Status Solidi (a)*, 218(13), 2000732.

Jurick II, W.M. and Adaskaveg, J.E. (2024). Postharvest diseases. In *Agrios' Plant Pathology* (pp. 317-330). Academic Press.

Kandeepan, G., & Tahseen, A. (2022). Modified atmosphere packaging (map) of meat and meat products: A review. *Journal of Packaging Technology and Research, 6*(3), 137-148.

Kohli, D., Loushigam, G., Richa, R. and Bajad, R. (2024). Recent Advances in Food Packaging for Shelf Life of Food. *Food Production, Diversity, and Safety Under Climate Change*, 273-284.

Lacey, J. (1989). Pre-and post- harvest ecology of fungi causing spoilage of foods and other stored products. *Journal of Applied Bacteriology*, 67, 11s-25s.

López-Dıaz, T.M., Santos, J.A., Garcia-López, M.L. and Otero, A. (2001). Surface mycoflora of a Spanish fermented meat sausage and toxigenicity of Penicillium isolates. *International Journal of Food Microbiology*, 68(1-2), 69-74.

Ludemann, V., Greco, M., Rodríguez, M.P., Basílico, J.C. and Pardo, A.G. (2010). Conidial production by *Penicillium nalgiovense* for use as starter cultures in dry fermented sausages by solid state fermentation. *LWT-Food Science and Technology*, 43(2), 315-318.

Magan, N. and Aldred, D. (2007) Why do fungi produce mycotoxins? In: Dijksterhuis J., Samson R.A. (eds) Food mycology: a multifaceted approach to fungi and food (pp. 121-133). CRC Press, Boca Raton.

Martín, I., Gálvez, L., Guasch, L. and Palmero, D. (2022). Fungal pathogens and seed storage in the dry state. *Plants*, 11(22), 3167.

Morris, D.D. (2024). *Post-Harvest Detection and Pathogen Control of Foodborne Pathogens in the Fresh Produce Industry* (Master's thesis, Michigan State University).

Nnadi, N.E. and Carter, D.A. (2021). Climate change and the emergence of fungal pathogens. *PLoS Pathogens*, 17(4), e1009503.

Okori, F., Cherotich, S., Baidhe, E., Komakech, A.J. and Banadda, N. (2022). Grain hermetic storage and post-harvest loss reduction in sub-saharan Africa: Effects on grain damage, weight loss, germination, insect infestation, and mold and mycotoxin contamination. *Journal of Biosystems Engineering*, 47(1), 48-68.

Oraon, S., Padamini, R., Shahni, Y.S., Das, N., Sinha, D., Sujatha, G.S. and Karanwal, R. (2024). Impact of Emerging Pathogens in Crop Production. *Microbiology Research Journal International*, 34(7), 80-92.

Pakdel, M., Olsen, A. and Bar, E.M.S. (2023). A review of food contaminants and their pathways within food processing facilities using open food processing equipment. *Journal of Food Protection*, 86(12), 100184.

Pfaller, M.A. (1994). Epidemiology and control of fungal infections. *Clinical Infectious Diseases*, 19(1), S8-S13.

Pinto, C.A., Moreira, S.A., Fidalgo, L.G., Inácio, R.S., Barba, F.J. and Saraiva, J.A. (2020). Effects of high-pressure processing on fungi spores: Factors affecting spore germination and inactivation and impact on ultrastructure. *Comprehensive Reviews in Food Science and Food Safety*, 19(2), 553-573.

Pitt, J.I. and Hocking, A.D. (2009). Fungi and food spoilage. New York: Springer-Verlag.

Possas, A. and Pérez-Rodríguez, F. (2023). New insights into cross-contamination of fresh-produce. *Current Opinion in Food Science*, 49, 100954.

Pouris, J., Kolyva, F., Bratakou, S., Argyro Vogiatzo, C., Chaniotis, D. and Beloukas, A. (2024). The role of fungi in food production and processing. *Applied Sciences*, 14(12), 5046.

Sagar, N.A. and Pareek, S. (2020). Safe storage and preservation techniques in commercialized agriculture. In *Natural Remedies for Pest, Disease and Weed Control* (pp. 221-234). Academic Press.

Samson, R.A., Visagie, C.M., Houbraken, J., Hong, S.B., Hubka, V., Klaassen, C.H. W., Perrone, G., Seifert, K.A., Susca, A., Tanney, J.B., Varga, J., Kocsubé, S., Szigeti, G., Yaguchi, T. and Frisvad, J.C. (2014). Phylogeny, identification and nomenclature of the genus Aspergillus. Studies in Mycology, 78, 141-173.

Seib, T., Fischer, K., Sturm, A.M. and Stephan, D. (2023). Investigation on the influence of production and incubation temperature on the growth, virulence, germination, and conidial size of Metarhizium brunneum for Granule development. Journal of Fungi, 9(6), 668.

Singh, K. and Kumari, A. (2022). *Mycotoxins and Mycotoxicoses* (pp. 129-136). Singapore: Springer.

Sweeney, M.J. and Dobson, A.D. (1998). Mycotoxin production by *Aspergillus*, *Fusarium* and *Penicillium* species. *International Journal of Food Microbiology*, 43(3), 141-158.

Tang, W., Kuehn, T.H. and Simcik, M.F. (2015). Effects of temperature, humidity and air flow on fungal growth rate on loaded ventilation filters. *Journal of Occupational and Environmental Hygiene,* 12, 525–537.

Vermelho, A.B., Moreira, J.V., Junior, A.N., da Silva, C.R., Cardoso, V.D.S. and Akamine, I.T. (2024). Microbial preservation and contamination control in the baking industry. *Fermentation*, 10(5), 231.

Chapter 5

Historical Perspectives Concerning the Detection of Fungal Pathogens and Mycotoxins

Rajbala Yadav[1*], Pooja[2] and Richa Kumari[3]

[1]Department of Microbiology, [2]Department of Plant Pathology, CCS HAU, Hisar-125004

[3]Department of Biotechnology, CURAJ, Ajmer, Rajasthan-305817

rajbalaydv335@gmail.com

Abstract

Mycotoxins, toxic secondary metabolites from various fungal pathogens, threaten food safety and public health. The detection and management of these compounds have evolved alongside our understanding of their health implications, including carcinogenicity, immunosuppression and endocrine disruption. Aflatoxins, ochratoxin A, and fumonisins are notable mycotoxins linked to severe health outcomes, particularly in regions reliant on staple crops susceptible to fungal contamination. Conventional detection methods, such as high-performance liquid chromatography (HPLC) and mass spectrometry, while effective, are often impractical for rapid field testing due to their complexity and cost. To address this, innovative methodologies have emerged, including enzyme-linked immunosorbent assay (ELISA) lateral flow assays enabling quicker and more accessible detection of mycotoxins. The historical perspective on mycotoxin mitigation has transitioned from conventional physical and chemical methods to more advanced strategies, such as the application of cold atmospheric plasma (CAP), natural essential oils (NEOs), and nanoparticles.

Keywords: Mycotoxins, Fungal pathogens, Food safety, Detection methods, ELISA, Biosensors

Introduction

The word "mycotoxin" was first introduced in 1962 after a London Veterinary emergency involving roughly 1,00,000 turkeys perished due to what was initially termed "turkey X-disease". This condition was linked later to contaminated feed containing aflatoxins from *Aspergillus flavus* (Allcroft and Carnaghan, 1962). Since then, mycotoxins have been recognised as a global concern affecting various agricultural products. Fungal infection can occur before and after harvesting, including drying, storage, and consumption (Streit et al., 2012). Exposure of mycotoxin is particularly problematic in developing countries, where food safety measures are often inadequate. In contrast, developed nations typically have established monitoring systems to ensure food quality (Stoev, 2013). The health effects of mycotoxin exposure can range from acute symptoms to chronic diseases, including rapid death in severe cases, tumour formation, immunosuppression, and endocrine disorders (Marin et al., 2013). Despite the identification of approximately 18,000 fungal metabolites until the 1960s, only a few have garnered significant interest in science (Cole and Cox, 1981; Science, 2003). Among most of the critical mycotoxins are aflatoxin, ochratoxins, trichothecenes, zearalenone, and fumonisins, which have substantial public health and economic implications. Aflatoxins are particularly notorious, being produced by common fungal species like *Aspergillus* and *Penicillium*. Notably, the primary producers of aflatoxins are *A. flavus, A. parasiticus*, and *P. puberulum* including aflatoxin B1 (AFB1), B2 (AFB2), G1 (AFG1) and G2 (AFG2). AFB1 is recognized as one of the most potent hepatocarcinogens and has also been shown to possess genotoxic properties. In dairy cattle, AFB1 and AFB2 can be converted into hydroxylated metabolites, known as aflatoxins M1 and M2 (AFM1 and AFM2), which are detectable in milk and dairy products from livestock that have consumed contaminated feed (Boudra et al., 2007). The WHO-International Agency for Research on Cancer has also recognised other mycotoxins known as ochratoxin A, trichothecenes, fumonisins, and zearalenone, as carcinogenic (WHO-IARC, 1993). A lethal metabolite of *Aspergillus ochraceus* Ochratoxin A has been demonstrated as teratogenic, hepatotoxic, immunotoxic, and carcinogenic effects in the experimental studies. Trichothecenes, primarily generated by *Fusarium* species, may lead to immunosuppression, neurotoxicity and renal toxicity (Richard, 2007).

Zearalenone (ZEA) is an estrogenic mycotoxin synthesized from various *Fusarium* sp., predominantly found in cereals, particularly maize. This fungal mycotoxin exhibits distinct effects on different sexes; in females, it can lead to precocious puberty while in males it may result in infertility and exhibit anabolic properties (Goertz et al., 2010). Fumonisins is another group of mycotoxins, are typically present in the maize which are contaminated and the other cereal grains, produced primarily by *Fusarium verticillioides* and *F. proliferatum* in a humid climate with high temperature. These mycotoxins frequently co-occur with aflatoxins, especially in corn (Sun et al., 2011). Fumonisins are known to be nephrotoxic, hepatotoxic, and potentially carcinogenic to humans (Alizadeh et al., 2012; Humans et al., 2002). Ergot alkaloids, which encompass several mycotoxins, are produced by the fungus *Claviceps purpurea* and are responsible for ergotism and historically referred to as St. Anthony's fire. This condition is recognized as one of the ancient foodborne diseases in humans and caused hallucinatory symptoms and numerous fatalities in France and other European nations during the Middle Ages (Betina, 1989). Despite being a widely overlooked issue, the Food and Agriculture Organization of the United Nations (FAO) estimates that around 25% of global cereals and crops are contaminated with molds and mycotoxins (Rice and Ross, 1994). Thus, it is essential to develop strategies to effectively manage the challenges posed by mycotoxins, thereby enhancing food security. Mycotoxicoses can be classified as either acute or chronic in humans. Acute toxicity presents a rapid onset and noticeable toxic response, while chronic toxicity results from prolonged low-dose exposure that can lead to cancer and other generally reversible effects (James and Zikankuba, 2018). Aflatoxin has been linked to serious conditions such as Kwashiorkor and Reye's syndrome in children (Blunden et al., 1991) and immunosuppression (Turner et al., 2003). The ruminants are generally less affected than the non-ruminant animals, but long-term consumption of mycotoxin-contaminated feed can negatively impact their production (milk, beef, or wool), reproduction, and growth (Hussein and Brasel, 2001). Aflatoxins are known to be potent hepatotoxins, immunosuppressants, mutagens, and carcinogens affecting companion animals, livestock, poultry, and humans (Eaton and Gallagher, 1994). Zearalenone significantly impacts reproduction by causing hyperestrogenism, particularly in prepubertal swine. Clinical signs include vulvar swelling, increased uterine size and secretions, mammary gland hyperplasia, prolonged estrus, anestrus, infertility, and complications like prolapses (Gupta et al., 2018). Fumonisins B1 and B2, commonly found in corn, have been associated with porcine pulmonary edema (PPE) (Colvin et al., 1993; Harrison et al., 1990; Osweiler et al., 1993) and equine leukoencephalomalacia (ELEM) (Wilson et al., 1995). They have been shown to cause liver damage across multiple species including pigs, horses, cattle, rabbits, and primates (Gumprecht et al., 1995; Haschek et al., 1992; Jaskiewicz et al., 1987; Osweiler et al., 1993; Voss et al., 1996), with specific organ toxicity observed in pigs (lungs), horses (brain), rats/rabbits/sheep (kidneys), and rats/pigs (esophagus) (Edrington et al., 1995; Gumprecht et al., 1995; Casteel et al., 1994; Lim et al., 1996). Epidemiological studies have linked the consumption of corn infected with *F. verticillioides* to human esophageal cancer (Rheeder et al., 1992) while fumonisins have demonstrated hepatocarcinogenic properties in rats and mice (Gelderblom et al., 1988; Howard et al., 2001). Ochratoxin A (OTA) and citrinin are recognised to cause nephropathy in animals and have been implicated in Balkan endemic nephropathy in humans. OTA is also carcinogenic to rodents (Creppy et al., 1985) and possesses teratogenic (Arora et al., 1983), immunotoxic (Størmer & Lea, 1995), neurotoxic (Bruinink & Sidler, 1997; Sava et al., 2006), mutagenic (Stetina & Votava, 1986), and genotoxic properties (Meisne et al., 1983). In comparison to OTA, ochratoxin B is less toxic and significantly less commonly found. Therefore, the ingestion of clover hay containing slaframine can induce salivary episodes lasting several hours to over three days in ruminants and horses (Gupta et al., 2018).

Table 1: Mycotoxin and its detection methods

Mycotoxin	*Source Organism*	*Health Effects*	*Detection Methods*	*References*
Aflatoxins (AFs)	*Aspergillus* sp.	Carcinogenic, hepatotoxic	HPLC, LC-MS/MS, ELISA	Andrade et al. (2013)
Ochratoxin A (OTA)	*Aspergillus* and *Penicillium* sp.	Nephrotoxic, possibly carcinogenic	HPLC, LC-MS/MS, ELISA	Huertas-Pérez et al. (2017)
Zearalenone (ZEA)	*Fusarium* sp.	Estrogenic effects, reproductive toxicity	HPLC, LC-MS/MS, ELISA	Han et al. (2017)
Patulin (PAT)	*Penicillium* sp.	Gastrointestinal toxicity, immunotoxicity	HPLC, LC-MS/MS, ELISA	Welke et al. (2009)
Fumonisins (FUMs)	*Fusarium* sp.	Carcinogenic, neurotoxic, affects sphingolipid metabolism	HPLC, LC-MS/MS, ELISA	Ottoboni et al. (2018)
Trichothecenes (TCs, e.g., DON, T-2)	*Fusarium* sp.	Immunosuppressive, gastrointestinal toxicity	HPLC, LC-MS/MS, ELISA	Pereira et al. (2015)

Decontamination of mycotoxin-contaminated feeds and foods

- Physical strategies, such as sorting, washing, crushing, and de-hulling of maize grains, have been effective in removing aflatoxins and fumonisins in Benin (Fandohan et al., 2005).
- Chemical strategies involve the fungicides application like prochloraz, propiconazole, epoxiconazole, tebuconazole, cyproconazole, Oltipraz, chlorophyllin, and azoxystrobin to reduce fumonisin and aflatoxin infection (Haidukowski et al., 2005; Hayes et al., 1998; Ni & Streett, 2005).
- Biological strategies rely on emergence of atoxigenic fungi that compete with toxigenic fungi in environment; the introducing atoxigenic strains of *A. flavus* and *A. parasiticus* to the soil of developing crops resulted in a 74.3% to 99.9% reduction in aflatoxin contamination of peanuts in the USA (Dorner, Cole, & Blankenship, 1998).

Conventional techniques used for detection and analysis of mycotoxins and fungal pathogens

Ever since the first mycotoxin was discovered, researchers have explored various methods to detect their presence in food and feed (Vasseghian et al., 2021). Chromatographic techniques have become the most prominent, utilizing different forms of chromatography like thin layer chromatography (TLC) and high-performance liquid chromatography (HPLC) which are often paired with the detectors such as diode array, fluorescence, and UV. Also, liquid chromatography-tandem mass spectrometry (LC-MS/MS) and gas chromatography-tandem mass spectrometry (GC-MS/MS) are widely used for the detection of mycotoxin (Zhang et al., 2018). For situations requiring quick analysis, immunoassay methods, including enzyme-linked immunosorbent assay (ELISA) and lateral flow immunoassay (LFIA), play a vital role (Lattanzio et al., 2019). Furthermore, biosensors are emerging as valuable tools for identifying mycotoxins in food products (Lattanzio et al., 2019).

Chromatography technique for mycotoxins

Thin Layer Chromatography

Thin-layer chromatography (TLC) has emerged as a popular and cost-effective method for screening large numbers of samples for the presence of mycotoxins. This technique involves a stationary phase which typically composed of alumina, silica, or cellulose that immobilized on an inert material like plastic or glass (Wacoo et al., 2014). The sample is carried through the stationary phase by a mobile phase consisting of a mixture of methanol, acetonitrile and water. TLC has proven to be an effective tool for detecting mycotoxins and fungal pathogens. Its low cost, simplicity, and ability to produce fluorescent spots under UV light make it an essential technique for analyzing various mycotoxins and fungi. TLC has been developed for both qualitative and quantitative evaluation of mycotoxins (Singh and Mehta, 2020). However, TLC has some limitations. It lacks sensitivity and accuracy, making quantification challenging. Also, proper sample preparation and clean-up procedures are crucial as they depend on specific properties and the type of mycotoxin being analyzed. Despite these drawbacks, TLC remains a valuable technique for economical screening and the detection of mycotoxins in a wide range of samples (Yang et al., 2014).

Liquid chromatography

To address the constraints of thin-layer chromatography (TLC) such as issues with plate height and environmental factors like temperature and humidity, liquid chromatography (LC) methods have been developed (Singh and Mehta, 2020). LC allows for the concurrently detection of multiple mycotoxins, despite their chemical structure or biological activity. This technique utilizes an analytical column along with a mobile phase to separate analytes from matrix components and make it particularly effective for high-polarity, non-volatile and thermally sensitive mycotoxins (Yang et al., 2020).

High-performance liquid chromatography (HPLC) is the most often used form of LC in mycotoxin and fungal pathogen analysis. The choice of adsorbent in HPLC depends on the physical and chemical characteristics of the specific mycotoxin being analyzed. Most protocols for mycotoxin detection are quite similar and the primary detectors used in HPLC include UV-visible (UV) and fluorescence (FLD) detectors as well as mass spectrometry (MS) options. Some mycotoxins, like aflatoxins and ochratoxin A, naturally fluoresce and can be detected directly using HPLC-FLD which is particularly effective for analyzing ochratoxin A in various food matrices, such as rice (Zinedine et al., 2007). For mycotoxins like fumonisin B1, which lack chromophores derivatization is necessary to introduce fluorescent moieties either before or after the chromatographic separation. Despite its advantages, HPLC has limitations, including portability issues and challenges related to matrix effects, sample types and calibration. The use of HPLC in mycotoxin analysis is well-documented in numerous studies.

In recent years, the application of LC-MS/MS has significantly increased for detecting low molecular weight contaminants and residues at trace levels. The combination of mass spectrometry with LC enhances reliability and sensitivity and making LC-MS/MS a standard tool for tackling analytical challenges in food and feed safety (Turner et al., 2009). This method offers a high selectivity and sensitivity ensuring better identification of the analytes across a wider range of matrices compared to traditional methods. Many mycotoxigenic fungi can produce multiple mycotoxins simultaneously, LC-MS/MS is especially valuable for the reliable and sensitive analysis of multiple mycotoxins in agricultural commodities. Studies have shown that the LC-MS/MS provides some of the most dependable results for the simultaneous mycotoxin detection (Bessaire et al., 2019).

Gas Chromatography

Gas chromatography (GC) relies on the differential partitioning of analytes between the mobile phase (carrier gas) and the stationary phase within the GC column. As the sample components distribute themselves between these two phases, they separate based on their relative affinities. After the separation process, the volatile products are detected using various detectors such as mass spectrometry (MS), electron capture detector (ECD), or flame ionization detector (FID) (Singh and Mehta, 2020).

However, GC is not commonly used for mycotoxin analysis due to the high polarity and low volatility of these compounds. To overcome this limitation, a derivatization step is often required to convert mycotoxins into more volatile derivatives. Despite these challenges, GC-MS/MS has been successfully employed for detecting mycotoxins in milled grain-based products and wheat, and semolina (Rodríguez-Carrasco et al., 2012). This technique offers high sensitivity and specificity for mycotoxin analysis when the compounds are properly derivatized to increase their volatility (Alshannaq et al., 2017). While GC-MS/MS can be a powerful tool for mycotoxin detection but it also faces some challenges. These include potential column blockage, drifting responses, cross-contamination from previous samples, and nonlinearity of calibration curves with certain detectors. Researchers must carefully optimize the derivatization process and address these technical issues to obtain reliable results when using GC-MS/MS for mycotoxin analysis.

Rapid Technologies for mycotoxin detection and fungal pathogens

Enzyme-Linked Immunosorbent Assay

Immunochemical methods like the Enzyme-Linked Immunosorbent Assay (ELISA) offer a rapid and straightforward approach for on-site mycotoxin and fungal pathogen analysis, complementing more complex chromatographic techniques (Al-Jaal et al., 2019). ELISA is user-friendly, enabling simultaneous testing of multiple samples with high accuracy and requiring only small sample volumes, which simplifies the process compared to methods like High-Performance Liquid Chromatography (HPLC) (Oplatowska-Stachowiak et al., 2018). This test relies on the interaction between an antigen-antibody complex and chromogenic substrates with results measured via spectrophotometry. ELISA has been successfully employed to detect mycotoxins in various food products, such as aflatoxin B1 in chicken liver samples (Solcan et al., 2013). However, it has limitations, including potential cross-reactivity that can lead to inaccurate results due to similar compounds binding to antibodies. The effectiveness of ELISA also depends on proper validation for specific matrices; inadequate validation restricts its applicability (Thway et al., 2014). Therefore, comprehensive studies are needed to assess ELISA's accuracy across a wider range of food commodities.

Lateral Flow Immunoassay

Lateral Flow Immunoassay (LFIA), also known as the immunochromatographic strip test, is a membrane-based immunoassay that utilizes a competitive method with a labelled antibody as a signalling agent (Song et al., 2014). In this test, capillary beds resembling porous paper facilitate the movement of the analyte, allowing the specific recognition elements to bind to immobilized moieties on the membrane surface. The accuracy of LFIA largely depends on the signal labels used, with the gold nanoparticles (GNPs) being the most common (Anfossi et al., 2013). Other materials like magnetic nanoparticles (MNPs), carbon nanoparticles (CNPs), and quantum dots (QDs) have also been employed to enhance detection effectiveness (Wang et al., 2009). LFIA offers simplicity, rapid results, and low cost, making it ideal for the large-scale on-site screening. It eliminates the need for sample clean-up, streamlining the testing process. However, limitations include potential interferences that can affect result accuracy and challenges in detecting trace analytes in complex matrices (Krska et al., 2009).

Biosensor

Biosensors integrate biological or biologically derived sensing elements to detect specific bio-analytes, coupled with a transducer to convert biological signals into electrical signals (Perumal et al., 2014). Various transducers can be used for the mycotoxin detection, including optical (surface plasmon resonance-SPR and fluorescence), electrochemical (potentiometric, amperometric, impedimetric, and piezoelectric (quartz crystal microbalance-QCM) methods (Oliveira et al., 2019). Common sensing materials include peptides, nucleic acid, antibodies, enzymes, cells, and the bioinspired elements like recombinant antibodies, aptamers, and molecularly imprinted polymers (MIPs) (Malekzad et al., 2017). To enhance sensitivity, metal nanoparticles, carbon nanotubes (CNTs), nanofibers, and quantum dots (QDs) are often incorporated due to their favourable properties (Doria et al., 2012). Quartz crystal microbalance transducers detect mass changes and vibrations at the electrode surface when molecular recognition occurs (Oliveira et al., 2019). Rapid mycotoxin analysis methods offer advantages such as speed, low cost, simplicity, and portability, allowing for on-site testing in food production environments without delays from laboratory shipping (Zheng et al., 2006). However, challenges remain, including antibody cross-reactivity, matrix interference and the need for proper confirmation of testing matrices.

Historical Approaches to Mitigate mycotoxins and fungal pathogens

Historically, mycotoxin and fungal pathogens mitigation has relied on preventive measures aimed at controlling fungal growth during crop production and storage (Ji et al., 2019). Traditional strategies included proper harvesting techniques, drying crops to moisture levels below 14%, and maintaining controlled storage conditions to hinder fungal proliferation. Physical methods such as sorting, sieving, and washing have been employed to lower mycotoxin levels in food products. These approaches focus on maintaining a "clean substrate" to avoid contamination, which is crucial for ensuring food safety (Owolabi et al., 2022). Physical methods include temperature and humidity control, irradiation, and various processing techniques like milling and peeling, which can help eliminate contaminated portions of food. Chemical methods often involve the use of agents such as ammonia or acid treatments that can degrade mycotoxins or develop them less toxic (Furlong et al., 2024). The Food and Agriculture Organization (FAO) has also created guidelines for the procedures of decontamination, emphasizing the importance of preserving the nutritional and technological attributes of food while minimizing the toxic residues (Owolabi et al., 2022).

One such innovation is cold atmospheric plasma (CAP) technology which has shown promise in decreasing the amount of mycotoxin without the need for high temperatures, which can affect the quality of food (Owolabi et al., 2022). This technology is based on generating reactive species that can degrade mycotoxins, although further research is needed to optimize its application across different food types. Natural essential oils (NEOs) also present a viable alternative for mycotoxin mitigation. Oils derived from plants, such as clove and turmeric, have demonstrated antifungal properties that may supress the growth of mycotoxin-producing fungi and reduce toxin levels in the food products. Studies have indicated that NEOs can effectively suppress aflatoxin production, making them a valuable tool in food safety. Additionally, nanoparticles have emerged as a promising solution for mycotoxin adsorption. Research indicates that magnetic nanoparticles and other nanomaterials can effectively bind to mycotoxins, facilitating their removal from contaminated food (Furlong et al., 2024). However, the safe and efficient implementation of these materials in food systems requires further investigation to address potential toxicity and environmental concerns.

Cold Atmospheric Plasma (CAP)

Cold atmospheric plasma (CAP) represents a distinct state of matter formed when a neutral gas is exposed to an electric field, leading to the ionization of specific atoms or molecules in the gas and the subsequent generation of plasma. When the voltage surpasses the breakdown threshold, reactive species of oxygen and nitrogen are produced, enabling the creation of CAP at room temperature and atmospheric pressure. The utilization of CAP includes the treatment of materials such as water, buffer solutions, and acids through plasma discharge, resulting in the production of sanitizing agents that are effective for cleaning and disinfection purposes (Ali et al., 2022). The antimicrobial efficacy of cold atmospheric plasma (CAP) and its ability to decompose intricate biochemical compounds stem from a combination of reactive chemical species. This includes superoxide (O^{-2}), ozone (O_3), hydrogen peroxide (H_2O_2), hydroxyl radicals (OH), peroxynitrite ($ONOO^-$), nitrite (NO^{-2}), nitrate (NO^{3-}), ultraviolet (UV) radiation, and the electrical voltage applied. Collectively, these factors sustain temperatures akin to ambient levels, thereby enhancing the nonthermal properties of CAP. The capacity of CAP to degrade mycotoxins is affected by the molecular configuration of the toxins and the particular conditions under which the treatment occurs (Esua et al., 2021). Numerous factors influence the degradation efficiency of deoxynivalenol (DON) in wheat-based products, including the type of gas used, the duration of processing, the characteristics of the matrix, the moisture content, and the plasma voltage. Nitrogen-based cold atmospheric plasma (CAP) generates reactive nitrogen species (RNS) such as nitric oxide (NO) and nitrogen dioxide (NO_2), which exhibit antimicrobial properties beneficial for food safety, sterilization, and surface disinfection. Conversely, oxygen-based CAP produces reactive oxygen species (ROS) like ozone (O_3), singlet oxygen ($^1O_2$), and atomic oxygen (O), which are also effective in decontaminating surfaces from mycotoxins. Inert gases, such as argon or helium, can act as carriers for these reactive species generated during electrical discharge; although they are less reactive, inert gas-based CAP remains effective for applications such as the modification of material surfaces. Chen et al. (2017) demonstrated that treating samples with 50 kV atmospheric air plasma for a duration of 8 minutes led to a 25.82% decrease in deoxynivalenol (DON) levels. Additionally, an increase in moisture content from 8% to 20% further amplified this reduction by 36.10%. The efficacy of cold atmospheric plasma (CAP) in diminishing T-2 and HT-2 toxins in oatmeal is affected by various factors, such as relative humidity, exposure duration, and the types of gases utilized in the production of CAP. Kiš et al. (2021) observed that CAP, when applied with nitrogen (N_2), resulted in reductions of 42.24% for T-2 and 37.53% for HT-2 toxins. Similarly, corona discharge plasma has been shown to effectively degrade aflatoxin B1 (AFB1) in rice and wheat samples, achieving reductions between 45% and 56% after only 30 minutes, as reported by Puligundla et al. (2020). The observed degradation is linked to the lack of a C8–C9 double bond in the furofuran rings, as well as alterations in lactone, cyclopentanone, and methoxyl groups due to oxidative effects from various chemicals. Notably, cold atmospheric plasma (CAP) has demonstrated superior efficacy compared to gamma irradiation, achieving a reduction of 70–73% in aflatoxin B1 (AFB1) levels and 15–47% in aflatoxin B2 (AFB2) levels in hazelnuts, while having a minimal impact on sensory attributes. Research by Wielogorska et al. (2019) indicated that CAP effectively generates safe products during the detoxification of mycotoxins, with reductions in AFB1, ochratoxin A (OTA), fumonisin B1 (FB1), deoxynivalenol (DON), zearalenone (ZEN), and enniatin B (ENB) following a first-order kinetic model. Their findings revealed that ENB and DON exhibited half-lives of 1.1 minutes and 74 minutes, respectively. A 10-minute exposure to CAP resulted in a substantial decrease in FB1 and AFB1 levels, with reductions of 64% and 65%, respectively. However, the identification of degradation products, particularly AFB1 and ZEN, was limited and did not affect the cytotoxicity towards liver cancer cells. Various mechanisms have been proposed for the degradation of mycotoxins *via* CAP, which are influenced by plasma characteristics such as oxygen concentration, hydroxyl radicals, photons, and UV light. The species generated by CAP interact with different regions within fungal cells, leading to structural damage, inactivation, and eventual cell death. Furthermore, these species facilitate the degradation of toxins through multiple processes, resulting in the formation of less toxic byproducts. According to Gavahian and Cullen et al. (2020), the reactivity generated by the plasma can disrupt fungal cell walls, leading to cytoplasmic leakage and cell inactivation. Lee et al. (2015) proposed that the spores of *Cordyceps bassiana* could be rendered inactive by modifying their structure through reactive species produced by plasma. Subsequently, Wu et al. (2020) noted that cold atmospheric plasma (CAP) detoxifies aflatoxins by compromising the integrity of fungal cell walls, cytoplasm, and metabolic functions. In summary, the reactive compounds generated by plasma can deactivate cells by impacting their membranes. CAP increases the permeability of cell membranes through oxidation, which results in damage to cell walls, leading to DNA fragmentation, leakage of intracellular materials, and protein degradation. These mechanisms may also induce cellular apoptosis and the contraction of fungal spores (Chen et al., 2022).

Polyphenolic Compounds

Numerous plant-based foods are rich in a category of naturally occurring compounds known as polyphenols, which have garnered significant interest due to their numerous beneficial attributes. Their established antiviral, antibacterial, and anti-inflammatory effects have made these compounds the focus of extensive research. Polyphenols demonstrate various mechanisms of action against mycotoxins, including antioxidant activity, lipophilicity, suppression of genes associated with mycotoxin production, and inhibition of enzymatic processes related to mycotoxin synthesis (Mehany et al., 2021).

Hamad et al. (2021) demonstrated that extracts from plants and polyphenolic compounds exhibit inhibitory effects on fungal growth and can significantly lower mycotoxin concentrations. Sanzani et al. (2014) indicated that polyphenolic compounds, including umbelliferone and quercetin, can diminish the expression of genes associated with patulin (PAT) biosynthesis, leading to a decrease in the accumulation of this mycotoxin. In a similar vein, chlorogenic and gallic acids have been found to suppress the production of aflatoxin B1 (AFB1) in beans. Furthermore, Salas et al. (2012) revealed that flavanones derived from citrus fruits, such as hesperidin and naringin, can achieve a remarkable 95% reduction in PAT production. Additionally, a nanosponge formulated with β-cyclodextrin and enriched with bioactive phenols has demonstrated the ability to thwart fungal infestations and detoxify mycotoxins.

Natural essential oils

Natural essential oils (NEOs) serve as environmentally sustainable additives capable of inhibiting fungal proliferation and reducing mycotoxin concentrations in a range of food products. These substances interfere with critical enzymes that play a role in carbohydrate breakdown, mycotoxin synthesis, and the structural integrity of fungal cells (Tian et al., 2011). Additionally, they influence fungal gene expression and disrupt cell membranes *via* polyphenols, leading to the destruction of these membranes (Cai et al., 2022). Studies have demonstrated that NEOs can significantly suppress mycotoxin production.

Essential oil derived from *Origanum majorana* L. has demonstrated antifungal activity in laboratory studies, effectively inhibiting the growth of *Aspergillus flavus* and the production of aflatoxins. This indicates its potential as a novel food-preserving agent to improve food safety (Chaudhari et al., 2020). Additional investigations have shown that capsaicin can inhibit the production of ochratoxin A (OTA) by *Aspergillus niger*. Furthermore, essential oils extracted from garlic, rosemary, sage, and mint have proven effective in preventing OTA production (Kollia et al., 2019). Likewise, turmeric oil has exhibited significant antifungal properties against *A. flavus*, thereby reducing aflatoxin contamination in corn. Research by Kedia et al. (2016) indicates that essential oils from *Mentha spicata* effectively target the plasma membrane to inhibit mycotoxin production by *A. flavus* in chickpea. Conversely, essential oils from lemon, grapefruit, and eucalyptus have been shown to effectively suppress the production of zearalenone (ZEN). Recent case studies have also indicated that essential oils from *Curcuma longa* can effectively inhibit the growth of *Fusarium graminearum* and the production of ZEN (Kumar et al., 2016).

NEOs have demonstrated potential in reducing mycotoxins in food; however, it is crucial to recognize their limitations. For instance, the efficacy of NEOs may be inferior to that of synthetic or chemical alternatives and can vary depending on the specific mycotoxin and environmental conditions. Additionally, the flavors and appealing aromas associated with these ingredients may influence the overall sensory attributes of processed foods. The absence of regulations and guidelines presents obstacles to achieving broad acceptance, while safety issues emerge from their instability and the potential for allergic reactions. Furthermore, the practicality of these food products is constrained by concerns related to their cost, residue, and adverse environmental effects during production (Reynolds et al. 2015).

Magnetic materials and nanoparticles

Recent studies have underscored the efficacy of magnetic materials and nanoparticles in addressing mycotoxin contamination in food products. These methods are not only environmentally friendly but also cost-effective and proficient in diminishing mycotoxin levels (Kumar et al., 2016). For instance, chitosan-coated magnetic particles (Fe_3O_4) have shown significant capability in adsorbing patulin (PAT) from fruit juices. It is crucial to consider, however, that these chitosan-coated particles may interact with other constituents in the juice, potentially affecting its taste, aroma, color, or texture, which could influence the sensory characteristics and overall quality of the juice. Additionally, nano-cellulose and retinoic acid have proven effective in adsorbing aflatoxin B1 (AFB1) from a variety of food sources (Jebali et al., 2015). These approaches are non-toxic and function optimally at specific pH levels (Kumar et al., 2016). A range of magnetic nanoparticles, such as surface-active maghemite (SAMNs), silver nanoparticles (SLN), copper nanoparticles (CN), selenium nanoparticles (SEN), and zinc oxide nanoparticles (ZON), have demonstrated their effectiveness in eliminating mycotoxins from food products. SEN, which is derived from *Trichoderma harzianum* JF309, has demonstrated a capacity to decrease fumonisin B1 (FB1) levels by 63% and deoxynivalenol (DON) levels by 76% (Hu et al., 2019). Consequently, SLN has been effective in inhibiting the proliferation of mycotoxin-producing molds, thereby reducing the production of aflatoxins and ochratoxin A (OTA) in corn-based substrates. Additionally, ZON has proven advantageous in mitigating the growth of *Aspergillus, Penicillium,* and *Fusarium* species (Hu et al., 2019).

Magnetic carbon nanocomposites derived from corn byproducts have effectively eliminated AFB1 in poultry, achieving an impressive 90% removal rate at pH 7 within a span of three hours. Additionally, a combination of chitosan and glutaraldehyde has exhibited remarkable adhesion capabilities for various mycotoxins (Zahoor et al., 2016). Moreover, nano-fungicides incorporating phytochemicals such as tannins, phloretin, catechol, eugenol, and thymol have demonstrated significant antimicrobial efficacy without endangering human or animal health (Miu et al., 2022). The mechanisms by which nanoparticle-mediated mitigation of mycotoxin risks occurs involve molecular interactions that induce oxidative stress (Yadav et al., 2024), engage nucleic acids, and provoke inflammatory responses, effectively eliminating microbes in both plant and animal systems (Horky et al., 2018). Interactions with cellular membranes can initiate apoptosis, produce reactive oxygen species (ROS), inhibit mitochondrial functions, and promote lipid peroxidation or autophagy. For instance, ZON induces ROS production that compromises fungal cell membranes, as evidenced by observations made through scanning electron microscopy (SEM) (Pietrzak et al., 2015). A recent study indicated that chitosan nanoparticles loaded with quercetin (Q) can initiate a hepato-protective cascade that enhances antioxidant defense mechanisms. This cascade commences with the activation of the nuclear factor erythroid 2-related factor 2 (NRF2), which subsequently promotes the synthesis of heme oxygenase-1 (HO-1). Additionally, quercetin inhibits the activity of nitric oxide synthase (NOS) and diminishes the production of nitric oxide (NO) induced by lipopolysaccharide, acting through IκB kinase (IKK) and p38 mitogen-activated protein kinases (p38MAPK). The antioxidant response element (ARE) within detoxification promoter genes facilitates this transcriptional process (Horky et al., 2018).

Challenges in mycotoxin management

Mycotoxins pose a significant threat to global food and feed safety, necessitating fast, reliable, and economical testing methods (https://food.r-biopharm.com). Climate change and shifts in agricultural practices are complicating this issue, as some mycotoxins are spreading to new regions that lack effective management strategies (Carvajal-Moreno et al., 2022). While modern LC–tandem MS (LC-MS/MS) instruments provide accurate results, their high cost and need for qualified personnel make them impractical for farms and smaller processing facilities. Consequently, the mycotoxin community is focusing on developing antibody-based and spectrophotometric methods for on-site rapid testing. Innovative technologies, such as miniaturized MS, may offer detailed information on toxin mixtures at costs comparable to quicker methods like ELISA. Although initial capital costs might be higher, these technologies could lower per-sample testing costs and time requirements while maintaining laboratory-level accuracy. Effective sampling remains a major challenge in the testing process, often being the most costly and time-consuming aspect (Zhang et al., 2024). Altomare et al. (2021) proposed using grain dust for sampling Fusarium toxins, potentially enhancing accuracy and reducing costs. The overall costs of mycotoxin testing include equipment, facilities, personnel, consumables, and sample shipping. Balancing cost per sample with the number of samples needed for reliable measurements is crucial to avoid misreporting toxin levels, which can lead to regulatory issues and health risks (Mukhtar et al., 2023). Despite significant innovations over the past four decades, the industry still requires "rapid, accurate, and precise methods." A shift towards higher capital investments may be necessary to achieve lower per-sample costs, enabling more tests and better management of commodities.

Conclusion

The detection and management of mycotoxins and fungal pathogens have evolved significantly over time with advancements in analytical techniques and a growing understanding of their impact on human and animal health. While conventional techniques like HPLC and mass spectrometry remain effective, rapid, cost-effective and accessible testing led to the development of innovative approaches such as ELISA, LFAs, and biosensors. Similarly, the mitigation strategies from traditional physical and chemical methods to more sophisticated techniques involving CAP, NEOs and nanoparticles. However, challenges persist in the terms of sampling, matrix effects and the need for comprehensive validation across diverse food commodities. Continued research and collaboration are crucial to address these issues and ensure the safety and security of the global food supply.

References

Ali, M., Sun, D. W., Cheng, J. H., & Esua, O. J. (2022). Effects of combined treatment of plasma activated liquid and ultrasound for degradation of chlorothalonil fungicide residues in tomato. *Food chemistry, 371*, 131162.

Alizadeh A.M. (2012). Fumonisin B1 contamination of cereals and risk of esophageal cancer in a high-risk area in northeastern Iran. *Asian Pacific journal of cancer prevention*. 13 (6): 2625-2628.

Al-Jaal, B., Salama, S., Al-Qasmi, N., & Jaganjac, M. (2019). Mycotoxin contamination of food and feed in the Gulf Cooperation Council countries and its detection. *Toxicon*, 171, 43-50.

Allcroft R & Carnaghan R (1962). Groundnut toxicity. Aspergillus flavus toxin (aflatoxin) in animal products: preliminary communication. *Veterinary record*, 74: 863-864.

Alshannaq, A., & Yu, J. H. (2017). Occurrence, toxicity, and analysis of major mycotoxins in food. *International journal of environmental research and public health*, 14(6), 632.

Altomare, C., Logrieco, A. F., & Gallo, A. (2021). Mycotoxins and mycotoxigenic fungi: risk and management. A challenge for future global food safety and security. *Food Safety, Volume 1*, Pages 64-93.

Andrade, P. D., da Silva, J. L. G., & Caldas, E. D. (2013). Simultaneous analysis of aflatoxins B1, B2, G1, G2, M1 and ochratoxin A in breast milk by high-performance liquid chromatography/fluorescence after liquid–liquid extraction with low temperature purification (LLE–LTP). *Journal of Chromatography* A, 1304, 61-68.

Anfossi, L., Baggiani, C., Giovannoli, C., D'Arco, G., & Giraudi, G. (2013). Lateral-flow immunoassays for mycotoxins and phycotoxins: a review. *Analytical and bioanalytical chemistry*, 405, 467-480.

Arora, R. G., Frölén, H., & Fellner-Feldegg, H. (1983). Inhibition of ochratoxin A teratogenesis by zearalenone and diethylstilboestrol. *Food and Chemical Toxicology, 21*(6), 779-783.

Bessaire, T., Mujahid, C., Mottier, P., & Desmarchelier, A. (2019). Multiple mycotoxins determination in food by LC-MS/MS: An international collaborative study. *Toxins*, 11(11), 658.

Betina V 1989. Mycotoxins. Chemical, biological and environmental aspects. Elsevier.

Blunden, G., Roch, O. G., Rogers, D. J., Coker, R. D., Bradburn, N., & John, A. E. (1991). Mycotoxins in food. *Medical Laboratory Sciences, 48*(4), 271-282.

Boudra H, Barnouin J, Dragacci S & Morgavi D 2007. Aflatoxin M 1 and ochratoxin A in raw bulk milk from French dairy herds. Journal of dairy science. 90 (7): 3197-3201.

Bruinink, A., & Sidler, C. (1997). The neurotoxic effects of ochratoxin-A are reduced by protein binding but are not affected byl-phenylalanine. *Toxicology and applied pharmacology*, *146*(2), 173-179.

Cai, J., Yan, R., Shi, J., Chen, J., Long, M., Wu, W., & Kuca, K. (2022). Antifungal and mycotoxin detoxification ability of essential oils: A review. *Phytotherapy Research*, *36*(1), 62-72.

Carvajal-Moreno, M. (2022). Mycotoxin challenges in maize production and possible control methods in the 21st century. *Journal of Cereal Science*, 103, 103293.

Casteel, S. W., Turk, J. R., & Rottinghaus, G. E. (1994). Chronic effects of dietary fumonisin on the heart and pulmonary vasculature of swine. *Fundamental and Applied Toxicology*, *23*(4), 518-524.

Chaudhari, A. K., Singh, V. K., Das, S., Prasad, J., Dwivedy, A. K., & Dubey, N. K. (2020). Improvement of in vitro and in situ antifungal, AFB1 inhibitory and antioxidant activity of Origanum majorana L. essential oil through nanoemulsion and recommending as novel food preservative. *Food and Chemical Toxicology*, *143*, 111536.

Chen, P., Liu, T., Jiang, S., Yang, Z., Huang, L., & Liu, F. (2017). Effects of purified zearalenone on selected immunological and histopathologic measurements of spleen in post-weanling gilts. *Animal Nutrition*, *3*(3), 212-218.

Chen, X., Qiu, Y., Zhang, J., Guo, Y., Ding, Y., & Lyu, F. (2022). Degradation efficiency and products of deoxynivalenol treated by cold plasma and its application in wheat. *Food Control*, *136*, 108874.

Cole R & Cox R 1981. Handbook of Toxic Fungal Metabolites, 1981. *Academic Press* INC.

Colvin, B. M., Cooley, A. J., & Beaver, R. W. (1993). Fumonisin toxicosis in swine: clinical and pathologic findings. *Journal of Veterinary Diagnostic Investigation*, *5*(2), 232-241.

Creppy, E. E., Kane, A., Dirheimer, G., Lafarge-Frayssinet, C., Mousset, S., & Frayssinet, C. (1985). Genotoxicity of ochratoxin A in mice: DNA single-strand break evaluation in spleen, liver and kidney. *Toxicology Letters*, *28*(1), 29-35.

Doria, G., Conde, J., Veigas, B., Giestas, L., Almeida, C., Assunção, M., ... & Baptista, P. V. (2012). Noble metal nanoparticles for biosensing applications. *Sensors*, 12(2), 1657-1687.

Dorner, J. W., Cole, R. J., & Blankenship, P. D. (1998). Effect of inoculum rate of biological control agents on preharvest aflatoxin contamination of peanuts. *Biological control*, *12*(3), 171-176.

Eaton, D. L., & Gallagher, E. P. (1994). Mechanisms ofaflatoxin carcinogenesis. *Annual review of pharmacology and toxicology, 34*, 135-172.

Edrington, T. S., Harvey, R. B., & Kubena, L. F. (1995). Toxic Effects of Aflatoxin B~ 1 and Ochratoxin A, Alone and in Combination, on Chicken Embryos. *Bulletin of environmental contamination and toxicology, 54*(3), 331-336.

Esua, O. J., Cheng, J. H., & Sun, D. W. (2021). Functionalization of water as a nonthermal approach for ensuring safety and quality of meat and seafood products. *Critical reviews in food science and nutrition, 61*(3), 431-449.

Fandohan, P., Gnonlonfin, B., Hell, K., Marasas, W. F. O., & Wingfield, M. J. (2005). Natural occurrence of Fusarium and subsequent fumonisin contamination in preharvest and stored maize in Benin, West Africa. *International Journal of Food Microbiology, 99*(2), 173-183.

Furlong, E. B., Buffon, J. G., Cerqueira, M. B., & Kupski, L. (2024). Mitigation of Mycotoxins in Food-Is It Possible?. *Foods* (Basel, Switzerland), 13(7), 1112. https://doi.org/10.3390/foods13071112

Gavahian, M., & Cullen, P. J. (2020). Cold plasma as an emerging technique for mycotoxin-free food: Efficacy, mechanisms, and trends. *Food Reviews International, 36*(2), 193-214.

Gelderblom, W. C., Jaskiewicz, K., Marasas, W. F., Thiel, P. G., Horak, R. M., Vleggaar, R., & Kriek, N. (1988). Fumonisins--novel mycotoxins with cancer-promoting activity produced by Fusarium moniliforme. *Applied and environmental microbiology, 54*(7), 1806-1811.

Goertz A, et al. 2010. Fusarium species and mycotoxin profiles on commercial maize hybrids in Germany. *European journal of plant pathology*. 128 (1): 101-111.

Gumprecht, L. A., Marcucci, A., Weigel, R. M., Vesonder, R. F., Riley, R. T., Showker, J. L., ... & Haschek, W. M. (1995). Effects of intravenous fumonisin B1 in rabbits: Nephrotoxicityxs and sphingolipid alterations. *Natural Toxins, 3*(5), 395-403.

Gupta, R. C., Mostrom, M. S., & Evans, T. J. (2018). Zearalenone. In *Veterinary Toxicology* (pp. 1055-1063). *Academic Press.*

Haidukowski, M., Pascale, M., Perrone, G., Pancaldi, D., Campagna, C., & Visconti, A. (2005). Effect of fungicides on the development of Fusarium head blight, yield and deoxynivalenol accumulation in wheat inoculated under field conditions with Fusarium graminearum and Fusarium culmorum. *Journal of the Science of Food and Agriculture, 85*(2), 191-198.

Hamad, G. M., Mohdaly, A. A. A., El-Nogoumy, B. A., Ramadan, M. F., Hassan, S. A., & Zeitoun, A. M. (2021). Detoxification of aflatoxin B1 and ochratoxin A using Salvia farinacea and Azadirachta indica water extract and application in meat products. *Applied Biochemistry and Biotechnology, 193*(10), 3098-3120.

Han, Z., Jiang, K., Fan, Z., Di Mavungu, J. D., Dong, M., Guo, W., ... & Wu, Y. (2017). Multi-walled carbon nanotubes-based magnetic solid-phase extraction for the determination of zearalenone and its derivatives in maize by ultra-high performance liquid chromatography-tandem mass spectrometry. *Food Control*, 79, 177-184.

Harrison, L. R., Colvin, B. M., Greene, J. T., Newman, L. E., & Cole Jr, J. R. (1990). Pulmonary edema and hydrothorax in swine produced by fumonisin B1, a toxic metabolite of Fusarium moniliforme. *Journal of Veterinary Diagnostic Investigation, 2*(3), 217-221.

Haschek, W. M., Motelin, G., Ness, D. K., Harlin, K. S., Hall, W. F., Vesonder, R. F., ... & Beasley, V. R. (1992). Characterization of fumonisin toxicity in orally and intravenously dosed swine. *Mycopathologia, 117*, 83-96.

Hayes, J. A., & Wall, T. N. (1998). What influences clinicians' responsibility attributions? The role of problem type, theoretical orientation, and client attribution. *Journal of Social and Clinical Psychology, 17*(1), 69-74.

Horky, P., Skalickova, S., Baholet, D., & Skladanka, J. (2018). Nanoparticles as a solution for eliminating the risk of mycotoxins. *Nanomaterials, 8*(9), 727.

Howard, D. V., & Howard, J. H. (2001). When it does hurt to try: Adult age differences in the effects of instructions on implicit pattern learning. *Psychonomic bulletin & review, 8*, 798-805.

Hu, D., Yu, S., Yu, D., Liu, N., Tang, Y., Fan, Y., ... & Wu, A. (2019). Biogenic Trichoderma harzianum-derived selenium nanoparticles with control functionalities originating from diverse recognition metabolites against phytopathogens and mycotoxins. *Food Control, 106*, 106748.

Huertas-Pérez, J. F., Arroyo-Manzanares, N., García-Campaña, A. M., & Gámiz-Gracia, L. (2017). Solid phase extraction as sample treatment for the determination of Ochratoxin A in foods: A review. *Critical reviews in food science and nutrition*, 57(16), 3405-3420.

Humans IWGotEoCRt, Organization WH & Cancer IAfRo 2002. Some traditional herbal medicines, some mycotoxins, naphthalene and styrene. *World Health Organization.*

Hussein, H. S., & Brasel, J. M. (2001). Toxicity, metabolism, and impact of mycotoxins on humans and animals. *Toxicology, 167*(2), 101-134.

James, A., & Zikankuba, V. L. (2018). Mycotoxins contamination in maize alarms food safety in sub-Sahara Africa. *Food Control, 90*, 372-381.

Jaskiewicz, K., Marasas, W. F. O., & Taljaard, J. J. F. (1987). Hepatitis in vervet monkeys caused by Fusarium moniliforme. *Journal of comparative pathology, 97*(3), 281-291.

Jebali, A., Yasini Ardakani, S. A., Sedighi, N., & Hekmatimoghaddam, S. (2015). Nanocellulose conjugated with retinoic acid: its capability to adsorb aflatoxin B1. *Cellulose, 22*, 363-372.

Ji, F., He, D., Olaniran, A. O., Mokoena, M. P., Xu, J., & Shi, J. (2019). Occurrence, toxicity, production and detection of Fusarium mycotoxin: A review. *Food Production, Processing and Nutrition*, 1(1), 1-14.

Kedia, A., Dwivedy, A. K., Jha, D. K., & Dubey, N. K. (2016). Efficacy of Mentha spicata essential oil in suppression of Aspergillus flavus and aflatoxin contamination in chickpea with particular emphasis to mode of antifungal action. *Protoplasma, 253*, 647-653.

Kiš, M., Vulić, A., Kudumija, N., Šarkanj, B., Jaki Tkalec, V., Aladić, K., ... & Pleadin, J. (2021). A two-year occurrence of Fusarium T-2 and HT-2 toxin in Croatian cereals relative of the regional weather. *Toxins, 13*(1), 39.

Kollia, E., Proestos, C., Zoumpoulakis, P., & Markaki, P. (2019). Capsaicin, an inhibitor of Ochratoxin A production by Aspergillus section Nigri strains in grapes (Vitis vinifera L.). *Food Additives & Contaminants: Part A, 36*(11), 1709-1721.

Krska, R., & Molinelli, A. (2009). Rapid test strips for analysis of mycotoxins in food and feed. *Analytical and bioanalytical chemistry*, 393, 67-71.

Kumar, K. N., Venkataramana, M., Allen, J. A., Chandranayaka, S., Murali, H. S., & Batra, H. V. (2016). Role of Curcuma longa L. essential oil in controlling the growth and zearalenone production of Fusarium graminearum. *LWT-Food Science and Technology, 69*, 522-528.

Lattanzio, V. M., von Holst, C., Lippolis, V., De Girolamo, A., Logrieco, A. F., Mol, H. G., & Pascale, M. (2019). Evaluation of mycotoxin screening tests in a verification study involving first time users. *Toxins*, 11(2), 129.

Lee, G. J., Sim, G. B., Choi, E. H., Kwon, Y. W., Kim, J. Y., Jang, S., & Kim, S. H. (2015). Optical and structural properties of plasma-treated Cordyceps bassiana spores as studied by circular dichroism, absorption, and fluorescence spectroscopy. *Journal of Applied Physics, 117*(2).

Lim, C. W., Parker, H. M., Vesonder, R. F., & Haschek, W. M. (1996). Intravenous fumonisin B1 induces cell proliferation and apoptosis in the rat. *Natural Toxins, 4*(1), 34-41.

Malekzad, H., Sahandi Zangabad, P., Mirshekari, H., Karimi, M., & Hamblin, M. R. (2017). Noble metal nanoparticles in biosensors: recent studies and applications. *Nanotechnology reviews*, 6(3), 301-329.

Marin S, Ramos A, Cano-Sancho G & Sanchis V 2013. Mycotoxins: occurrence, toxicology, and exposure assessment. *Food and chemical toxicology*. 60: 218-237.

Mehany, T., Khalifa, I., Barakat, H., Althwab, S. A., Alharbi, Y. M., & El-Sohaimy, S. (2021). Polyphenols as promising biologically active substances for preventing SARS-CoV-2: A review with research evidence and underlying mechanisms. *Food Bioscience, 40*, 100891.

Meisner, H., Cimbala, M. A., & Hanson, R. W. (1983). Decrease of renal phosphoenolpyruvate carboxykinase RNA and poly (A)+ RNA level by ochratoxin A. *Archives of Biochemistry and Biophysics, 223*(1), 264-270.

Miu, B. A., & Dinischiotu, A. (2022). New green approaches in nanoparticles synthesis: An overview. *Molecules, 27*(19), 6472.

Mukhtar, K., Nabi, B. G., Ansar, S., Bhat, Z. F., Aadil, R. M., & Khaneghah, A. M. (2023). Mycotoxins and consumers' awareness: Recent progress and future challenges. *Toxicon*, 107227.

Ni, X., & Streett, D. A. (2005). Modulation of water activity on fungicide effect on Aspergillus niger growth in Sabouraud dextrose agar medium. *Letters in applied microbiology*, *41*(5), 428-433.

Oliveira, I. S., da Silva Junior, A. G., de Andrade, C. A. S., & Oliveira, M. D. L. (2019). Biosensors for early detection of fungi spoilage and toxigenic and mycotoxins in food. *Current Opinion in Food Science*, 29, 64-79.

Oplatowska-Stachowiak, M., Reiring, C., Sajic, N., Haasnoot, W., Brabet, C., Campbell, K., ... & Salden, M. (2018). Development and in-house validation of a rapid and simple to use ELISA for the detection and measurement of the mycotoxin sterigmatocystin. *Analytical and bioanalytical chemistry*, 410, 3017-3023.

Osweiler, G. D., Kehrli, M. E., Stabel, J. R., Thurston, J. R., Ross, P. F., & Wilson, T. M. (1993). Effects of fumonisin-contaminated corn screenings on growth and health of feeder calves. *Journal of animal science, 71*(2), 459-466.

Ottoboni, M., Pinotti, L., Tretola, M., Giromini, C., Fusi, E., Rebucci, R., ... & Cheli, F. (2018). Combining E-nose and lateral flow immunoassays (LFIAs) for rapid occurrence/co-occurrence aflatoxin and fumonisin detection in maize. *Toxins*, 10(10), 416.

Owolabi, I. O., Kolawole, O., Jantarabut, P., Elliott, C. T., & Petchkongkaew, A. (2022). The importance and mitigation of mycotoxins and plant toxins in Southeast Asian fermented foods. npj *Science of Food*, 6(1), 39.

Pereira, V. L., Fernandes, J. O., & Cunha, S. C. (2015). Comparative assessment of three cleanup procedures after QuEChERS extraction for determination of trichothecenes (type A and type B) in processed cereal-based baby foods by GC–MS. *Food chemistry*, 182, 143-149.

Perumal, V., & Hashim, U. (2014). Advances in biosensors: Principle, architecture and applications. *Journal of applied biomedicine*, 12(1), 1-15.

Pietrzak, K., Twarużek, M., Czyżowska, A., Kosicki, R., & Gutarowska, B. (2015). Influence of silver nanoparticles on metabolism and toxicity of moulds. *Acta Biochimica Polonica, 62*(4), 851-857.

Puligundla, P., Lee, T., & Mok, C. (2020). Effect of corona discharge plasma jet treatment on the degradation of aflatoxin B1 on glass slides and in spiked food commodities. *Lwt, 124*, 108333.

Reynolds, T. W., Waddington, S. R., Anderson, C. L., Chew, A., True, Z., & Cullen, A. (2015). Environmental impacts and constraints associated with the production of major food crops in Sub-Saharan Africa and South Asia. *Food Security, 7,* 795-822.

Rheeder, J. P., Marasas, W. F. O., Thiel, P. G., Sydenham, E. W., Shephard, G. S., & Van Schalkwyk, D. J. (1992). Fusarium moniliforme and fumonisins in corn in relation to human esophageal cancer in Transkei. *Phytopathology, 82*(3), 353-357.

Rice LG & Ross PF 1994. Methods for detection and quantitation of fumonisins in corn, cereal products and animal excreta. *Journal of food protection.* 57 (6): 536-540.

Richard JL 2007. Some major mycotoxins and their mycotoxicoses—An overview. *International journal of food microbiology*. 119 (1): 3-10.

Rodríguez-Carrasco, Y., Berrada, H., Font, G., & Mañes, J. (2012). Multi-mycotoxin analysis in wheat semolina using an acetonitrile-based extraction procedure and gas chromatography–tandem mass spectrometry. *Journal of Chromatography* A, 1270, 28-40.

Salas, M. P., Reynoso, C. M., Céliz, G., Daz, M., & Resnik, S. L. (2012). Efficacy of flavanones obtained from citrus residues to prevent patulin contamination. *Food Research International, 48*(2), 930-934.

Sanzani, S. M., Schena, L., & Ippolito, A. (2014). Effectiveness of phenolic compounds against citrus green mould. *Molecules, 19*(8), 12500-12508.

Sava, V., Reunova, O., Velasquez, A., & Sanchez-Ramos, J. (2006). Can low level exposure to ochratoxin-A cause parkinsonism?. *Journal of the neurological sciences, 249*(1), 68-75.

Science CfA (2003). Mycotoxins: risks in plant, animal, and human systems. *Council for Agricultural.*

Singh, J., & Mehta, A. (2020). Rapid and sensitive detection of mycotoxins by advanced and emerging analytical methods: A review. *Food science & nutrition,* 8(5), 2183-2204.

Solcan, C., Gogu, M., Floristean, V., Oprisan, B., & Solcan, G. (2013). The hepatoprotective effect of sea buckthorn (Hippophae rhamnoides) berries on induced aflatoxin B1 poisoning in chickens. *Poultry Science,* 92(4), 966-974.

Song, S., Liu, N., Zhao, Z., Njumbe Ediage, E., Wu, S., Sun, C., ... & Wu, A. (2014). Multiplex lateral flow immunoassay for mycotoxin determination. *Analytical chemistry,* 86(10), 4995-5001.

Stětina, R., & Votava, M. (1986). Induction of DNA single-strand breaks and DNA synthesis inhibition by patulin, ochratoxin A, citrinin, and aflatoxin B1 in cell lines CHO and AWRF. *Folia biologica, 32*(2), 128-144.

Stoev SD 2013. Food safety and increasing hazard of mycotoxin occurrence in foods and feeds. *Critical reviews in food science and nutrition*. 53 (9): 887-901.

Størmer, F. C., & Lea, T. (1995). Effects of ochratoxin A upon early and late events in human T-cell proliferation. *Toxicology, 95*(1-3), 45-50.

Streit E, et al. 2012. Current situation of mycotoxin contamination and co-occurrence in animal feed—Focus on Europe. *Toxins*. 4 (10): 788-809.

Sun, G., Wang, S., Hu, X., Su, J., Zhang, Y., Xie, Y., ... & Wang, J. S. (2011). Co-contamination of aflatoxin B1 and fumonisin B1 in food and human dietary exposure in three areas of China. *Food additives and contaminants, 28*(4), 461-470.

Thway, T., & Salimi-Moosavi, H. (2014). Evaluating the impact of matrix effects on biomarker assay sensitivity. *Bioanalysis*, 6(8), 1081-1091.

Tian, J., Ban, X., Zeng, H., He, J., Huang, B., & Wang, Y. (2011). Chemical composition and antifungal activity of essential oil from Cicuta virosa L. var. latisecta Celak. *International Journal of Food Microbiology, 145*(2-3), 464-470.

Turner, N. W., Subrahmanyam, S., & Piletsky, S. A. (2009). Analytical methods for determination of mycotoxins: a review. *Analytica chimica acta*, 632(2), 168-180.

Turner, P. C., Moore, S. E., Hall, A. J., Prentice, A. M., & Wild, C. P. (2003). Modification of immune function through exposure to dietary aflatoxin in Gambian children. *Environmental health perspectives, 111*(2), 217-220.

United Nations Environmental Program (UNEP) GRID Adrenal, 2013. *Food Demand and Need.* http://www.grida.no/ publications/ rr/ food-crisis/ page/ 3559. aspx

Vasseghian, Y., Dragoi, E. N., Moradi, M., & Khaneghah, A. M. (2021). A review on graphene-based electrochemical sensor for mycotoxins detection. *Food and Chemical Toxicology*, 148, 111931.

Voss, K. A., Bacon, C. W., Norred, W. P., Chapin, R. E., Chamberlain, W. J., Plattner, R. D., & Meredith, F. I. (1996). Studies on the reproductive effects of Fusarium moniliforme culture material in rats and the biodistribution of [14C] fumonisin B1 in pregnant rats. *Natural Toxins, 4*(1), 24-33.

Wacoo, A. P., Wendiro, D., Vuzi, P. C., & Hawumba, J. F. (2014). Methods for detection of aflatoxins in agricultural food crops. *Journal of applied chemistry*, 2014(1), 706291.

Wang, Y., Xu, H., Wei, M., Gu, H., Xu, Q., & Zhu, W. (2009). Study of superparamagnetic nanoparticles as labels in the quantitative lateral flow immunoassay. *Materials Science and Engineering*: C, 29(3), 714-718.

Welke, J. E., Hoeltz, M., Dottori, H. A., & Noll, I. B. (2009). Quantitative analysis of patulin in apple juice by thin-layer chromatography using a charge coupled device detector. *Food Additives and Contaminants*, 26(5), 754-758.

Wielogorska, E., Ahmed, Y., Meneely, J., Graham, W. G., Elliott, C. T., & Gilmore, B. F. (2019). A holistic study to understand the detoxification of mycotoxins in maize and impact on its molecular integrity using cold atmospheric plasma treatment. *Food chemistry, 301*, 125281.

Wilson, J. P., Casper, H. H., & Wilson, D. M. (1995). Effect of delayed harvest on contamination of pearl millet grain with mycotoxin-producing fungi and mycotoxins. *Mycopathologia, 132*, 27-30.

World Health Organization International Agency for Research on Cancer (WHO-IARC), 1993b. IARC Monographs on the Evaluation of Carcinogenic Risks to Humans.

Wu, Z., He, D., Cui, B., Jin, Z., Xu, E., Yuan, C., ... & Chai, Q. (2020). Trimer-based aptasensor for simultaneous determination of multiple mycotoxins using SERS and fluorimetry. *Microchimica Acta, 187*, 1-7.

Yadav, R., Kumar, A., Saharan, B. S., Gera, R., Rana, A., & Parshad, J. (2024). Arbuscular mycorrhizal fungi as biostimulant for plant growth and abiotic stress amelioration. In *Microbial Biostimulants for Plant Growth and Abiotic Stress Amelioration* (pp. 81-95). Elsevier.

Yang, J., Li, J., Jiang, Y., Duan, X., Qu, H., Yang, B., ... & Sivakumar, D. (2014). Natural occurrence, analysis, and prevention of mycotoxins in fruits and their processed products. *Critical reviews in food science and nutrition*, 54(1), 64-83.

Yang, Y., Li, G., Wu, D., Liu, J., Li, X., Luo, P., ... & Wu, Y. (2020). Recent advances on toxicity and determination methods of mycotoxins in foodstuffs. *Trends in food science & Technology*, 96, 233-252.

Zahoor, M., & Ali Khan, F. (2016). Aflatoxin B1 detoxification by magnetic carbon nanostructures prepared from maize straw. *Desalination and water treatment, 57*(25), 11893-11903.

Zhang, K., Flannery, B., & Zhang, L. (2024). Challenges and Future State for Mycotoxin Analysis: A Review From a Regulatory Perspective. *Journal of agricultural and food chemistry*, 72(15), 8380–8388. https://doi.org/10.1021/acs.jafc.4c01746

Zhang, L., Dou, X. W., Zhang, C., Logrieco, A. F., & Yang, M. H. (2018). A review of current methods for analysis of mycotoxins in herbal medicines. *Toxins*, 10(2), 65.

Zheng, M. Z., Richard, J. L., & Binder, J. (2006). A review of rapid methods for the analysis of mycotoxins. *Mycopathologia*, 161(5), 261-273.

Zinedine, A., Soriano, J. M., Juan, C., Mojemmi, B., Molto, J. C., Bouklouze, A., ... & Manes, J. (2007). Incidence of ochratoxin A in rice and dried fruits from Rabat and Salé area, Morocco. *Food Additives and Contaminants*, 24(3), 285-291.

Chapter 6

Detection and Quantification of Mycotoxins and their Genetic Regulation

M. Chaithra[1*] *and Rashmi E. R.*[2]

[1]*ICAR-Central Plantation Crop Research Institute, Research Center, Kahikuchi, Guwahati, Assam, India-781017*

[2]*ICAR-Centeral Institute of Temperate Horticulture, Regional Station, Mukteshwar, Uttarakhand, India- 263138*

**Corresponding author: M. Chaithra; chaithram06@gmail.com*

Abstract

Mycotoxins, secondary metabolites produced by various fungal species, pose significant threats to food safety, public health, and agricultural productivity. Accurate detection and quantification of mycotoxins are essential to mitigate their risks in food and feed products. This chapter explores the advanced techniques used for detecting and quantifying mycotoxins, including chromatographic (HPLC, GC-MS), immunoassay-based methods (ELISA), and emerging molecular and biosensor-based approaches. Furthermore, the chapter elucidates the genetic and molecular regulation of mycotoxin biosynthesis in key fungal genera such as *Aspergillus*, *Fusarium*, and *Penicillium*. Special emphasis is given to the role of regulatory genes, including transcription factors, cluster-specific regulators, and environmental signals that control mycotoxin production. Additionally, the chapter discusses the integration of "omics" technologies—genomics, transcriptomics, and metabolomics—for understanding the complex regulatory pathways governing mycotoxin biosynthesis and their application for the development of effective control strategies. This information provides a robust foundation for innovative approaches to mycotoxin detection, prevention, and management.

Keywords: Bioinformatics; Chromatographic techniques; Enzyme-Linked Immunosorbent Assay; Omics technologies, Transcriptomics

Introduction

Mycotoxins have been detected not only in fruits and vegetables but also in their derivative products. Detecting these toxins at low concentrations is crucial due to their significant impact on health and the economy. Given the very low levels of mycotoxins in fruits and vegetables, there are high technical demands for detection accuracy and sensitivity. The complexity of food matrices and the trace amounts of mycotoxins present further complicate direct analysis. Several methods, ranging from traditional to modern, have been developed for the detection of mycotoxins in post-harvest fruits, vegetables, and their products (Table 1). The choice of analytical technique is primarily related to the nature of the mycotoxin. The primary techniques include chromatographic methods, immuno-based methods, and biosensors. Chromatographic techniques, such as high-performance liquid chromatography (HPLC) coupled with various detectors like fluorescence, diode array, UV, liquid chromatography-mass spectrometry (LC-MS), and liquid chromatography-tandem mass spectrometry (LC-MS/MS), are powerful tools for analyzing and detecting major mycotoxins (Stroka and Anklam, 2002). Recent advancements in rapid immune-affinity-based detection techniques, such as immunoassays and biosensors, have shown promise (Maragos, 2004). Additionally, emerging technologies like proteomic and genomic methods, molecular techniques, electronic noses, aggregation-induced emission dyes, quantitative NMR, and hyperspectral imaging are being explored to detect mycotoxins in food (Sanzani et al., 2016).

(A) Chromatographical Technique

Chromatography is widely accepted for mycotoxin detection due to its accuracy and reliability. Major organizations such as ISO, AOAC, and CEN often choose chromatography as the standard method. Among chromatographic techniques, thin-layer chromatography (TLC) is the oldest, used for rapid screening of mycotoxins and approved for aflatoxin (AF) detection for over half a century. TLC has been developed into high-performance TLC (HPTLC), achieving more precise and accurate results for mycotoxins like AFs and ochratoxin A (OTA). Despite being low-cost, TLC is not very accurate or sensitive (Zollner and Mayer-Helm, 2006). Gas chromatography (GC) is rarely used for mycotoxins because most are nonvolatile and polar, requiring a derivation step. However, GC is advantageous for volatile mycotoxins. When used, single mass spectrometry (MS) detectors are common, along with electron capture (ECD) and flame ionization (FID) detectors. Ion traps, quadrupoles, and time-of-flight (TOF) analyzers are also used, with the GC–QqQ-MS/MS technique featuring a triple quadrupole (QqQ) analyzer for trichothecenes (TCs), patulin (PAT), and zearalenone (ZEA) analysis (Turner et al., 2009). Despite GC's advantages, its derivatization requirement has led to a preference for liquid chromatography (LC) methods. LC offers exceptional development in mycotoxin detection, allowing simultaneous determination of several mycotoxins regardless of their chemical structure (Rios and Freitas-Silva, 2014). In 1993, ISO issued a standard method for detecting PAT in apple juice using HPLC-UV. CEN set standards for PAT in apple products and OTA in grape products and dried fruits, using liquid-liquid distribution, solid-phase extraction, or immuno affinity columns combined with HPLC (ISO, 1993). AOAC has formulated three PAT test standards, including TLC, HPLC, and IAC-HPLC for OTA determination (AOAC, 1998). While accurate, chromatography is time-consuming, requires professional testers, and involves expensive equipment. Liquid chromatography–tandem mass spectrometry (LC–MS/MS) combines chromatographic separation with mass spectrometry identification and has become the main method for multi-mycotoxin analysis (Spanjer et al., 2008). A rapid solid-phase extraction (SPE) method coupled with LC–MS/MS has been developed to determine nine mycotoxins in betel nut products, achieving an LOQ of 0.25–10 μg/L, an LOD of 5–20 μg/L, and recovery rates of 70.1–113.9% (Lin et al., 2021). High-performance liquid chromatography (HPLC) with fluorescence (FLD) or UV detectors is also used for chemically related mycotoxins. HPLC–FLD has been widely reported for analyzing AFs, ZEA, and deoxynivalenol (DON) (Skrbic et al., 2013). The presence of multiple mycotoxins in agricultural products has driven the development of techniques capable of simultaneously identifying many mycotoxins. LC–MS/MS is now the most modern and widely used technique for mycotoxin analysis at trace levels due to its sensitivity, specificity, and reliability compared to HPLC (Songsermsakul and Razzazi-Fazeli, 2008).

Table 1: Different techniques used for the detection of major Mycotoxins

Mycotoxins	Detection Techniques	References
1. Aflatoxin	AgraStrip®	Abbas and Abbas , 2021
	Immunodipstick	Luppa et al. 2001
	ELISA	Omar et al. 2020
	Quartz crystal microbalance (QCMs)	Van der Gaag 2003
	Electrochemical	Linting et al. 2012; Masoomi et al. 2013
	Fourier-transform infrared spectroscopy (FTIR)	Pearson et al. 2001
	Radio-immunoassay (RIA)	Romero-Sanchez et al. 2022
	High/Ultrahigh Performance Liquid Chromatography (HPLC/UHPLC)	Rahmani et al. 2009
	Fluorometer	Malone et al. 2000
	Thin layer chromatography (TLC)	Salisu et al. 2021
	Liquid chromatography (LC)	Kokkonen et al. 2005
	HPLC-MS/MS	Romero-Sanchez et al. 2022
	LC-MS/MS	Zhou et al. 2021
	Lateral Flow Immunoassay	Shen et al. 2022
	Direct analysis in real time-mass spectrometry (DART-MS)	Vaclavik et al. 2010
	MALDI-ToF	Casarova et al. 2017

2. Ochratoxin	ELISA	Casarova et al. 2017
	TLC	Braicu et al. 2008
	Lateral Flow Immunoassay	Shen et al. 2022
	UHPLC-MS/MS	Braicu et al. 2008
	HPLC-FLR	Tangni et al. 2002
	LC-MS/MS	Soleimany et al. 2012; Deng et al. 2020
	LC-ESI-qTOF	Carballo et al. 2021
	GC-MS	Olsson et al. 2002
	MALDI-ToF	Hickert et al 2016
3. Patulin	ELISA	Przybylska et al. 2021
	HPLC	Sadhasivam et al. 2021
	GC-MS	Monteagudo et al 2023
	Lateral Flow Immunoassay	Tang et al. 2022
	LC-MS/MS	Radicevic et al. 2021
	UPLC–MS/MS	Wang et al. 2016
	MALDI-ToF	Chalupova et al. 2014
4. Fumonisins	Lateral flow immunoassay	Tran et al. 2009
	Immunochromatographic test strips (ICSs)	Huang et al. 2020
	TLC	Radicevic et al. 2021
	HPLC/UHPLC	Ren et al. 2011
	UHPLC-MS/MS	Pantano et al. 2021
	GCMS	Borowik et al. 2014
	Electronic Nose device	Borowik et al. 2014
	MALDI-ToF	Hickert et al. 2016
5. Zearalenone	ELISA	Pleadin et al. 2012
	LC-ESI-qTOF	Carballo et al. 2021
	Lateral flow immunoassay (LFIA)	Zhang et al. 2023
	LC-MS/MS	Belakova et al., 2014
	GC-MS	Zhang et al. 2023
	MALDI-ToF	Casarova et al 2017

6. Deoxynivalenol	ELISA	Pleadin et al. 2012
	TLC	Vujanovic et al. 2012
	Dual-label time-resolved fluoroimmunoassay (TRFIA)	Wang et al. 2016
	Multiplex flow cytometric microsphere immunoassay	Jeroen et al. 2010
	Immunochromatographic test strips (ICSs)	Huang et al. 2020
	Immunochip	Wang et al. 2016
	Immuno-rotary biosensor (IRB)	Zhao et al. 2012
	Electronic nose (EN)	Campagnoli et al. 2011
	HPLC	Muscarella et al. 2012
	LC-MS/MS	Soleimany et al. 2012; Zou et al. 2012
	LC-ESI-qTOF	Carballo et al. 2021
	Lateral flow immunoassays (LFIA)-MS	Geballa-Koukoula et al 2022
	GC-MS	Olsson et al. 2002
	MALDI-ToF	Casarova et al. 2017
7. Citrinin	HPLC	Sadhasivam et al. 2021
	ELISA	Singh et al. 2019
	TLC	Casarova et al 2017
	LC-MS	Sadhasivam et al. 2021
	UPLC–MS/MS	Wang et al. 2016
	MALDI-ToF	Casarova et al 2017

(B) Immunoassay-based Methods:

Immunochemical-based methods for mycotoxin detection rely on the recognition of these toxins as antigens by specific antibodies. Detection is typically facilitated by marker compounds that react with an enzyme, non-enzymatic labels conjugated to the antibodies, or by leveraging the natural fluorescence of some mycotoxins (Goryacheva et al., 2007). Immunoassay kits, which are portable and rapid tests, provide results quickly, making them suitable for on-site testing and decision-making. These kits often include variations of enzyme-linked immunosorbent assay (ELISA), such as lateral flow immunoassay (LFIA) or flow-through ELISA, and fluorometric assays like fluorescent polarization immunoassays or immunoaffinity columns coupled with fluorometric assays.

ELISAs, commonly performed in microtiter plates or well strips, combine antibody specificity with the sensitivity of simple enzyme assays for detecting specific analytes. They offer fast, reliable, and simultaneous analysis of many samples, and are the most validated immunochemical methods for monitoring mycotoxins. ELISAs are highly specific, have relatively low limits of detection (LOD), and support high sample throughput (Sheng et al., 2012). However, they can be affected by the complexity of sample matrices. For example, Omar et al. used ELISA to detect AFB1, AFB2, AFG1, and AFG2 in wheat samples, OTA and ZEA in various types of tea, and AFB1, OTA, and ZEA in corn samples (Weaver et al., 2020).

Lateral flow immunoassay (LFIA), conducted on a nitrocellulose membrane imprinted with specific antibodies conjugated with non-enzymatic labels like gold colloid nanoparticles (AuNPs), is a low-cost, simple, paper-based test for the rapid identification and quantification of various analytes (Anfossi et al., 2013). LFIAs, also known as lateral flow devices (LFDs) or dipstick tests, often use a competitive ELISA format. They are easy to perform, suitable for in situ analysis, and available for multi-mycotoxin detection in food and feed (Xing et al., 2020). Validated LFDs can detect OTA and aflatoxins in wine and grape products, among other matrices (Nolan et al., 2019). Flow-through immunoassays, which use a gel-based column of microbeads, are typically in a direct competitive format like ELISAs. These kits are rapid, easy to use, and suitable for multiplexing, generating qualitative results that can be semi-quantified using portable readers (Beloglazova et al., 2017). Jin and co-worker detected diacetoxyscirpenol in rice using this method, while Li and their co-worker identified ZEA, T-2, and DON in maize (Chen et al., 2019).

Biosensors are analytical devices comprising a detection layer with a biological element, a transducer that converts the biological element's physical change into a measurable signal, and an output system that processes and displays the signal (Rovina et al., 2019). Recently, biosensors such as electrochemical, optical, and piezoelectric types have been used to identify mycotoxins in foodstuffs (Pohanka and Skladal, 2008). These biosensors offer a low-cost, fast alternative to traditional chromatographic methods for mycotoxin analysis (Vidal et al., 2013). Various biosensors have been developed for detecting mycotoxins in different food and feed commodities, each with specific advantages and limitations in sensitivity, selectivity, cost efficiency, and ease of in situ use (Shrivastava and Sharma, 2022). Many have been tested on different kinds of wine, primarily for detecting OTA, AFB1, and AFM1 (Stratakou and van der Fels-Klerx, 2010). For instance, Wei and co-workers developed a simple and sensitive method for detecting OTA toxins in beverages such as wine and beer (Chen et al., 2023).

(C) Genomic Methods:

Genomic methods for detecting mycotoxins begin with the extraction of fungus DNA or RNA from food samples. These methods include polymerase chain reaction (PCR), quantitative PCR (qPCR), loop-mediated isothermal amplification (LAMP), reverse transcription PCR (RT-PCR), fluorescent in situ hybridization (FISH), and

DNA barcoding (Magan and Olsen, 2004). While these techniques effectively detect fungal isolates, they do not quantify them (El Sheikha, 2019). Characterizing and classifying mycotoxigenic fungi, particularly those isolated from grape microflora, often involves multilocus sequence analysis (MLSA) and restriction or amplified fragment length polymorphism of PCR amplicons (RFLP, AFLP) (Kizis et al., 2021). An advantageous alternative to traditional PCR assays is LAMP, which can be directly applied to grape samples (Edwards et al., 2002). For instance, a LAMP assay has successfully detected P. oxalicum in grape samples (Mellikeche et al., 2022). Next-generation sequencing (NGS) provides comprehensive analysis of fungal genomes, enabling the identification of mycotoxin-producing species and their genetic variants (Asthana and Kumar, 2018). It offers detailed insights into the genetic diversity and potential for mycotoxin production (Janik et al., 2020). Microarray technology uses DNA probes immobilized on a solid surface to detect and quantify multiple mycotoxin-producing genes simultaneously, allowing for high-throughput screening and providing valuable data on the presence and expression levels of target genes (Rasooly and Herold, 2008). CRISPR technology can be adapted to detect specific DNA sequences associated with mycotoxin production. CRISPR-based assays offer high specificity and can be designed for rapid, onsite testing.

(E) Proteomic Methods:

Proteomic methods for detecting mycotoxin production begin with the extraction of peptides and proteins from fungus present in food samples. These proteomic methods provide valuable insights into the proteins involved in mycotoxin production, aiding in the precise detection and identification of mycotoxins and their fungal sources. Key methods include:

1. Matrix-Assisted Laser Desorption/Ionization Time-of-Flight Mass Spectrometry (MALDI- TOF MS): MALDI-TOF MS is a rapid technique for detecting and identifying fungal isolates and mycotoxins with high precision. It is an alternative to chromatographic techniques, analyzing proteins in the mass range of 2–20 k Da by calculating their mass-to-charge (m/z) values. However, the lack of a public database for all in-house entries and the high cost are notable drawbacks for the food industry.
2. Electrospray Ionization (ESI) MS: ESI MS, similar to MALDI-TOF, is used for the identification of proteins and peptides, particularly useful for analyzing complex protein mixtures. This technique is highly effective in identifying and characterizing the proteins involved in mycotoxin production.
3. Label-Free Quantitative Proteomics: This approach employs advanced mass spectrometry techniques to quantify proteins without labelling. It is highly accurate and capable of detecting subtle changes in protein expression associated with mycotoxin production, making it a valuable tool for monitoring and studying mycotoxin biosynthesis.

F. Sampling methods

Sampling is crucial for determining mycotoxin levels because mycotoxigenic fungi do not grow uniformly on substrates, making it difficult to obtain representative bulk samples. Mycotoxin contamination in natural samples is often heterogeneous. Therefore, extraction, purification, and chromatographic separation must be well-defined to minimize matrix effects. Traditional food sampling methods are generally unsuitable for mycotoxin analysis, except for liquid food samples like milk or highly processed foods such as peanut butter. To standardize sampling procedures, Commission Regulation (EC) No. 401/2006 specifies methods for official mycotoxin level control in foodstuffs.

Various methods are used for mycotoxin isolation, separation, and extraction, including

1. Liquid-Liquid Extraction (LLE): Utilizes different solubilities of toxins in organic and aqueous phases, commonly used for aflatoxins (AFs) and ochratoxin A (OTA) analysis.
2. Solid-Liquid Extraction (SLE): Involves homogenizing samples in an extraction solvent andagitating them, suitable for cereals.
3. Pressurized Liquid Extraction (PLE) / Accelerated Solvent Extraction (ASE): Employs high temperature and pressure to improve extraction efficiency, used for mycotoxins in tomato samples.
4. Supercritical Fluid Extraction (SFE): Uses supercritical CO2 to extract non-polar molecules like zearalenone (ZEA) in maize flour, minimizing organic solvent use.
5. Microwave-Assisted Extraction (MAE): A green technique using microwave energy for rapid extraction with minimal organic solvents.
6. QuEChERS (Quick, Easy, Cheap, Effective, Rugged, and Safe): Initially developed for pesticides but effective for mycotoxins using small amounts of acetonitrile.

After extraction, cleanup is necessary to increase selectivity and remove matrix interferences. Solid Phase Extraction (SPE) is popular for its speed and efficiency, using solid absorbents in cartridges to purify mycotoxins. Novel absorbents like carbon nano materials, including graphene oxide and multi-walled carbon nanotubes, offer high absorption capacities and unique properties. However, these new materials must be thoroughly evaluated for various mycotoxins to be considered appropriate alternatives. Efficient sample pretreatment techniques, focusing on eco-friendly solvents and rapid processes, are crucial before mycotoxin analysis, except for enzyme-linked immunosorbent assays, which generally do not require such extensive preparation.

GENE REGULATION OF MYCOTOXIN BIOSYNTHESIS IN FUNGI

1. Biosynthetic Gene Clusters for Mycotoxins

Biosynthetic gene clusters (BGCs) are tightly organized loci in fungal genomes containing structural genes, cluster-specific transcriptional regulators, and transporter genes (Bhatnagar et al., 2015).

- ✰ **Aflatoxins**: Produced by *Aspergillus flavus* and *A. parasiticus*, the aflatoxin gene cluster consists of 25 genes, with *aflR* as the key transcriptional activator (Macheleidt et al., 2016). The *aflR* gene regulates structural genes like *aflD*, *aflM*, and *aflP*, which encode enzymes for aflatoxin biosynthesis.
- ✰ **Trichothecenes**: In *Fusarium* spp., trichothecene biosynthesis is regulated by the *Tri* gene cluster, with *Tri6* and *Tri10* acting as transcriptional regulators (Phokane et al., 2018).
- ✰ **Fumonisins**: The *fum* gene cluster in *Fusarium verticillioides* controls fumonisin biosynthesis, with *Fum21* as the key cluster-specific regulator (Macheleidt et al., 2016).
- ✰ **Ochratoxins**: Produced by *Penicillium* and *Aspergillus* species, ochratoxin biosynthesis involves genes such as *otapks* (producing polyketide synthase) (Bhatnagar et al., 2015).

The identification and annotation of these BGCs have been facilitated by fungal genome sequencing projects.

2. Transcriptional Regulation of Mycotoxin Biosynthesis

Transcriptional regulation is a central mechanism controlling mycotoxin production. Cluster-specific transcription factors activate or suppress mycotoxin biosynthetic genes under appropriate conditions.

- ✰ **Aflatoxin Regulation**: *AflR* positively regulates aflatoxin production, while *AflS* assists in the process (Schmidt-Heydt et al., 2008). Deletion of *aflR* leads to complete loss of aflatoxin production.
- ✰ **Trichothecene Regulation**: *Tri6* binds to promoter regions of trichothecene genes to activate their expression. *Tri10* plays a role in cluster-wide regulation ((Macheleidt et al., 2016).
- ✰ **Global Regulators**: Regulatory pathways like the **LaeA**-mediated velvet complex (VeA, VelB, LaeA) modulate secondary metabolism and development in fungi (Bhatnagar et al., 2015).

Transcription factors integrate signals from environmental and epigenetic regulators to ensure optimal mycotoxin biosynthesis.

3. Environmental Regulation of Mycotoxin Genes

Environmental conditions significantly influence mycotoxin production.

- **Temperature and Humidity**: Aflatoxin production is enhanced at temperatures between 25-30°C and high humidity (Schmidt-Heydt et al., 2008).
- **Carbon and Nitrogen Sources**: The availability of glucose and nitrogen impacts gene cluster activation (Cairns et al., 2018).
- **Oxidative Stress**: Reactive oxygen species (ROS) activate signaling pathways that induce aflatoxin production (Schmidt-Heydt et al., 2008).
- **Light**: Light exposure modulates the velvet complex, which regulates mycotoxin biosynthesis and fungal development (Cairns et al., 2018).

The interplay between environmental signals and gene regulators enables fungi to produce mycotoxins in response to stress or favorable conditions.

4. Epigenetic Regulation of Mycotoxin Biosynthesis

Epigenetic mechanisms such as histone modifications and chromatin remodeling influence the activation of biosynthetic gene clusters.

- **Histone Acetylation/Methylation**: Modifications in histones H3K9ac and H3K4me regulate gene expression within mycotoxin clusters (Schmidt-Heydt et al., 2008).
- **Chromatin Remodeling**: The *LaeA* protein, a global regulator, functions as a histone methyltransferase, modulating secondary metabolism in *Aspergillus* spp. (Keller et al., 2005).

Epigenetic regulation provides an additional layer of control, ensuring mycotoxin genes are activated only under specific conditions.

5. Omics Technologies for Understanding Gene Regulation

Recent advancements in **omics technologies** have accelerated the study of mycotoxin gene regulation:

- **Genomics**: Genome sequencing identifies BGCs and regulatory genes (Phokane, 2018).
- **Transcriptomics**: RNA-Seq provides insights into gene expression patterns under varying conditions (Schmidt-Heydt et al., 2008).
- **Proteomics and Metabolomics**: Profiling of proteins and metabolites helps to connect gene expression with mycotoxin production (Keller et al., 2005).

Integrating multi-omics approaches reveals the complex regulatory networks governing mycotoxin biosynthesis.

6. Genetic Manipulation and CRISPR-Cas Applications

Functional genomics tools, including **CRISPR-Cas9**, have revolutionized studies on mycotoxin regulation.

- ✰ Targeted deletion of regulatory genes like *aflR* and *Tri6* confirms their roles in toxin production (Cairns et al., 2018).
- ✰ CRISPR-Cas9 enables precise genome editing, offering potential for disrupting biosynthetic pathways and reducing mycotoxin contamination.

Conclusion

In conclusion, the detection and quantification of mycotoxins, alongside understanding the genetic regulation of their biosynthesis, are critical for safeguarding food and feed safety. Advances in analytical techniques, particularly through chromatography and immunoassays, have enabled precise, rapid, and large-scale monitoring of mycotoxins in agricultural products. Additionally, newer approaches like biosensors and molecular tools are promising for on-site detection, offering real-time and sensitive solutions. The genetic regulation of mycotoxin biosynthesis, controlled by complex gene clusters and influenced by environmental and epigenetic factors, opens pathways for targeted intervention. By exploring transcriptional, epigenetic, and signaling mechanisms, researchers can develop novel methods to control or suppress mycotoxin production in pathogens. Genetic tools such as CRISPR-Cas9 and RNA interference are valuable for functional studies and may pave the way for future control strategies. Integrating these insights into agricultural practices and breeding programs holds promise for reducing mycotoxin contamination and mitigating the associated health and economic impacts, ultimately contributing to global food security.

References

1. Abbas, M. and Abbas, M. (2021). Chromatographic techniques for estimation of aflatoxins in food commodities. Aflatoxins-Occurrence, Detoxification, Determination and Health Risks.
2. Anfossi, L., Baggiani, C., Giovannoli, C., D'Arco, G. and Giraudi, G. (2013). Lateral-flow immunoassays for mycotoxins and phycotoxins: a review. Analytical and bio-analytical chemistry, 405, 467-480.
3. AOAC. (1998). Official methods of analysis for mycotoxins. Association of Official Analytical Chemists.
4. Asthana, M. and Kumar, A. (2018). Understanding the Diversity of Penicillium Using Next-Generation Sequencing. In New and Future Developments in Microbial Biotechnology and Bioengineering (pp. 19-43). Elsevier.
5. Belakova, S., Benesova, K., Caslavsky, J., Svoboda, Z. and Mikulíkova, R. (2014). The occurrence of the selected fusarium mycotoxins in Czech malting barley. Food Control, 37, 93–98. https://doi. org/10.1016/j.foodcont.2013.09.033
6. Beloglazova, N. V., Graniczkowska, K., Njumbe Ediage, E., Averkieva, O. and De Saeger, S. (2017). Sensitive flow-through immunoassay for rapid multiplex determination of cereal-borne mycotoxins in feed and feed ingredients. Journal of agricultural and food chemistry, 65(33), 7131-7137.

7. Bhatnagar-Mathur, P., Sunkara, S., Bhatnagar-Panwar, M., Waliyar, F. and Sharma, K. K. (2015). Biotechnological advances for combating Aspergillus flavus and aflatoxin contamination in crops. *Plant science*, 234, 119-132.

8. Borowik, P., Dyshko, V., Tkaczyk, M., Okorski, A., Polak-Sliwinska, M., Tarakowski, R. and Oszako, T. (2024). Analysis of Wheat Grain Infection by Fusarium Mycotoxin-Producing Fungi Using an Electronic Nose, GC-MS, and qPCR. Sensors, 24(2), 326.

9. Braicu, C., Puia, C., Bodoki, E. and Socaciu, C. (2008). Screening and quantification of aflatoxins and ochratoxin in different cereals cultivated in Romania using thin-layer chromatography-densitometry. Journal of Food Quality, 31(1), 108-120.

10. Cairns, T., Nai, C. and Meyer, V. (2018). How a fungus shapes biotechnology: 25 years of *Aspergillus niger* research. *Fungal Biology and Biotechnology,* 5, 1-13.

11. Campagnoli, A., Cheli, F., Polidori, C., Zaninelli, M., Zecca, O., Savoini, G., et al. (2011). Use of the electronic nose as a screening tool for the recognition of durum wheat naturally contaminated by deoxynivalenol: a preliminary approach. Sensors, 11(5), 4899e4916.

12. Carballo, D., Pallares, N., Ferrer, E., Barba, F. J. and Berrada, H. (2021). Assessment of human exposure to deoxynivalenol, ochratoxin A, zearalenone and their metabolites biomarker in urine samples using LC-ESI-qTOF. Toxins, 13(8), 530.

13. Casarova, M., Shariati, M. A. and Tana inova, D. (2017). Detection of mycotoxins using MALDI-TOF mass spectrometry. Journal of microbiology, biotechnology and food sciences, 7(2), 181-185.

14. Chalupova, J., Raus, M., Sedlarova, M. and Sebela, M. (2014). Identification of fungal microorganisms by MALDI-TOF mass spectrometry. Biotechnology advances, 32(1), 230-241.

15. Chen, X., Wu, H., Tang, X., Zhang, Z. and Li, P. (2023). Recent advances in electrochemical sensors for mycotoxin detection in food. Electroanalysis, 35(2), e202100223.

16. Chen, Y., Kistler, H. C., & Ma, Z. (2019). Fusarium graminearum trichothecene mycotoxins: biosynthesis, regulation, and management. Annual Review of Phytopathology, 57(1), 15-39.

17. Deng, Y., Wang, Y., Deng, Q., Sun, L., Wang, R., Wang, X. and Gooneratne, R. (2020). Simultaneous quantification of aflatoxin B1, T-2 toxin, ochratoxin A and deoxynivalenol in dried seafood products by LC-MS/MS. *Toxins,* 12(8), 488.

18. Edwards, S. G., O'callaghan, J. and Dobson, A. D. (2002). PCR-based detection and quantification of mycotoxigenic fungi. Mycological Research, 106(9), 1005-1025.

19. El Sheikha, A. F. (2019). Molecular detection of mycotoxigenic fungi in foods: The case for using PCR-DGGE. Food Biotechnology, 33(1), 54-108.

20. Geballa-Koukoula, A., Gerssen, A. and Nielen, M. W. (2020). Direct analysis of lateral flow immunoassays for deoxynivalenol using electrospray ionization mass spectrometry. Analytical and Bioanalytical Chemistry, 412, 7547-7558.

21. Goryacheva, I. Y., Saeger, S. D., Eremin, S. A. and Peteghem, C. V. (2007). Immunochemical methods for rapid mycotoxin detection: Evolution from single to multiple analyte screening: A review. Food additives and contaminants, 24(10), 1169-1183.

22. Hickert, S., Cramer, B., Letzel, M. C. and Humpf, H. (2016). MALDI-imaging mass spectrometry of ochratoxin A and fumonisins in mold-infected food. Rapid Commun Mass Spectrom, 30(23), 2508-2516.

23. Huang, X., Huang, X., Xie, J., Li, X. and Huang, Z. (2020). Rapid simultaneous detection of fumonisin B1 and deoxynivalenol in grain by immune chromatographic test strip. Analytical biochemistry, 606, 113878.

24. ISO. (1993). Standard method for determining patulin in apple juice by HPLC-UV. International Organization for Standardization.

25. Janik, E., Niemcewicz, M., Ceremuga, M., Stela, M., Saluk-Bijak, J., Siadkowski, A. and Bijak, M. (2020). Molecular aspects of mycotoxins—A serious problem for human health. International journal of molecular sciences, 21(21), 8187.

26. Jeroen, P., Bienenmann-Ploum, M., Rijk, T. D. and Willem, H. (2010). Development of a multiplex flow cytometric microsphere immunoassay for mycotoxins and evaluation of its application in feed. Mycotoxin Research, 27(1), 63e72.

27. Keller, N. P., Turner, G. and Bennett, J. W. (2005). Fungal secondary metabolism—from biochemistry to genomics. *Nature Reviews Microbiology, 3*, 937-947.

28. Kizis, D., Vichou, A. E. and Natskoulis, P. I. (2021). Recent advances in mycotoxin analysis and detection of mycotoxigenic fungi in grapes and derived products. Sustainability, 13(5), 2537.

29. Kokkonen, M., Jestoi, M. and Rizzo, A. (2005). Determination of selected mycotoxins in mould cheeses with liquid chromatography coupled to tandem with mass spectrometry. Food additives and contaminants, 22(5), 449-456.

30. Lin, H. Y., Agrawal, D. C., Yang, W. G. and Chien, W. J. (2021). A simple HPLC-MS/MS method for the analysis of multi-mycotoxins in betel nut. *International Journal of Applied Science and Engineering*, 18(5), 1-7.

31. Linting, Z., Ruiyi, L., Zaijun, L., Qianfang, X., Yinjun, F. and Junkang, L. (2012). An immune sensor for ultrasensitive detection of aflatoxin B1 with an enhanced electrochemical performance based on graphene/conducting polymer/gold nanoparticles/the ionic liquid composite film on modified gold electrode with electrodeposition. Sensors and actuators B: chemical, 174, 359-365.

32. Luppa, P. B., Sokoll, L. J. and Chan, D. W. (2001). Immuno sensors—principles and applications to clinical chemistry. Clinica chimica acta, 314(1-2), 1-26.

33. Macheleidt, J., Mattern, D. J., Fischer, J., Netzker, T., Weber, J., Schroeckh, V. and Brakhage, A. A. (2016). Regulation and role of fungal secondary metabolites. *Annual review of genetics*, 50 (1), 371-392.

34. Magan, N. and Olsen, M. (Eds.). (2004). Mycotoxins in food: detection and control. Woodhead Publishing.

35. Malone, B. R., Humphrey, C. W., Romer, T. R. and Richard, J. L. (2000). Determination of aflatoxins in grains and raw peanuts by a rapid procedure with fluorometric analysis. Journal of aoac International, 83(1), 95-98.

36. Maragos, C. M. (2004). Emerging technologies for mycotoxin detection. *Journal of Toxicology: Toxin Reviews*, 23(2-3), 317-344.

37. Masoomi, L., Sadeghi, O., Banitaba, M. H., Shahrjerdi, A. and Davarani, S. S. H. (2013). A non-enzymatic nanomagnetic electro-immunosensor for determination of Aflatoxin B1 as a model antigen. Sensors and actuators B: chemical, 177, 1122-1127.

38. Mellikeche, W., Casini, G., Ricelli, A., Colelli, G., Gallo, M. and D'onghia, A. M. (2022). Detection of post-harvest pathogens by loop-mediated isothermal amplification: a review. Phytopathologia Mediterranea, 61(3), 531-547.

39. Monteagudo, A. A., Torres, A., Chamorro, F., Prieto, M. A., Perez-Gregorio, R., Simal-Gandara, J. and Otero, P. (2023, July). Rapid Identification of the Mycotoxin Patulin by Gas Chromatography–Mass Spectrometry. In *Biology and Life Sciences Forum* (Vol. 24, No. 1, p. 6). MDPI.

40. Mortimer, D. N., Shepherd, M. J., Gilbert, J. and Clark, C. (1988). Enzyme-linked immunosorbent (ELISA) determination of aflatoxin B1 in peanut butter: Collaborative trial. Food Additives & Contaminants, 5(4), 601-608.

41. Muscarella, M., Iammarino, M., Nardiello, D., Palermo, C. and Centonze, D. (2012). Determination of deoxynivalenol and nivalenol by liquid chromatography and fluorimetric detection with on-line chemical post-column derivatization. Talanta, 97, 145e149.

42. Nolan, P., Auer, S., Spehar, A., Elliott, C. T. and Campbell, K. (2019). Current trends in rapid tests for mycotoxins. Food Additives & Contaminants: Part A, 36(5), 800-814.

43. Olsson, J., Borjesson, T., Lundstedt, T. and Schnurer, J. (2002). Detection and quantification of ochratoxin A and deoxynivalenol in barley grains by GC-MS and electronic nose. International Journal of Food Microbiology, 72(3), 203-214.

44. Omar, S. S., Haddad, M. A. and Parisi, S. (2020). Validation of HPLC and Enzyme-Linked Immunosorbent Assay (ELISA) techniques for detection and quantification of aflatoxins in different food samples. Foods, 9(5), 661.

45. Pantano, L., La Scala, L., Olibrio, F., Galluzzo, F. G., Bongiorno, C., Buscemi, M. D. and Vella, A. (2021). QuEChERS LC–MS/MS screening method for mycotoxin detection in cereal products and spices. International Journal of Environmental Research and Public Health, 18(7), 3774.

46. Pearson, T. C., Wicklow, D. T., Maghirang, E. B., Xie, F. and Dowell, F. E. (2001). Detecting aflatoxin in single corn kernels by transmittance and reflectance spectroscopy. Transactions of the ASAE, 44(5), 1247.

47. Phokane, S. (2018). *Mycotoxin contamination of maize and groundnut produced by subsistence farmers in northern KwaZulu-Natal* (Doctoral dissertation, Stellenbosch: Stellenbosch University).

48. Pleadin, J., Persi, N., Zadravec, M., Sokolović, M., Vulic, A., Jaki, V. and Mitak, M. (2012). Correlation of deoxynivalenol and fumonisin concentration determined in maize by ELISA methods. Journal of Immunoassay and Immunochemistry, 33(4), 414-421.

49. Pohanka, M. and Skladal, P. (2008). Electrochemical biosensors--principles and applications. Journal of applied biomedicine, 6(2).

50. Przybylska, A., Chrustek, A., Olszewska-Słonina, D., Koba, M. and Kruszewski, S. (2021). Determination of patulin in products containing dried fruits by Enzyme-Linked Immunosorbent Assay technique Patulin in dried fruits. Food Science & Nutrition, 9(8), 4211-4220.

51. Radicevic, T., Jankovic, S., Simunovic, S., Stefanovic, S., Petrovic, Z., & Jelusic, V. (2021, October). Determination of patulin in apple juice by liquid chromatography-electrospray tandem mass spectrometry. In *IOP Conference Series: Earth and Environmental Science* (Vol. 854, No. 1, p. 012075). IOP Publishing.

52. Rahmani, A., Jinap, S. and Soleimany, F. (2009). Qualitative and quantitative analysis of mycotoxins. Comprehensive reviews in food science and food safety, 8(3), 202-251.

53. Rasooly, A. and Herold, K. E. (2008). Food microbial pathogen detection and analysis using DNA microarray technologies. Foodborne pathogens and disease, 5(4), 531-550.

54. Ren, Y., Zhang, Y., Han, S., Han, Z. and Wu, Y. (2011). Simultaneous determination of fumonisins B1, B2 and B3 contaminants in maize by ultra-high-performance liquid chromatography tandem mass spectrometry. Analytica chimica acta, 692(1-2), 138-145.

55. Rios, L. C. and Freitas-Silva, O. (2014). Application of liquid chromatography in food analysis. In Tech, 13, 100-130.

56. Romero-Sanchez, I., Ramirez-García, L., Gracia-Lor, E. and Madrid-Albarrán, Y. (2022). Simultaneous determination of aflatoxins B1, B2, G1 and G2 in commercial rices using immunoaffinity column clean-up and HPLC-MS/MS. Food chemistry, 395, 133611.

57. Rovina, K., Shaeera, S. N., Vonnie, J. M. and Yi, S. X. (2019). Recent biosensors technologies for detection of mycotoxin in food products. Mycotoxins and food safety.

58. Sadhasivam, S., Barda, O., Zakin, V., Reifen, R. and Sionov, E. (2021). Rapid detection and quantification of patulin and citrinin contamination in fruits. Molecules, 26(15), 4545.

59. Salisu, B., Anua, S. M., Ishak, W. R. W. and Mazlan, N. (2021). Development and validation of quantitative thin layer chromatographic technique for determination of total aflatoxins in poultry feed and food grains without sample clean-up. Journal of advanced veterinary and animal research, 8(4), 656.

60. Sanzani, S. M., Reverberi, M. and Geisen, R. (2016). Mycotoxins in harvested fruits and vegetables: Insights in producing fungi, biological role, conducive conditions, and tools to manage postharvest contamination. Postharvest Biology and Technology, 122, 95-105.

61. Schmidt-Heydt, M., Magan, N. and Geisen, R. (2008). Stress induction of mycotoxin biosynthesis genes by abiotic factors. *FEMS Microbiology Letters, 284*(2), 142-149.

62. Shen, K., Hu, X., Sun, L., Han, C. and Yang, J. (2022). Development of a lateral flow strip with a positive readout for the on-site detection of aflatoxin B1. Molecules, 27(15), 4949.

63. Sheng, Y., Jiang, W., De Saeger, S., Shen, J., Zhang, S. and Wang, Z. (2012). Development of a sensitive enzyme-linked immunosorbent assay for the detection of fumonisin B1 in maize. Toxicon, 60(7), 1245-1250.

64. Shrivastava, A. and Sharma, R. K. (2022). Biosensors for the detection of mycotoxins. Toxin Reviews, 41(2), 618-638.

65. Singh, G., Velasquez, L., Huet, A. C., Delahaut, P., Gillard, N. and Koerner, T. (2019). Development of a sensitive polyclonal antibody-based competitive indirect ELISA for determination of citrinin in grain-based foods. Food Additives & Contaminants: Part A, 36(10), 1567-1573.

66. Skrbic, B., Koprivica, S. and Godula, M. (2013). Validation of a method for determination of mycotoxins subjected to the EU regulations in spices: The UHPLC–HESI–MS/MS analysis of the crude extracts. Food control, 31(2), 461-466.

67. Soleimany, F., Jinap, S. and Abas, F. (2012). Determination of mycotoxins in cereals by liquid chromatography tandem mass spectrometry. Food Chemistry, 130(4), 1055e1060.

68. Songsermsakul, P. and Razzazi-Fazeli, E. (2008). A review of recent trends in applications of liquid chromatography-mass spectrometry for determination of mycotoxins. Journal of Liquid Chromatography & Related Technologies®, 31(11-12), 1641-1686.

69. Spanjer, M. C., Rensen, P. M. and Scholten, J. M. (2008). LC–MS/MS multi-method for mycotoxins after single extraction, with validation data for peanut, pistachio, wheat, maize, cornflakes, raisins and figs. Food Additives and Contaminants, 25(4), 472-489.

70. Stratakou, I. and van der Fels-Klerx, H. (2010). Mycotoxins in grapes and wine in Europe: occurrence, factors affecting the occurrence and related toxicological effects. World Mycotoxin Journal, 3(3), 283-300.

71. Stroka, J. and Anklam, E. (2002). New strategies for the screening and determination of aflatoxins and the detection of aflatoxin-producing moulds in food and feed. TrAC Trends in Analytical Chemistry, 21(2), 90-95.

72. Tangni, E. K., Ponchaut, S., Maudoux, M., Rozenberg, R. and Larondelle, Y. (2002). Ochratoxin A in domestic and imported beers in Belgium: occurrence and exposure assessment. Food Additives & Contaminants, 19(12), 1169–1179. https://doi.org/10.1080/02652030210007859

73. Tran, T. V., Do, B. N., Nguyen, T. P. T., Tran, T. T., Tran, S. C., Van Nguyen, B. and Le, H. Q. (2019). Development of an IgY-based lateral flow immunoassay for detection of fumonisin B in maize. F1000Research, 8.

74. Turner, N. W., Subrahmanyam, S. and Piletsky, S. A. (2009). Analytical methods for determination of mycotoxins: a review. Analytica chimica acta, 632(2), 168-180.

75. VaclavikCasarova , L., Zachariasova, M., Hrbek, V. and Hajslova, J. (2010). Analysis of multiple mycotoxins in cereals under ambient conditions using direct analysis in real time (DART) ionization coupled to high resolution mass spectrometry. Talanta, 82(5), 1950-1957.

76. Van der Gaag, B., Spath, S., Dietrich, H., Stigter, E., Boonzaaijer, G., van Osenbruggen, T. and Koopal, K. (2003). Biosensors and multiple mycotoxin analysis. Food Control, 14(4), 251-254.

77. Vidal, J. C., Bonel, L., Ezquerra, A., Hernández, S., Bertolín, J. R., Cubel, C. and Castillo, J. R. (2013). Electrochemical affinity biosensors for detection of mycotoxins: A review. Biosensors and Bioelectronics, 49, 146-158.

78. Vujanovic, V., Goh, Y. K. and Daida, P. (2012). Heat- and cold-shock responses in Fusarium graminearum 3 acetyl- and 15 acetyl-deoxynivalenol chemotypes. The Journal of Microbiology, 50(1), 97e102.

79. Wang, M., Jiang, N., Xian, H., Wei, D., Shi, L. and Feng, X. (2016). A single-step solid phase extraction for the simultaneous determination of 8 mycotoxins in fruits by ultra-high performance liquid chromatography tandem mass spectrometry. Journal of Chromatography A, 1429, 22-29.

80. Weaver, A. C., Adams, N. and Yiannikouris, A. (2020). Invited Review: Use of technology to assess and monitor multimycotoxin and emerging mycotoxin challenges in feedstuffs. Applied Animal Science, 36(1), 19-25.

81. Xing, K. Y., Shan, S., Liu, D. F. and Lai, W. H. (2020). Recent advances of lateral flow immunoassay for mycotoxins detection. TrAC Trends in Analytical Chemistry, 133, 116087.

82. Zhang, Y., Wang, Z., Qu, X., Zhou, J., Yang, H., Wang, W. and Yang, C. (2023). A photothermal lateral flow immunoassay for zearalenone with high sensitivity and wide detection range. Sensors and Actuators B: Chemical, 390, 133909.

83. Zhao, Y., Wang, P., Wang, F., Zhou, H., Li, W., Yue, J., et al. (2012). A novel biosensor regulated by the rotator of F0F1eATPase to detect deoxynivalenol rapidly. Biochemical & Biophysical Research Communications, 423(1), 195e199.

84. Zhou, H., Liu, N., Yan, Z., Yu, D., Wang, L., Wang, K. and Wu, A. (2021). Development and validation of the one-step purification method coupled to LC-MS/MS for simultaneous determination of four aflatoxins in fermented tea. Food Chemistry, 354, 129497.

85. Zollner, P. and Mayer-Helm, B. (2006). Trace mycotoxin analysis in complex biological and food matrices by liquid chromatography–atmospheric pressure ionisation mass spectrometry. Journal of Chromatography A, 1136(2), 123-169.

86. Zou, Z., He, Z., Li, H., Han, P., Tang, J., Xi, C., et al. (2012). Development and application of a method for the analysis of two trichothecenes: deoxynivalenol and T-2 toxin in meat in China by HPLCeMS/MS. Meat Science, 90(3), 613e617.

Chapter 7

Molecular Techniques for Detection and Diagnosis of Fungal Pathogens during Post-Harvest Phase

Shubham[*1], *Pankaj Yadav*[3], *Ankit Kumar*[2], *Abhishek Kumar*[3] *and Ankit Phoughat*[4]

[1]*Department of Entomology,* [2]*Centre of Food Science and Technology,* [3]*Department of Plant Pathology,* [4]*Department of Agriculture Meteorology, CCS HAU, Hisar-125004*

Abstract

Fungal pathogens threaten the agricultural produce even after the harvest which is a critical period in the supply chain. Early detection as well as diagnosis of these pathogens is crucial for preventing significant losses of the produce. Use of traditional detection methods such as visual inspection, culture-based techniques, and microscopy are limited due to their lack of sensitivity, specificity, and time efficiency. Advancements in molecular biology have revolutionized pathogen detection, with techniques such as Polymerase Chain Reaction (PCR) and its variants (e.g., Nested PCR, Real-Time PCR, Multiplex PCR) offering significant improvements in speed, sensitivity, and specificity in identifying fungal pathogens such as *Botrytis cinerea*, *Aspergillus spp.*, and *Fusarium spp.*, etc. Other techniques such as Enzyme-Linked Immunosorbent Assay (ELISA), Next-Generation Sequencing (NGS), *etc.* are also becoming increasingly popular due to their ability to detect low levels of fungal DNA or RNA, identify multiple pathogens simultaneously, and even visualize pathogen spatial distribution within tissues. Application of these techniques enable accurate pathogen identification, differentiation of closely related species, and detection of latent infections or antimicrobial resistance genes. To facilitate accurate as well as quick detection, future emphasise should be laid on the integration of multiple techniques for better pathogen identification and technological advancements such as AI-driven diagnostics, nanotechnology-based biosensors, and portable field devices for real-time detection for comprehensive pathogen tracking and enhanced food safety.

Introduction

The post-harvest phase is a critical period in the agricultural supply chain, where harvested produce is susceptible to various pathogens, particularly fungi, which can cause significant economic losses. Fungal pathogens such as *Botrytis cinerea*, *Penicillium* spp., *Monilinia* spp., and *Alternaria alternata* are among the primary culprits responsible for post-harvest diseases in fruits and vegetables. *B. cinerea* is often regarded as the second most significant plant pathogenic fungus, following only *Magnaporthe oryzae*. This is due to its ability to cause gray mold disease in a wide range of horticultural crops (Zhang *et al.*, 2021). These pathogens can lead to rotting, spoilage, and a substantial reduction in the quality and marketability of produce. This chapter delves into the intricate biochemical and molecular methods employed in identifying and diagnosing fungal pathogens that pose a significant threat to post-harvest crops. The need for effective detection and accurate diagnosis of these fungal pathogens as early as possible is paramount to mitigate post-harvest losses, ensure food safety, and minimize their impact. This chapter explores the principles, applications, and limitations of these biochemical and molecular techniques in the context of post-harvest pathogen detection.

Traditional methods for detecting and diagnosing fungal pathogens include visual inspection, culture-based techniques, and microscopic examination. While these methods have been fundamental in plant pathology, they often lack the sensitivity and specificity required for early and accurate detection. Culture-based techniques are time-consuming and may not detect non-culturable or slow-growing pathogens. Moreover, visual and microscopic inspections are subjective and may not differentiate between closely related fungal species.

Recent advancements in molecular biology have also revolutionized the detection and diagnosis of fungal pathogens. DNA-based methods, particularly Polymerase Chain Reaction (PCR) and its variants, have become popular due to their high sensitivity (ability to detect low levels of fungal DNA), specificity, speed, and rapid turnaround time (Hariharan and Prasannath, 2021). Polymerase chain reaction (PCR) and its variants, such as quantitative PCR (qPCR) and reverse transcription PCR (RT-PCR), have become indispensable in pathogen diagnostics. These techniques amplify specific DNA or RNA sequences, enabling the identification of fungal species with high accuracy. This chapter discusses various molecular approaches, such as; Conventional PCR, Nested PCR, Real-Time PCR or Quantitative PCR, Multiplex PCR, Next-generation Sequencing (NGS), Loop-Mediated Isothermal Amplification (LAMP) and Fluorescence *In Situ* Hybridization (FISH), etc.

Enzyme-linked immunosorbent assay (ELISA)

ELISA is a widely used technique for detecting and quantifying specific antigens or antibodies in post-harvest pathogen detection.

Principle of ELISA

The ELISA process involves the immobilizing of the target antigen onto a 96-well microplate, blocking of the non-specific binding sites, and adding a specific primary antibody to bind to the antigen. Afterwards, a secondary enzyme-linked antibody is then introduced to bind to the primary antibody and then a substrate for the enzyme is added, which the enzyme converts into a detectable signal (e.g., a color change). Finally, the signal's intensity is measured via spectrophotometry to quantify the antigen or antibody in the sample (Alhajj *et al.*, 2023)

Applications in Post-Harvest Pathogen Detection

ELISA can detect fungi like *Aspergillus flavus* and *A. parasiticus*, which produce aflatoxins in crops such as peanuts, maize, and rice with high sensitivity and specificity due to the specific antigen-antibody interaction. It can also detect fungi such as *Botrytis cinerea*, *Alternaria, etc.* that cause spoilage in post-harvest in many fruits and vegetables. It has also been used to quantify mycotoxins produced by fungi in agricultural products such as grains and nuts as well as provide quick screening of large numbers of samples. This capability is crucial for quality control in the food industry and it can also be adapted for a wide range of pathogens (Wang *et al.*, 2017; Quintela *et al.*, 2022; Youssef *et al.*, 2023).

Conventional Polymerase Chain Reaction (PCR)

Polymerase Chain Reaction (PCR) is a molecular biology technique used to amplify specific DNA sequences, making millions of copies from a small initial sample with rapid results, simplicity, and high sensitivity (Valones *et al.*, 2009). PCR is a powerful technique that has found significant applications in various fields, including the detection of post-harvest pathogens.

Principle of PCR

In the first step called denaturation, the double-stranded DNA is subjected to heat at around 94-98°C, which leads to its separation into two single strands. Then in the annealing process, the temperature is lowered to 50-65°C, allowing primers (short DNA sequences complementary to the target region) to bind (anneal) to the single-stranded DNA. Finally, extension phase is carried out in which the temperature is raised to around 72°C, where Taq polymerase (a heat-stable DNA polymerase) extends the primers, synthesizing new DNA strands complementary to the target sequence. By repeating these steps 20-40 times, PCR exponentially amplifies the target DNA region, making it easier to detect even small amounts of pathogen DNA (Valones *et al.*, 2009).

Applications in Post-Harvest Pathogen Detection

PCR can identify specific pathogens like fungi, bacteria, or viruses in harvested crops simultaneously in a single PCR reaction which is crucial for preventing the spread of diseases during storage and transportation. For e.g., fungal pathogens causing postharvest rot such as *Alternaria iridiaustralis*, *Aspergillus niger*, Penicillium, Trichoderma, and *Colletotrichum fioriniae* can be easily identified by analyzing their ITS (Internal Transcribed Spacer) regions using PCR, providing precise genetic characterization. It can also distinguish between closely related pathogen species or strains and can detect even small amounts of pathogen DNA as in case of *Lasiodiplodia theobromae*, *Neofusicoccum parvum*, and *Fusarium solani* which cause stem-end rot in avocados. Also, the identification of many fungal pathogens can be carried out simultaneously in just a few hours allowing for rapid detection such as Botrytis spp., Colletotrichum spp., and Verticillium spp. which cause postharvest loss in strawberries (Wanjiku *et al.*, 2020; Youssef *et al.*, 2023; Ling *et al.*, 2023).

Nested PCR

Nested Polymerase Chain Reaction (PCR) is a modified advanced version of the standard PCR technique that aims to increase the sensitivity and specificity of DNA amplification. It is particularly useful in detecting low-abundance pathogens or in situations where non-specific amplification may be a concern, such as in the detection of post-harvest pathogens in agricultural products.

Principle

Nested PCR is a two-step amplification technique that enhances specificity and sensitivity by using two sets of primers in successive rounds of PCR. In the first round, external primers amplify a DNA fragment containing the target sequence, but non-specific products may also be generated. The second round uses the product from the first PCR as a template and employs internal (nested) primers that bind within the initial amplified region, ensuring only the specific target DNA is amplified. This approach minimizes non-specific amplification, as non-target products from the first round are unlikely to contain binding sites for the internal primers, resulting in highly specific and sensitive detection of the target sequence (Green & Sambrook, 2019).

Applications in Post-Harvest Pathogen Detection

Nested PCR can detect pathogens present in very low quantities, (as little as 100 fg of target DNA, corresponding to just a few copies of the target sequence), which is often the case in post-harvest samples where pathogens may not be evenly distributed. The increased sensitivity and specificity also allows for the early as well as simultaneous detection and differentiation between closely related pathogen strains leading to more accurate identification like *Penicillium marneffei*, *P. granati* in pomegranate fruits. It can also detect fungal pathogens at latent infection stages when fungal load is very low and undetectable by conventional methods. For example, it has been used to identify *Colletotrichum* spp. in strawberries during their biotrophic phase (Poritz *et al.*, 2011; Chung *et al.*, 2022).

Real-Time PCR or Quantitative PCR

Real-Time PCR, also known as quantitative PCR (qPCR), is an advanced form of PCR based on the standard PCR technique but it allows for the quantification and monitoring of DNA in real-time during the amplification process. This technique is widely used in various fields, including the detection of post-harvest pathogens in agricultural products, due to its ability to provide both qualitative and quantitative data (Nehra *et al.*, 2022).

Principle

Real-time PCR combines amplification and detection in a single step, eliminating the need for post-PCR processing. As in standard PCR, it involves denaturation, annealing, and extension cycles, but a fluorescent reporter dye or probe is included for real-time fluorescence detection. Common methods include SYBR Green, which binds to double-stranded DNA and fluoresces upon binding, and TaqMan probes, which use sequence-specific probes with a fluorescent reporter and quencher (suppresses the fluorescence of the reporter dye when the probe is intact) in which the probe is cleaved, releasing fluorescence. Fluorescence intensity is measured at each cycle which is directly proportional to the amount of DNA generated, enabling quantification. The threshold cycle (Ct value) marks the cycle where fluorescence surpasses a set threshold and inversely correlates with the initial DNA quantity in which lower Ct values indicate higher starting DNA levels (Nehra *et al.*, 2022).

Applications in Post-Harvest Pathogen Detection

Real-Time PCR allows for rapid, simultaneous and sensitive detection of foodborne pathogens like Aspergillus species (e.g. *A. flavus, A. parasiticus, A. niger*), Penicillium species (e.g. *P. expansum, P. digitatum*), *Botrytis cinerea*, etc. which enables timely interventions to prevent significant losses. It can also identify specific strains or serotypes of pathogens which allows for precise identification and differentiation between pathogen species or strains, e.g. *Diaporthe batatas* and *Diaporthe destruens* causing dry rot and foot rot in sweet potato; *Colletotrichum acutatum, C. gloeosporioides,* and *C. siamense* causing bitter rot in apple (Sanzani *et al.*, 2014; Fujiwara *et al.*, 2021; McHenry and Aćimović, 2024).

Multiplex PCR

Multiplex Polymerase Chain Reaction (Multiplex PCR) is a variant of PCR that allows for the simultaneous amplification of multiple DNA targets in a single reaction. This technique is particularly useful in situations where multiple pathogens need to be detected or when different genetic markers of a pathogen must be identified in a single assay. In the post-harvest pathogen detection, Multiplex PCR offers a highly efficient and cost-effective approach (Al-Zaban *et al.*, 2023).

Principle

The principle of Multiplex PCR is similar to that of standard PCR, but with the key difference that multiple sets of primers specific to different target sequences are used in the same reaction mix to facilitate simultaneous amplifications. Each primer set is designed to amplify a specific DNA target corresponding to different pathogens or different regions of a single pathogen's genome, resulting in multiple PCR products of different sizes that can be distinguished and identified by gel electrophoresis or other detection methods such as capillary electrophoresis, or Real-Time PCR with fluorescent probes, depending on the nature of the assay. The primers are carefully designed to ensure that they work efficiently at the same annealing temperature and do not interfere with each other (no primer-dimer formation) (Al-Zaban *et al.*, 2023).

Applications in Post-Harvest Pathogen Detection

Multiplex PCR can identify and screen several fungal pathogens in a single reaction, saving time and resources with high sensitivity and specificity, e.g. *Fusarium verticillioides* and *Fusarium subglutinans* in maize. It can also detect multiple pathogens at once, e.g. *Neofabraea alba, N. perennans*, and *N. keinholzii* in apple which allows for faster screening of food products, which is crucial in post-harvest quality control. Multiplex PCR enhances the ability to detect and differentiate between various pathogens and their different strains or species that may cause similar symptoms or contaminate the same food products as it amplifies strain-specific markers, providing detailed information about the pathogens present in post-harvest products (Zhou *et al.*, 2013; Hariharan and Prasannath, 2021).

Next-generation Sequencing (NGS)

Next-generation sequencing (NGS) is a powerful technology that has revolutionized genomic research, including pathogen detection in post-harvest settings. NGS is a high-throughput technology that allows for the rapid sequencing of entire genomes or targeted regions of DNA by massively parallel sequencing of DNA or RNA fragments, providing detailed insights into the genetic composition of samples. NGS is increasingly used in the detection and study of post-harvest pathogens due to its ability to identify multiple organisms, including those that are difficult to culture or detect by traditional methods.

Principle

The genetic material is extracted from the sample contaminated with pathogens. Afterwards, the DNA or RNA is extracted, fragmented and then converted into a library of smaller sequences, each tagged with unique adapters for identification. The clonal amplification of fragments is also carried out which is carried out by simultaneous parallel sequencing of millions of fragments which leads to generation of vast amounts of data. The data generated is subjected to bioinformatic processing to assemble and analyze sequence data, align them to reference genomes, and identify the presence of pathogens (Jiang *et al.*, 2022).

Applications in Post-Harvest Pathogen Detection

NGS allows for comprehensive profiling of microbial communities in samples, identifying all potential pathogens present. It enables the sequencing of entire genomes of pathogens, providing detailed information on their genetic makeup E.g. *Alternaria alternata* causing black spot on various fruits. It can also provide high-resolution data for distinguishing between closely related pathogen strains. It can also identify mixed infections in a sample, where multiple pathogens are present simultaneously, which may not be detected by other methods, e.g. *Colletotrichum* species (e.g., *C. truncatum* and *C. brevisporum*) which causes anthracnose in fruits like passion fruit and mangoes. It can detect genes associated with pathogenicity and antimicrobial resistance, e.g. *Pestalotiopsis microspora* in kiwi fruit rot. NGS can also be used to detect genetic markers associated with fungicide resistance in post-harvest fungal pathogens, informing disease management strategies, e.g. *Penicillium digitatum* and *Penicillium italicum* responsible for green and blue mold in citrus fruits. Transcriptomic NGS approaches allow for monitoring expression of mycotoxin biosynthesis genes in fungi like *Alternaria alternata* contaminating stored products which provides information on toxin production potential (Rizwan *et al.*, 2021; Deng *et al.*, 2024)

Loop-Mediated Isothermal Amplification (LAMP)

Loop-Mediated Isothermal Amplification (LAMP) is a molecular DNA amplification technique that has gained popularity for detecting post-harvest pathogens due to its simplicity, speed, and sensitivity. Unlike PCR, LAMP is performed at a constant temperature, eliminating the need for thermal cycling.

Principle

LAMP uses 4-6 specially designed primers and a DNA polymerase with high strand displacement activity to amplify target DNA sequences at a constant temperature (typically 60-65°C) called Isothermal amplification (no thermal cycling is required). The process leads to formation of cauliflower-like DNA structures with multiple loop structures that allow rapid, exponential, and specific amplification without the need for thermal cycling equipment. LAMP is highly specific due to the use of 4-6 primers that recognize 6-8 distinct regions on the target gene. Amplification can often be detected visually, as the reaction produces a large amount of DNA, leading to a visible turbidity or color change when specific dyes are used and the use of multiple primers enhances the specificity of the reaction (Notomi *et al.*, 2000).

Applications of LAMP in Post-Harvest Pathogen Detection

LAMP enables detection with high sensitivity of a variety of pathogens such as fungal pathogens, like *Botrytis cinerea*, *Penicillium* species, and *Fusarium* species affecting stored produce. LAMP assays to detect *Talaromyces flavus* in strawberries and soil samples have also been developed with high specificity, achieving a detection limit of 1 femtogram (fg) of genomic DNA or 64 ascospores per gram of sample Early detection of mycotoxin-producing fungi present in post-harvest produce, e.g. *Aspergillus flavus* and *Aspergillus carbonarius*, which contaminate nuts, grains, and other food products. It can also provide rapid on-site detection of pathogens in storage facilities provides results within 30-60 minutes, which is much faster than traditional PCR methods. This method is also relatively inexpensive, requiring only basic equipment like a water bath or a simple heating block, and it can also be easily coupled with colorimetric or fluorescent dyes, allowing for real-time detection and easy interpretation of results without the need for sophisticated instruments making it accessible for use in resource-limited settings and practical for routine monitoring (Niessen, 2015; Panek & Frac, 2019; Mellikeche *et al.*, 2024).

Fluorescence In Situ Hybridization (FISH)

Fluorescence In Situ Hybridization (FISH) is a powerful molecular technique used for detecting and localizing specific short, single-stranded DNA or RNA sequences within cells or tissues using fluorescently labelled probes. In post-harvest pathogen detection, FISH is particularly useful for identifying and visualizing the presence of pathogens directly in plant tissues or on surfaces where they may reside.

Principle

Fluorescence in situ hybridization (FISH) is a molecular technique that involves the hybridization of fluorescently labelled nucleic acid probes to complementary sequences in a target organism's DNA or RNA. The process includes fixation of the sample to preserve cellular structure, permeabilization to allow probe entry (DNA or RNA probes that are complementary to specific sequences of the target pathogen's DNA or RNA). Then, the fluorescent probe are hybridization to the target sequence and washing is carried out to remove unbound probes. The visualization is carried out using fluorescence microscopy where the fluorescence emitted at specific locations which indicates where the probes have bound which shows the presence of the target pathogen (Shakoori, 2017).

Applications in Post-Harvest Pathogen Detection

FISH can specifically detect and visualize particular fungal pathogens in post-harvest products based on unique genetic sequences, e.g. *Alternaria tenuis* and *Botrytis cinerea* which cause post-harvest diseases in peaches. It can also detect pathogens in food samples within hours, much faster than traditional culturing methods. For example fungi like Penicillium spp., Fusarium spp., and Aspergillus spp. which cause post-harvest spoilage of fruits, vegetables, and grains. It can also be used to differentiate between closely related species, allowing for precise identification of pathogens such as *Aspergillus tenuis and A. niger* which cause postharvest disease in mature peach fruits. Similarly, this technique can also detect pathogens in a Viable but non-culturable (VBNC)_state, which is crucial for food safety such as *Aspergillus flavus* and *Fusarium verticillioides*. Multiple pathogens can also be detected simultaneously using probes with different fluorophores. It also allows for visualization of pathogens directly within food matrices, providing information on their spatial distribution (where pathogens are colonizing or infecting the product) (Namulawa *et al.*, 2020; Zhang *et al.*, 2019).

Conclusion

The detection and diagnosis of post-harvest pathogens are critical aspects of ensuring food safety and quality in the agricultural supply chain. This chapter highlights a range of molecular techniques that have revolutionized the field of pathogen detection in post-harvest crops. Molecular techniques have significantly advanced the pathogen detection. Polymerase Chain Reaction (PCR) and its variants (Nested PCR, Real-Time PCR, Multiplex PCR, etc.) offer rapid, sensitive, and specific detection of pathogen DNA. Next-Generation Sequencing (NGS) provides comprehensive genomic analysis, enabling the detection of novel pathogens and strain-level identification. Loop-Mediated Isothermal Amplification (LAMP) offers a rapid, cost-effective alternative for on-site diagnostics. Advanced imaging and spectroscopy techniques have also contributed to pathogen detection. Fluorescence microscopy and Fluorescence In Situ Hybridization (FISH) allow for visual localization of pathogens within host tissues. The most effective strategies for post-harvest pathogen detection involve integrating multiple techniques. For instance, combining ELISA or LAMP with traditional visual inspection for initial pathogen screening, followed by Real-Time PCR for confirmation and quantification. NGS can be employed for in-depth genomic analysis of complex pathogen populations or novel pathogens. Portable spectroscopy techniques integrated with LAMP offer the potential for real-time, in-field pathogen detection.

The future of detection and diagnosis will focus on developing automated systems which will integrate multiple detection methods with artificial intelligence for rapid, accurate pathogen identification. The incorporation of nanomaterials in biosensors for ultra-sensitive pathogen detection and advancements in miniaturized, field-deployable devices that combine multiple detection technologies are also promising areas. Increased use of NGS and bioinformatics for comprehensive pathogen profiling and microbiome analysis in post-harvest environments is expected. Additionally, implementing blockchain technology for transparent tracking of pathogen testing throughout the supply chain could enhance food safety measures.

References

Alhajj, M., Zubair, M., & Farhana, A. (2023). Enzyme linked immunosorbent assay. *StatPearls*.

Al-Zaban, M. I., Alrokban, A. H., & Mahmoud, M. A. (2023). Development of a real-time PCR and multiplex PCR assay for the detection and identification of mycotoxigenic fungi in stored maize grains. *Mycology, 14*(3), 227-238.

Chung, P. C., Wu, H. Y., Chen, Y. C., Hung, T. H., & Chung, C. L. (2022). Development of a nested PCR assay for detecting *Colletotrichum siamense* and *Colletotrichum fructicola* on symptomless strawberry plants. *Plos one, 17*(6), e0270687.

Deng, L., Qiu, X., Su, Q., Pan, H., Wang, Z., Qian, G., & Li, L. (2024). Complete genome sequence analysis of *Pestalotiopsis microspora,* a fungal pathogen causing kiwifruit postharvest rots. *BMC genomics, 25*(1), 839.

Fujiwara, K., Kobayashi, Y. O., Usui, M., Nishioka, K., Nakamura, M., Kawano, S., & Inoue, H. (2021). Real-Time PCR Assay for the Diagnosis and Quantification of Co-infections by *Diaporthe batatas* and *Diaporthe destruens* in Sweet Potato. *Frontiers in Plant Science, 12*, 694053.

Green, M. R., & Sambrook, J. (2019). Nested polymerase chain reaction (PCR). *Cold Spring Harbor Protocols, 2019*(2), pdb-prot095182.

Hariharan, G., & Prasannath, K. (2021). Recent advances in molecular diagnostics of fungal plant pathogens: a mini review. *Frontiers in Cellular and Infection Microbiology, 10*, 600234.

Jiang, S., Chen, Y., Han, S., Lv, L., & Li, L. (2022). Next-generation sequencing applications for the study of fungal pathogens. *Microorganisms, 10*(10), 1882.

Ling, L., Luo, H., Zhao, Y., Yang, C., Cheng, W., & Pang, M. (2023). Fungal pathogens causing postharvest fruit rot of wolfberry and inhibitory effect of 2, 3-butanedione. *Frontiers in Microbiology, 13*, 1068144.

McHenry, D. J., & Aćimović, S. G. (2024). New Species-Specific Real-Time PCR Assays for Colletotrichum Species Causing Bitter Rot of Apple. *Microorganisms, 12*(5), 878.

Mellikeche, W., Ricelli, A., Casini, G., Gallo, M., Baser, N., Colelli, G., & D'Onghia, A. M. (2024). Development of Loop-Mediated Isothermal Amplification (LAMP) Assays for the Rapid Detection of Toxigenic *Aspergillus flavus* and *A. carbonarius* in Nuts. *International Journal of Molecular Sciences, 25*(7), 3809.

Namulawa, V. T., Mutiga, S., Musimbi, F., Akello, S., Ngángá, F., Kago, L., & Ghimire, S. (2020). Assessment of fungal contamination in fish feed from the Lake Victoria Basin, Uganda. *Toxins, 12*(4), 233.

Nehra, M., Kumar, V., Kumar, R., Dilbaghi, N., & Kumar, S. (2022). Current scenario of pathogen detection techniques in agro-food sector. *Biosensors, 12*(7), 489.

Niessen, L. (2015). Current state and future perspectives of loop-mediated isothermal amplification (LAMP)-based diagnosis of filamentous fungi and yeasts. *Applied microbiology and biotechnology, 99*, 553-574.

Notomi, T., Okayama, H., Masubuchi, H., Yonekawa, T., Watanabe, K., Amino, N., & Hase, T. (2000). Loop-mediated isothermal amplification of DNA. *Nucleic acids research, 28*(12), e63-e63.

Panek, J., & Frąc, M. (2019). Loop-mediated isothermal amplification (LAMP) approach for detection of heat-resistant *Talaromyces flavus* species. *Scientific Reports, 9*(1), 5846.

Poritz, M. A., Blaschke, A. J., Byington, C. L., Allen, L., Nilsson, K., Jones, D. E., Thatcher, S. A., Robbins, T., Lingenfelter, B., Amiott, E., Herbener, A & Ririe, K. M. (2011). FilmArray, an automated nested multiplex PCR system for multi-pathogen detection: development and application to respiratory tract infection. *PloS one, 6*(10), e26047.

Quintela, I. A., Vasse, T., Lin, C. S., & Wu, V. C. (2022). Advances, applications, and limitations of portable and rapid detection technologies for routinely encountered foodborne pathogens. *Frontiers in Microbiology, 13,* 1054782.

Rizwan, H. M., Zhimin, L., Harsonowati, W., Waheed, A., Qiang, Y., Yousef, A. F., & Chen, F. (2021). Identification of fungal pathogens to control postharvest passion fruit (Passiflora edulis) decays and multi-omics comparative pathway analysis reveals purple is more resistant to pathogens than a yellow cultivar. *Journal of Fungi, 7*(10), 879.

Rizwan, H. M., Zhimin, L., Harsonowati, W., Waheed, A., Qiang, Y., Yousef, A. F., Munir, N., Wei, X., Scholz, S. S., Reichelt, M., Oelmüller, R., & Chen, F. (2021). Identification of fungal pathogens to control postharvest passion fruit (*Passiflora edulis*) decays and multi-omics comparative pathway analysis reveals purple is more resistant to pathogens than a yellow cultivar. *Journal of Fungi, 7*(10), 879.

Sanzani, S. M., Li Destri Nicosia, M. G., Faedda, R., Cacciola, S. O., & Schena, L. (2014). Use of quantitative PCR detection methods to study biocontrol agents and phytopathogenic fungi and oomycetes in environmental samples. *Journal of Phytopathology, 162*(1), 1-13.

Shakoori, A. R. (2017). Fluorescence in situ hybridization (FISH) and its applications. *Chromosome structure and aberrations,* 343-367.

Valones, M. A. A., Guimarães, R. L., Brandão, L. A. C., Souza, P. R. E. D., Carvalho, A. D. A. T., & Crovela, S. (2009). Principles and applications of polymerase chain reaction in medical diagnostic fields: a review. *Brazilian Journal of Microbiology, 40,* 1-11.

Wang, T., Li, P., Zhang, Q., Zhang, W., Zhang, Z., Wang, T., & He, T. (2017). Determination of Aspergillus pathogens in agricultural products by a specific nanobody-polyclonal antibody sandwich ELISA. *Scientific Reports, 7*(1), 4348.

Wanjiku, E. K., Waceke, J. W., Wanjala, B. W., & Mbaka, J. N. (2020). Identification and pathogenicity of fungal pathogens associated with stem end rots of avocado fruits in Kenya. *International Journal of Microbiology, 2020*(1), 4063697.

Youssef, K., Roberto, S. R., Ippolito, A., & Abo-Elyousr, K. A. (2023). Advances in detection and control of post-harvest pathogens. *Frontiers in Microbiology, 14,* 1184039.

Zhang, S., Zheng, Q., Xu, B., & Liu, J. (2019). Identification of the fungal pathogens of postharvest disease on peach fruits and the control mechanisms of *Bacillus subtilis* JK-14. *Toxins, 11*(6), 322.

Zhang, Z. Q., Chen, T., Li, B. Q., Qin, G. Z., & Tian, S. P. (2021). Molecular basis of pathogenesis of postharvest pathogenic fungi and control strategy in fruits: progress and prospect. *Molecular horticulture, 1*, 1-10.

Zhou, B., Xiao, J., Liu, S., Yang, J., Wang, Y., Nie, F., Zhou, Q., Li, Y., & Zhao, G. (2013). Simultaneous detection of six food-borne pathogens by multiplex PCR with a GeXP analyzer. *Food Control, 32*(1), 198-204.

Chapter 8

Imaging Techniques for Fungal Pathogen Detection and Diagnosis

Kumar Sathiyaseelan[1], Emmadi Venu[2], Akurathi Ramya[3], Andonissamy Daniel[4], Vimalkumar C[5*]

[1,2] Division of Plant Pathology,
Indian Agricultural Research Institute, Pusa, New Delhi.
[3] Department of Plant Pathology,
Junagadh Agricultural University, Gujarat.
[4] Division of Plant Physiology,
Indian Agricultural Research Institute, Pusa, New Delhi.
[5] CISH-Regional Research Station, Malda, West Bengal
[*]Corresponding author: vimalkumar.c@icar.gov.in

Abstract

Fungal infections are a major threat to global agriculture, causing significant crop losses and economic consequences. Rapid and accurate diagnosis to improve the management of the disease is still a challenging concern. In this chapter we are going to examine advanced imaging techniques for the detection and diagnosis of fungal pathogens in plants. It begins with the limitations of traditional techniques such as visual inspection and laboratory tests that are time-consuming, subjective, and require special skills. This chapter examines contemporary imaging techniques, among others and optical, hyperspectral, fluorescence, and thermal imaging. The principles, advantages, and applications of any method of fungal infection diagnosis are discussed. For example, hyperspectral imaging can detect minor physiological changes in plants. Artificial intelligence (AI) and machine learning combined with these imaging techniques are also analyzed, showing how AI increases accuracy and speed in image data analysis overcoming challenges such as the need for high quality imaging equipment, complex image analysis, and environmental change management. The chapter concludes by highlighting the transformational impact of advanced imaging techniques in fungal pathogen detection, offering effective, accurate and scalable solutions to improve plant health management and agricultural outcomes. Future research directions and the importance of interdisciplinary collaboration are emphasized.

Keywords: Hyperspectral, Artificial intelligence, Thermal imaging, Image processing, Precision agriculture.

Introduction

Fungal infections represent a major threat to global agriculture, being one of the most significant biotic factors causing crop diseases. With over 8,000 species of fungi and oomycetes identified as plant pathogens, the impact on agricultural productivity is substantial (Horst, 2008; Fisher et al., 2020). These pathogens can infect plants at various stages of development-from seedlings to mature plants-either independently or in combination with other pathogens, complicating disease management (Narayanasamy, 2011). Among the most prevalent fungal diseases are blight, blast, damping-off, dieback, leaf spot, powdery mildew, downy mildew, smut, rust, root rot, scab, and wilt. These diseases not only reduce crop yields but also deteriorate the quality of produce, which has significant economic repercussions across diverse agricultural sectors (Hussain and Usman, 2019; Jain et al., 2019). The global agricultural landscape is deeply affected by these diseases, impacting major crops, horticultural plants, and forest species alike (Agrios, 2009; Wenneker and Thomma, 2020; Marciulynas et al., 2020).

The complexity of accurately detecting and identifying fungal pathogens is heightened by the vast diversity of fungal species and their strains, which necessitates sophisticated diagnostic methods. Traditional diagnostic practices often rely on the visual inspection of disease symptoms and signs, such as the presence of fungal structures like conidia, sclerotia, or mycelia on plant surfaces (Tör and Woods-Tör, 2017). This method, however, can be labour-intensive, time-consuming, and occasionally prone to inaccuracies. Consequently, there is an increasing requirement for advanced diagnostic approaches that offer rapid, accurate, and efficient disease detection, particularly at early stages of infection (Thind, 2015).

Recent advancements in non-invasive diagnostic techniques have revolutionized plant disease detection. These include spectroscopy-based methods, imaging techniques, and remote sensing (RS) technologies, which have proven effective for real-time and large-scale disease monitoring. Various imaging sensors, such as digital, fluorescence, thermal, multispectral, and hyperspectral sensors, have been extensively studied for detecting both symptomatic and asymptomatic diseases (Zhan et al., 2022). Hyperspectral imaging, in particular, stands out due to its ability to capture a wide spectrum of light and analyze each pixel's spectral signature. This technique has been further enhanced when combined with microscopy, providing high-resolution images that enable detailed analysis at the genotypic level of plant tissues (Rumpf et al., 2010; Zhan et al., 2022). For instance, recent studies have demonstrated that integration of hyperspectral imaging with machine learning algorithms can improve the detection of early-stage infections and differentiate between disease types with high accuracy (García-Vera et al.,2024).

Artificial Intelligence (AI) has also made significant strides in plant disease detection, with machine vision-based methods becoming increasingly prevalent. Machine Learning (ML) and Deep Learning (DL), specialized areas within AI, have demonstrated considerable promise in enhancing disease detection accuracy. Recent developments in DL architectures, such as Convolutional Neural Networks (CNNs) and Recurrent Neural Networks (RNNs), have shown remarkable improvements in identifying and classifying plant disease symptoms. These algorithms can analyze large datasets of plant images to detect patterns and anomalies that are indicative of disease (Kamilaris and Prenafeta-Boldúliu, 2018; He et al., 2021). For example, recent research has applied CNNs to hyperspectral and multispectral images, achieving high accuracy in detecting various fungal diseases, including powdery mildew and rust (Wang et al., 2023). Additionally, advancements in Transfer Learning have further improved the performance of DL models by leveraging pre-trained networks to enhance classification accuracy with limited data (Xu et al., 2022).

The integration of AI into agricultural practices offers several advantages, including reduced labour costs, decreased time inefficiencies, and enhanced crop quality and yield. AI technologies facilitate early and accurate disease detection, enabling timely intervention and effective disease management. This approach not only helps in mitigating disease spread but also supports sustainable agricultural practices by minimizing the need for excessive pesticide use (Sinha and Shekhawat, 2020). By harnessing early data on crop health and disease location, AI-driven tools allow farmers to implement targeted disease management strategies, improving overall crop health and productivity.

Conventional Approaches to Fungal Pathogen Detection and Diagnosis

Traditional methods for identifying plant fungal pathogens, heavily rely on visual inspection and culture-based techniques. Visual assessments involve identifying symptoms or signs on infected tissues, such as mildew on leaves, spots, blight, or fruit bodies following established guidelines and standards (Luchi et al., 2020). However, many diseases present symptoms that are visually indistinguishable from one another, complicating accurate pathogen diagnosis. The results of visual assessments are not always conclusive, as they depend heavily on the observer's experience, which introduces temporal variation that can affect outcomes. This variation leads to significant inconsistencies and reduced reproducibility (Nutter and Guan, 2001). Culture based identification is a critical technique involves collecting symptomatic plant tissue, which is sterilized on the surface to remove any external contaminants before being cultured on specialized media like Potato Dextrose Agar (PDA) or Malt Extract Agar (MEA) etc., and then employing microscopy to evaluate morphological traits i.e., spore shapes, sporulation patterns, and the presence of structures bearing spores for taxonomic classification (Narayanasamy, 2011). These techniques are time-consuming, require expertise, and are sometimes inconclusive. Isozyme analysis is a valuable technique for detecting, distinguishing, and identifying morphologically similar or closely related fungal species. This method is effective for revealing genetic diversity within fungal populations, as different isozyme patterns reflect underlying genetic variation (Micales et al., 1986). However, isozyme analysis has been successfully employed to identify several important fungal pathogens, including *Fusarium* spp. (Hossain et al., 2020), *Pythium* spp. (Dugan et al., 2021), *Phytophthora* spp. (Goss et al., 2020; Van den Bosch et al., 2017), *Rhizoctonia solani* (Zhao et al., 2019), *Fusarium solani* (Miller et al., 2021), *Fusarium oxysporum* (Mohamed et al., 2018), and *Phytophthora infestans* (Schaad et al., 2019) Isozyme analysis is noted for its reliability, cost-effectiveness, and robustness in assessing large populations and identifying fungal species (Guarro et al., 1999).

Non-invasive/indirect plant pathogen detection methods

Indirect detection methods involve identifying the plant's responses to pathogenic stress as a way to diagnose disease. These stress reactions can present as alterations in morphology or physiology, including shifts in their temperature, color, transpiration rates, and the buildup of certain metabolites and hormones. Major indirect techniques for identifying fungal pathogens include imaging-based methods such as visible/RGB (red, green, and blue) imaging, hyperspectral imaging, multispectral imaging, thermography, and chlorophyll fluorescence (fig 1). A qualitative comparison of these indirect methods is provided in Table 1.

Table 1: A qualitative comparison of different indirect methods

Imaging Technique	*Wavelength Range (nm)*	*Application*	*Advantage*	*Limitation*	*Reference*
Visible/RGB imaging	400-700	Visual symptom detection	Widely available, easy to use	Limited to visible symptoms, not sensitive to early-stage infections	Mahlein et al., 2012
Hyperspectral Imaging	400-2500	Early detection, stress analysis High spectral resolution,	High spectral resolution, early detection	early detection Expensive, complex data processing	Sankaran et al., 2010
Multispectral Imaging	400-1000	Disease monitoring, precision agriculture	Specific band selection, less complex than hyperspectral	Limited spectral information, may miss subtle symptoms	Zhang et al., 2019
Thermal Imaging	8000-15000	Stress detection, water content analysis	Non-invasive, detects physiological changes	Sensitive to environmental conditions	Jones, 2004
Chlorophyll fluorescence Imaging	400-700	Detection of fungal infections and stress	Highly sensitive, detects biochemical changes	Requires specific excitation sources	Bauriegel and Herppich, 2014.

Visible/RGB-imaging

An image is essentially a pictorial representation of an object or scene, composed of a matrix of pixels arranged in a grid, each containing specific color and intensity values. These values

Figure 1: Various imaging techniques used for fungal disease detection and diagnosis include the following, listed in a clockwise direction starting from the top: 1) RGB imaging, 2) Multispectral imaging, 3) Hyperspectral imaging, 4) Thermal imaging, and 5) Fluorescence imaging allow the reconstruction of the original scene captured. In digital imaging, pixels encode information about colors and brightness, which is crucial for analyzing visual data (Gonzalez & Woods, 2020). The optical properties of plants, such as light transmission through tissues, light absorption by pigments, sugars, and water, and light reflection, play a critical role in digital imaging. These properties make it possible to detect changes and abnormalities in plant tissues using digital images, which can be analyzed at the field or farm level (Mahlein, 2016).

RGB imaging, which uses red (550-750 nm), green (500-549 nm), and blue (400-499 nm) channels of light, has become a widely adopted technique for plant disease detection. This method is particularly valuable during the growing season, providing continuous monitoring of plant health. Recent advancements in camera technology have significantly enhanced the resolution and sensitivity of digital cameras, making RGB imaging more accessible and precise for agricultural applications. Modern RGB sensors, often embedded in smartphones and tablets, facilitate high-resolution imaging, enabling the early detection of disease symptoms in plants (Barbedo, 2019). RGB-color images enable the detection and identification of fungal pathogens by analyzing specific color spaces like LAB, YCbCr, and HSV. These color spaces help to highlight various features of plant diseases, allowing for more accurate diagnosis (Bock et al., 2010). Machine learning models, combined with RGB imaging, have further refined disease classification and identification. For instance, Behmann et al., (2014) emphasized the critical role of feature selection in enhancing classification accuracy, while explored deep learning techniques to improve disease detection using RGB images.

RGB imaging has also been utilized to assess the levels of metabolites like mycotoxins produced by plant pathogens, which are crucial for food safety and public health. For instance, this technique was employed to identify the toxin deoxynivalenol in plants infected with *Fusarium graminearum* (Cambaza et al., 2019). Furthermore, Goluguri and colleagues (2020) conducted a review on the use of image classifiers and deep learning models for detecting diseases in rice. In olive plants, RGB indexes were used to detect the initial symptoms of Verticillium wilt, revealing noticeable differences between untreated and treated plants (Sancho-Adamson et al., 2019). Advancements in image analysis software, such as ASSESS 2.0, Leaf Doctor, and custom-made modules, have further enhanced the ability to evaluate disease severity. These tools analyze color distribution, texture, and shape, which are crucial for accurately assessing plant health (Pethybridge & Nelson, 2015). However, achieving consistent and accurate results still depends on factors like image quality, lighting conditions, and standardized imaging techniques. Ongoing research and development are crucial to enhancing these methods and increasing their reliability for use in agricultural practices.

Multi and hyperspectral imaging

Spectral sensors, crucial in plant disease detection, are typically classified based on their spectral resolution (the number and width of the measured wavelengths), spatial scale, and type of detector used (whether imaging or non-imaging). The first imaging technique developed, multispectral imaging, captures spectral data across several relatively broad wavelength bands, including red, green, blue (RGB), and near-infrared (NIR) wavelengths. Recent advancements in hyperspectral imaging have significantly expanded the scope of data collection, covering wavelengths from 350 to 2,500 nm, with spectral resolutions often narrower than 1 nm, offering detailed spectral and spatial information (Steiner et al., 2008). Unlike non-imaging methods, hyperspectral sensors provide both spectral and spatial data, making them more suitable for precise plant health analysis. In hyperspectral imaging, the collected data is visualized as a three-dimensional cube, where the x and y axes represent spatial dimensions, and the z-axis represents spectral data as reflectance intensity per wavelength. The number of pixels in an image, which depends on the distance between the sensor and the target, is critical for detecting fungal diseases or interactions between plants and pathogens. Distant or airborne sensing systems typically capture fewer pixels than near-range systems, which can influence detection precision (Mahlein et al., 2012; West et al., 2003). For detecting soil-borne fungal pathogens in the field, airborne sensors are particularly effective due to their ability to cover large areas quickly. However, sensors with a spatial resolution around 1 meter are not suitable for detecting individual symptoms on infected plants, where proximal sensor platforms, which operate closer to the plant, are more appropriate (Hillnhütter et al., 2011; Oerke et al., 2014; West et al., 2003). Despite its potential, the application of hyperspectral imaging in studying plant pathogens and assessing disease severity remains in its infancy. While many studies have explored this technology, it is not yet widely adopted in agricultural practices (Bock et al., 2010). Plant interaction with light is complex, involving processes such as light transmission through the leaf, absorption by various leaf components (e.g., pigments, water, sugars, lignin), and reflectance within the leaf structure. These processes are influenced by multiple physiological and biochemical reactions, making light reflectance a critical indicator of plant health. Different wavelength ranges provide specific insights: the visible range (400 to 700 nm) is influenced by the content and structure of leaf pigments, the near-infrared range (700 to 1,100 nm) is affected by internal scattering and water absorption, and the short-wave infrared range (1,100 to 2,500 nm) reflects the chemical composition and water content of leaves (Carter & Knapp, 2001; Jacquemoud & Ustin, 2001). Changes in reflectance caused by pathogen attacks can be detected through hyperspectral imaging, often linked to alterations in leaf biology and chemical composition during infection. For instance, biotrophic fungi, such as downy mildews and rusts, minimally affect leaf structure and chlorophyll content in early infection stages. Conversely, necrotrophic pathogens, which cause leaf spots, degrade plant tissue through toxins or enzymes, leading to necrotic lesions. Fungi such as powdery mildew and rust form structures on the leaf surface that modify the plant's optical characteristics. The complex interactions between sugar beet leaf diseases and their pathogens, studied through techniques like raster electron microscopy, highlight the potential of spectral properties for pathogen detection (Mahlein et al., 2012). Recent advancements in hyperspectral imaging are pushing the boundaries of plant disease detection, with ongoing research focused on improving the accuracy and applicability of these techniques in agriculture (Zarco-Tejada et al., 2018; Rumpf et al., 2010).

Plant Disease Detection Using Hyperspectral Imaging

Hyperspectral imaging has garnered significant attention for its potential in assessing and forecasting agricultural productivity by detecting microbial contaminants and diagnosing various crop diseases. When combined with specialized sensors, hyperspectral imaging can measure the radiation reflected or emitted by plant tissues across multiple spectral bands, providing detailed insights into different plant components such as leaves, stems, and soil nutrient levels, as well as enabling the early detection of diseases (Dhingra et al., 2017). In cereal crops like wheat and rice, hyperspectral imaging has been utilized to monitor plant water status (Ravi et al., 2017) and assess nitrogen levels (Sharma et al., 2021). Recent advances include the development of deep learning-based tools for detecting diseases such as leaf rust in wheat, caused by *Puccinia triticina*. These tools have achieved prediction accuracies of up to 84.1% (Zhu et al., 2020). Recent studies have expanded the applications of hyperspectral imaging to various plant fungal pathogens. For instance, Powdery Mildew (*Erysiphe necator*) in grapevines has been effectively detected using hyperspectral imaging, which highlighted distinct spectral signatures associated with infection (Pérez-Roncal et al., 2020). Rust Diseases (*Puccinia* spp.) in crops like wheat have been monitored using hyperspectral imaging combined with machine learning algorithms to enhance detection accuracy and disease monitoring (Guo et al., 2020). Fusarium Head Blight (*Fusarium graminearum*) in wheat has also been studied using hyperspectral imaging. The approach integrates hyperspectral data with chlorophyll fluorescence imaging to assess disease severity and monitor pathogen impact, providing valuable information for both laboratory and field assessments (Bauriegel, & Herppich, 2014). In addition, Late Blight (*Phytophthora infestans*) in potatoes has been successfully detected using hyperspectral imaging, which allowed for early diagnosis and management of the disease (Kool & Evenhuis, 2023).

Thermal imaging

Infrared thermography (IRT) is an effective method in agriculture for monitoring plant temperature, which provides insights into plant water status (Jones et al., 2002), understanding the microclimatic conditions within crop canopies (Lenthe et al., 2007), and identifying changes in transpiration rates often associated with early stages of pathogen infection (Oerke et al., 2006). IRT works by detecting infrared radiation emitted in the thermal infrared spectrum (8 to 12 μm) using specialized thermographic and infrared cameras, creating false-color images where each pixel reflects the temperature of the object being measured. This technique is utilized across different spatial and temporal scales in agriculture, from aerial surveys to close-up inspections. However, the accuracy of IRT can be influenced by environmental factors (Oerke et al., 2011). Leaf temperature, which is directly related to plant transpiration, is affected by various factors, including fungal pathogens. Pathogens that target aerial parts of plants, like leaf spots or rusts, cause localized temperature changes, while root pathogens (e.g., *Rhizoctonia solani* or *Pythium* spp.) or systemic infections (e.g., *Fusarium* spp.) can have a broader impact, often disrupt the transpiration rate and water flow across the entire plant or specific parts of it. Temperature changes in plants due to pathogen infections or defense mechanisms have been extensively studied, particularly in plant-virus interactions, such as in tobacco, and in plant-fungal interactions like *Cercospora beticola* in sugar beet (Singh et al., 2020). Recent studies have shown that infrared thermography (IRT) can effectively detect early-stage infections by identifying subtle temperature variations in plants. For example, Oerke et al., (2016) used IRT to monitor cucumber plants infected with downy mildew and apple trees affected by scab. In the case of the apple-*Venturia inaequalis* pathosystem, IRT was able to distinguish between infected and healthy tissues before any visible symptoms appeared, with the fungal structures only detectable microscopically. IRT has also been successfully applied to study the pathogenesis of *Peronospora sparsa* in various Rosa cultivars, highlighting its versatility in different plant-pathogen systems (Gomez et al., 2014). Moreover, recent research has shown a strong correlation between disease severity and IRT measurements across various crops, including cereals, tubers, and solanaceous plants such as barley, wheat, oat, sugar beet, and potatoes (Li et al., 2022). Accurate analysis of IRT images now increasingly relies on advanced algorithms that differentiate temperature variations within and between leaves. The average temperature variation within single leaves, entire plants, and across crop stands is a crucial element in identifying and tracking the development of plant diseases.

Fluorescence imaging

Fluorescence imaging, particularly through chlorophyll fluorescence, is a sophisticated technique for assessing plant photosynthetic activity and health. This method utilizes dynamic sensors equipped with LED or laser light sources to measure the efficiency of the electron transport system in photosynthesis (Murchie & Niyogi, 2018). Chlorophyll fluorescence imaging has been applied to study variations in photosynthesis impacted by both biotic and abiotic stressors (Kalaji et al., 2017; Mishra et al., 2016). In recent years, fluorescence imaging has been instrumental in detecting fungal pathogens in plants. For instance, researchers have used chlorophyll fluorescence to study the effects of Powdery Mildew (*Erysiphe graminis*) on wheat. The technique revealed early declines in photosynthetic efficiency and changes in leaf fluorescence before visible symptoms appeared, allowing for earlier disease management (Shen et al., 2020). Similarly, rust disease (*Puccinia triticina*) in wheat has been monitored using fluorescence imaging, which identified physiological stress responses and disease progression before symptoms were evident (Li et al., 2021). One major drawback of chlorophyll fluorescence imaging is that the plant preparation needs a strict protocol which requires laboratory facilities and hence it cannot be maintained in greenhouse or field conditions. Consequently, research has been conducted to extract fluorescence parameters from solar reflectance in field settings, which could significantly influence disease assessment at both the canopy and field scales (Rossini et al., 2014). Various plant fungal diseases have been detected using different imaging techniques, as detailed in Table 2.

Table 2: Detection of Plant Fungal Diseases Using Various Imaging Techniques

Imaging Type	*Crop*	*Disease*	*Pathogen*	*Reference*
Visible/RGB imaging	Cotton	Ascochyta blight	(*Ascochyta gossypii)*	Camargo and Smith (2009)
	Sugar beet	Leaf spot	(*Cercospora beticola*)	Neumann *et al.* (2014)
		Rust	(*Uromyces betae*)	
	Apple	scab	*Venturia inequalis*	Wijekoon *et al.* (2008)
	Winter Wheat	Yellow rust	*Puccinia striiformis* f.sp. *tritici*	Zhou et al., 2014
	Poatato	Late Blight	*Phytophthora infestance*	Sugiura et al., 2016
	Olive	Leaf spot	*Spilocaea oleaginea*	Al-Tarawneh, M. S. (2013)
	Alfalfa	Root rot	*Phymatotrichopsis omnivora*	Mattupalli et al., 2018
Multispectral imaging	Soybean	Rust	*Phakopsora pachyrhizi*	Cui et al., 2009
	Spinach	Fungal disease	*Verticillium spp., Fusarium spp.,*	Olesen et al., 2011
	Barley	Powdery mildew	*Blumeria graminis f.sp. hordei*	Kuska et al., 2018
	Cucumber	Fungal Diseases	*Sphaerotheca fuliginea, Corynespora cassiicola,*	Feng et al., 2009
	Oilseed Rape	Leaf spot	*Pyrenopeziza brassicae*	Veys et al., 2019
Hyperspectral Imaging	Wheat	Leaf rust	*Puccinia triticina*	Ashourloo et al., 2014
		Fusarium head blight	*Fusarium graminarium*	Zhang et al., 2020
		Powdery mildew	*Blumeria graminis f.sp. tritici*	Zhao et al., 2020
	Barley	Powdery mildew	*Blumeria graminis f.sp. hordei*	Nagasubramanian et al., 2019
	Rice	Sheath blight	*Rhizoctonia solani*	Lin et al., 2020
	Banana	Black sigatoka	*Mycosphaerella fijiensis*	Ugarte et al., 2020
	Grapes	Powdery mildew	*Erysiphe necator*	Knauer et al.,2017
		Downey mildew	*Plasmopara viticola*	Oerke et al., 2016
	Bell pepper	Powdery mildew	*Leveillula Taurica*	Schor et al., 2017

Thermal imaging	Sugar beet	Leaf spot	*Cecospora beticola*	Chaerle *et al.*, (2004)
	Cucumber	Powdery mildew	*Pseudoperonospora cubensis*	Berdugo *et al.*, (2014)
		Downey mildew	*Podosphaera xanthii*	Oerke *et al.*, (2006),
	Apple	Scab	*Venturia inequalis*	Oerke *et al.*, (2011)
	Rose	Downey mildew	*Perenospora sparsa*	Gomez et al., (2014)
	Peanut	Early and late blight	*Mycosphaerella arachidis and Mycosphaerella berkeleyi*	Omran, E.S.E. (2017).
	Cucurbits	Downey mildew	*Pseudoperonospora cubensis*	Oerke et al., (2006)
	Wheat	Stripe rust	*Puccinia striiformis*	Yang et al., (2019)
Chlorophyll fluorescence imaging	Wheat	Leaf rust Powdery mildew	*Puccinia triticina, Blumeria graminis f.sp. tritici*	Burling *et al.* (2011)
		Septoria blotch	*Zymoseptoria tritici*	Romero et al., (2023)
	Grape	Downy mildew	*Plasmopara viticola*	Bauriegel et al, (2021)
	Tomato	Powdery mildew	*Oidium neolycopersici*	Lu et al., (2020)
	Apple	Scab	*Venturia inaequalis*	De Castro et al., (2021)
	Cucumber	Powdery mildew	*Podosphaera xanthii*	Yang et al., (2022)

Role of Image Processing in Detecting and Identifying fungal diseases

Image processing is essential for accurately detecting and identifying fungal diseases in plants, providing precision, efficiency, and the capability to handle large volumes of data. Advances in image processing and information mining enable us to

- Identify infected leaves and stems
- Quantify the affected regions
- Analyze the shape of the infected areas
- Assess the color characteristics of the infected regions

The subsequent sub-section outlines the systematic approach to image processing, including a comparative analysis of techniques such as image segmentation, feature extraction, feature selection, and classification, all of which are vital for diagnosing plant diseases.

Image acquisition

Image acquisition serves as a foundational step in image processing, providing the initial digital images that are essential for subsequent analysis. This stage encompasses pre-processing tasks, such as image scaling and normalization. The process of acquiring high-quality images can be quite challenging (Gonzalez & Woods, 2020). Researchers often utilize specialized datasets, such as those from the Plant Village database, the Kaggle Plant Pathology dataset, and other contemporary collections, which are typically captured under controlled environmental conditions. Effective image acquisition requires advanced equipment, such as high-resolution cameras or scanners, to ensure the quality and accuracy of the images (Fenu and malloci, 2021).

Image pre-processing

Image pre-processing is a critical step in image analysis that involves transforming raw images into a format suitable for further processing. This stage includes several operations designed to enhance the quality of the images or extract meaningful information from them. Essentially, it is a form of signal processing where the input is an image, whether a video frame or a photograph and the output can be either an improved image or features extracted from it (Rathore and prasad ,2021; Islam et al., 2021). Key components of image pre-processing include filtering to reduce noise, color conversion to standardize hues, and detail enhancement to improve the visibility of important features (Gonzalez & Woods, 2020). Recent advancements in pre-processing techniques have incorporated sophisticated algorithms and deep learning methods to further refine image quality and feature extraction (Hasan et al., 2023).

Image Segmentation

Image segmentation is a pivotal step in analyzing leaf images, as it involves partitioning an image into distinct regions or segments based on specific criteria. This process is crucial for isolating and analyzing the areas of interest, such as diseased portions of a leaf. The segmentation process can be challenging, particularly when the background contains numerous green elements or shadow transitions, which complicates the distinction between the leaf and its surroundings (Singh and Mishra., 2017; Wani et al., 2022). To address these challenges and enhance the quality of the segmented images, pre-processing techniques are often employed. These techniques improve the visibility of relevant features and facilitate more accurate segmentation. In recent research, clustering methods such as K-means and Fuzzy C-means have been widely utilized for plant disease detection. These methods categorize the image into clusters representing the background, diseased areas, and healthy portions of the leaf. Such approaches help in effectively distinguishing between the various regions of interest and improve the accuracy of disease detection (Wani et al., 2022).

Feature extraction

Feature extraction is a critical step in distinguishing between various classes of disease by analyzing the geometry and visual characteristics of an object. Prior to extraction, feature selection identifies relevant features using domain knowledge and optimization algorithms (Jaffery et al., 2013). This process is crucial for effective image classification, as the extracted features are the primary parameters used to categorize images. Texture extraction, which captures the arrangement and structure of an object's surface, is fundamental for analyzing image patterns and is widely applied in remote sensing, biomedical imaging, and object recognition (Yu et al., 2022). Color extraction, another vital factor, involves analyzing color information from digital images. Colors are often represented in RGB space, though models like HIS and CIE Lab are also used for their effectiveness in differentiating hues and shades, which are crucial for early disease detection (Vishnoi et al., 2022). Shape extraction involves descriptors such as object count, region area, and image dimensions, which help in defining the shape and extent of lesions (Vishnoi et al., 2022). Edge detection, which identifies boundaries and transitions in image brightness, is essential for segmenting and extracting relevant features in image processing and computer vision tasks (Jing et al., 2022). These techniques together enable precise analysis and classification of plant diseases, contributing to more accurate and efficient diagnostic processes.

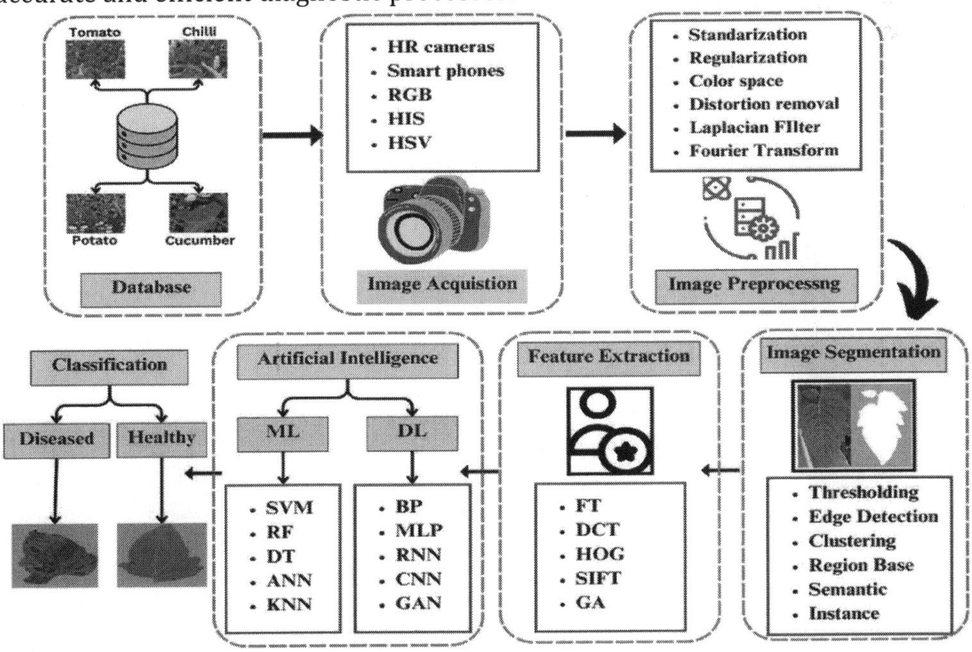

Figure 2: General frame work for plant disease detection (source: Jafar et al., 2024)

Classification: Classification is a crucial process in image analysis that involves categorizing objects or images into predefined categories based on their features. In the context of plants disease detection, classification differentiates between healthy and infected leaves by identifying and categorizing various disease symptoms. This process can be executed using machine learning or deep learning techniques. Machine learning algorithms, such as support vector machines or decision trees, require explicit feature extraction and a moderate amount of data to build effective models. Conversely, deep learning techniques, particularly convolutional neural networks (CNNs), are increasingly preferred due to their ability to automatically extract features from raw image data, offering greater efficiency and flexibility. These algorithms can handle large datasets and complex patterns, making them particularly suited for identifying subtle disease symptoms and variations (LeCun et al., 2015). General methodology of image processing is shown in Figure 2.

Artificial intelligence and machine learning in image processing and Fungal disease detection

Artificial Intelligence (AI) and Machine Learning (ML) have transformed image processing by automating complex tasks, enhancing accuracy, and efficiently managing large datasets. In plant disease detection, these technologies provide advanced solutions for analyzing and interpreting images, leading to more effective disease management. Recent advancements highlight the significant potential of AI and ML in fungal disease detection. Deep learning models, particularly Convolutional Neural Networks (CNNs), have demonstrated impressive

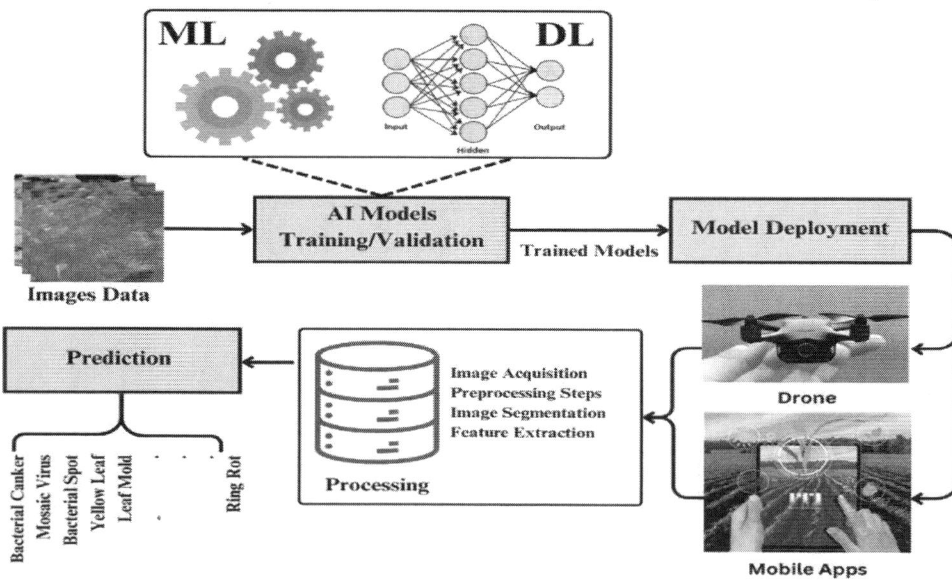

Figure 3: Schematic of Real-Time Plant Leaf Disease Detection Process (source: Jafar et al., 2024)

accuracy in classifying fungal infections by analyzing plant tissue images. Deployment of trained models on drones and smartphones for real-time identification of fungal plant diseases (fig.3). Studies reveal that CNN-based methods can effectively detect and differentiate between various fungal pathogens, such as Powdery Mildew and Rust Fungi, with high precision (Picon et al., 2021). Artificial intelligence and ML involve various approaches to categorizing images based on their features.

Supervised Classification: This method involves training a machine learning model using labeled data, where the model learns to differentiate between various classes based on predefined features. Algorithms like Support Vector Machines (SVM), Random Forest, and Decision Trees are commonly employed in traditional supervised classification. Recent progress includes the application of deep learning methods such as Convolutional Neural Networks (CNNs) and Recurrent Neural Networks (RNNs), which have significantly enhanced accuracy in identifying plant diseases (Hesamian et al., 2019).

Support Vector Machine (SVM): SVM is a differentiating classifier that identifies the optimal hyper-plane to distinguish between different classes. Recent advancements include incorporating kernel tricks and ensemble methods to enhance SVM's performance in complex disease detection tasks (Rashmi et al., 2020; Zhang et al., 2021). For example, in the detection of wheat powdery mildew, Zhao et al., (2020) utilized hyperspectral imaging to differentiate healthy leaves from those with varying infection levels, achieving a cross-validation accuracy of 93.33% using an SVM model. Furthermore, recent studies have demonstrated the efficacy of SVM in detecting Fusarium head blight in wheat, with an overall accuracy of 91.5% achieved when using multispectral imaging data (Rahimi et al., 2023). This highlights SVM's robustness in handling complex spectral data from plants affected by fungal diseases.

Random Forest: The Random Forest algorithm, an ensemble learning method, reduces the limitations of individual decision trees in classification problems by constructing multiple decision trees using random data subsets (Breiman, 2001). This approach has proven particularly effective in detecting plant diseases, including fungal infections. For instance, Patel et al. (2022) applied Random Forest to identify and classify Alternaria leaf spot in cotton using hyperspectral imaging, achieving an accuracy of 94.7%. The study underscores the model's ability to process high-dimensional data and effectively distinguish between healthy and diseased leaf regions.

Decision Tree: The Decision Tree algorithm, a monitored learning technique, is critical in classification tasks by assessing the relevance of various input features for predicting the target variable (Quinlan, 1986). In fungal plant disease detection, Decision Trees have been widely applied due to their straightforwardness and interpretability. Recent research by Ahmad et al. (2022) has demonstrated the effective classification of fungal diseases in crops like wheat and barley using Decision Trees. When paired with spectral imaging data, these models allow for the early and specific detection of fungal infections, which is vital for timely disease management. Moreover, Decision Tree-based models have been successfully used for classifying rice blast disease using drone-captured RGB images, achieving notable accuracy levels in large-scale field monitoring (Chen et al., 2021).

Convolutional Neural Networks (CNNs): CNNs have become a leading tool for image classification tasks due to their capacity to automatically learn spatial hierarchies of features. Recent innovations in CNNs, such as transfer learning and the use of pre-trained models, have significantly improved their accuracy in detecting fungal diseases across various crops (He et al., 2021; Xie et al., 2022). For example, Zhu et al. (2022) applied a CNN-based model with transfer learning to detect *Botrytis cinerea* in tomato plants using RGB images, achieving a remarkable accuracy of 96.2%. Another study by Wang et al. (2023) utilized a pre-trained ResNet model to classify grapevine diseases, including powdery mildew and downy mildew, with an accuracy of 94.8%, showcasing the effectiveness of deep learning approaches in challenging disease environments.

Unsupervised Classification: Unlike supervised methods, unsupervised classification does not work with labeled data. It works with clustering algorithms to group similar pixels based on their features. Recent improvements in clustering algorithms, such as Adaptive K-Means and Fuzzy C-Means, have enhanced the ability to segment and identify disease-affected regions in plant images (Mitra et al., 2021; Zhan et al., 2022).

Adaptive K-Means: Adaptive K-Means dynamically adjusts the number of clusters based on the data's complexity. This approach has been effectively used in the detection of Fusarium wilt in banana plants. For instance, recent research by Gupta et al. (2023) employed Adaptive K-Means to segment hyperspectral images of banana leaves. The study demonstrated that this method could successfully isolate the infected regions, enabling more accurate and timely diagnosis of Fusarium wilt, a fungal disease that causes significant crop losses.

Fuzzy C-Means (FCM): FCM is a popular clustering algorithm that enables data points to be associated with multiple clusters, each with different levels of membership. In the context of fungal plant disease detection, FCM has shown great potential in dealing with overlapping disease symptoms on plant leaves. For example, Zhan et al. (2022) applied FCM to segment disease-affected areas in rice crops infected with *Magnaporthe oryzae*, the pathogen responsible for rice blast disease. The study found that FCM was particularly effective in identifying early-stage infections, which are often difficult to detect with traditional methods.

Gaussian Mixture Models (GMM): In this, the data is retrieved from a pool of several Gaussian distributions. This method has been applied in the detection of powdery mildew in grapevines. Liu et al. (2022) demonstrated the effectiveness of GMM in identifying powdery mildew by analyzing the spectral features of infected grape leaves. The study reported that GMM could accurately differentiate between healthy and infected tissues, providing a robust tool for early detection.

Hierarchical Clustering: This method organizes data into a tree of clusters based on similarity. Chen et al. (2021) used hierarchical clustering to analyze hyperspectral images of wheat leaves infected with *Blumeria graminis*, the causative agent of wheat powdery mildew. The hierarchical approach allowed the researchers to identify subtle variations in the spectral data that corresponded to different infection stages, enabling more precise disease monitoring.

Self-Organizing Maps (SOMs): SOMs are neural networks that map high-dimensional data into lower-dimensional grids. Rahimi et al. (2023) applied SOMs to the classification of barley leaves infected with *Rhynchosporium commune*, the fungus responsible for scald disease. The study showed that SOMs could effectively visualize and interpret complex spectral data, offering insights into the progression of the disease. Different ML and DL methods used to identify the fungal diseases summarized in table 3.

Table 3: Different ML and DL methods used to identify fungal diseases.

Technique	*Host plant*	*Disease*	*Pathogen*	*Ref.*
SVM	Cucumber	Powdery mildew	*Sphaerotheca fuligina*	Youwen et al., (2008)
SVM	Grape	Scab and Rust Disease	*Elsinoe ampelina & Phakopsora euvitis*	Menukaewjinda et al., (2008)
SVM	Cucumber	Downey Mildew, Brown and Leaf spots	*Pseudoperonospora cubensis & Cercospora* sp., *& Alrernaria* sp.,	Jian and Wei., (2010)
SVM	Citrus	Anthracnose	*colletotrichum* sp.,	Gavhale et al., (2014)
SOM	Rice	Rice Blast, Brown Spot	*Magnaporthe grisea, Cochliobolus miyabeanus*	Phadikar and sil., (2008)
BPNN	Maize	Corn Sheath Blight, Southern Leaf Blight	*Rhizoctonia solani & Bipolaris maidis*	Song et al., (2011)
NN	Cotton	*Fusarium* and *Verticillium* Wilt	*Fusarium oxysporum f.sp., vasinfectum & Verticillium* sp.	Revathi and Hemalatha., (2012)
PNN	Rice	Rice Blast, Brown Spot	*Magnaporthe grisea, Cochliobolus miyabeanus*	Asfarian et al., (2013)
BPNN with LM optimization	Different hosts	*Cercospora* Leaf Spot,	*Cercospora* sp.	Hong and Huijie., (2014)
BPNN and GA	Tomato	Black mold	*Alternaria alternata*	Yadav and Verma., (2016

Neural Network	Watermelon	Anthracnose, Downey Mildew	*colletotrichum* sp., *Pseudoperonospora cubensis*	Kutty et al., (2013)
Neural Network	cotton	Red Spot	*Phymatotrichum* sp.	Bhong and Pawar, (2016)
KNN	Different hosts	*Alternaria* leaf spot, Anthracnose	*Alternaria Alternata, Colletotrichum* sp.	Hossain et al., (2019)
KNN, MLP	Avocado	Laurel Wilt, Phytophthora Root Rot,	*Raffaelea lauricola, Phytophthora* sp.	Abdhulridha et al., (2019)

SVM: Support Vector Machine; SOM: Self-Organizing Maps; BPNN: Backpropagation Neural Network; NN: Neural Networks; PNN: Probabilistic Neural Network; GA: Genetic Algorithm; LM: Levenberg-Marquardt Algorithm; MLP: Multi-Layer Perceptron; KNN: K-Nearest Neighbors

Conclusion

The early identification and treatment of plant diseases have advanced significantly as a result of the quick development of non-destructive imaging methods like hyperspectral and multispectral imaging. By capturing data across a broad range of the electromagnetic spectrum, including visible, ultraviolet, infrared, and near-infrared spectra, these technologies provide detailed insights into plant health that were previously unattainable. These imaging methods, when combined with artificial intelligence (AI), not only make it possible to identify diseases early on, but they also open the door to accurate disease control plans, which are crucial for sustainable farming methods. Timely intervention, which can significantly lower crop losses and stop the spread of disease, depends on early diagnosis. AI, especially through machine learning (ML) and deep learning (DL) techniques, improves this process by automating the analysis of the complex data produced by these imaging technologies. Utilizing algorithms such as Support Vector Machines (SVM) and Random Forest (RF) for disease classification, along with Convolutional Neural Networks (CNNs) for superior image analysis, enables researchers to achieve high precision in detecting diseases. This accuracy is vital in agriculture, where similar symptoms can obscure disease differentiation with traditional diagnostic methods.

Real-time implementation of disease detection technologies in agriculture faces challenges due to variability in environmental conditions such as lighting, temperature, and humidity, which can impact the accuracy of imaging systems and AI models. Developing robust algorithms that adapt to these dynamic conditions is essential for ensuring reliable disease detection across diverse settings. Future research should focus on enhancing the precision of imaging technologies and AI algorithms in quantifying disease severity, either through advanced imaging techniques or by refining models to interpret data more effectively. Integrating decision-support tools is also essential because these systems can offer farmers customized advice on how to manage diseases, such as sustainable practices or particular chemical treatments, increasing management efficacy while encouraging ecologically friendly behaviors and lowering dependency on chemical inputs. The continued evolution of these technologies promises significant benefits for researchers, pathologists, farmers, and the agricultural industry as a whole, paving the way for a future where early disease detection and precision management become standard practice in agriculture.

References

Abdhulridha, J., Ehsani, R., Abd-Elrahman, A., & Ampatzidis, Y. (2019). A remote sensing technique for detecting laurel wilt disease in avocado in presence of other biotic and abiotic stresses. *Computers and Electronics in Agriculture, 156,* 549–557. https://doi.org/10.1016/j.compag.2018.12.006

Agrios, G. N. (2009). Plant pathogens and disease: General introduction. In *Plant Pathology* (5th ed.). Elsevier Inc. https://doi.org/10.1016/B978-012373944-5.00344-8

Ahmad, H., Kumar, A., & Singh, R. (2022). Effective classification of fungal diseases in crops using decision trees. *International Journal of Plant Pathology, 14*(3), 245–259.

Al-Tarawneh, M. S. (2013). An empirical investigation of olive leaf spot disease using auto-cropping segmentation and fuzzy C-means classification. *World Applied Sciences Journal, 23*(9), 1207–1211.

Asfarian, A., Herdiyeni, Y., Rauf, A., & Mutaqin, K. H. (2013). Paddy diseases identification with texture analysis using fractal descriptors based on Fourier spectrum. *Proceedings of the International Conference on Computer Control Informatics and Its Applications (IC3INA)*, 77–81.

Ashourloo, D., Mobasheri, M. R., & Huete, A. (2014). Evaluating the effect of different wheat rust disease symptoms on vegetation indices using hyperspectral measurements. *Remote Sensing, 6*(6), 5107–5123.

Barbedo, J. G. A. (2019). Plant disease identification from individual lesions and spots using deep learning. *Biosystems Engineering, 180*, 96–107.

Bauriegel, E., & Herppich, W. B. (2014). Hyperspectral and chlorophyll fluorescence imaging for early detection of plant diseases, with special reference to *Fusarium* spec. infections on wheat. *Agriculture, 4*(1), 32–57. https://doi.org/10.3390/agriculture4010032

Bauriegel, E., Brabandt, H., Gärber, U., & Herppich, W. B. (2014). Chlorophyll fluorescence imaging to facilitate breeding of *Bremia lactucae*-resistant lettuce cultivars. *Computers and Electronics in Agriculture, 105*, 74–82.

Bauriegel, E., Grosse, I., & Schmidhalter, U. (2021). Downy mildew in grape (*Plasmopara viticola*). *Journal of Plant Diseases and Protection, 128*(3), 205–216.

Behmann, J., Mahlein, A.-K., Rumpf, T., Römer, C., & Plümer, L. (2014). A review of advanced machine learning methods for the detection of biotic stress in precision crop protection. *Precision Agriculture, 16*(3), 239–260. https://doi.org/10.1007/s11119-014-9372-7

Berdugo, C., Zito, R., Paulus, S., & Mahlein, A.-K. (2014). Fusion of sensor data for the detection and differentiation of plant diseases in cucumber. *Plant Pathology, 63*(6), 1344–1356. https://doi.org/10.1111/ppa.12204

Bhong, B. V., & Pawar, S. (2016). Study and analysis of cotton leaf disease detection using image processing. *International Journal of Advanced Research in Science, Engineering and Technology, 3*(2), 1447–1454.

Bock, C. H., Poole, G. H., Parker, P. E., & Gottwald, T. R. (2010). Plant disease severity estimated visually, by digital photography and image analysis, and by hyperspectral imaging. *Critical Reviews in Plant Sciences, 29*(2), 59–107. https://doi.org/10.1080/07352681003617285

Breiman, L. (2001). Random forests. *Machine Learning, 45*(1), 5–32. https://doi.org/10.1023/A:1010933404324

Burling, J., Han, L., & Zhang, L. (2011). Leaf rust and powdery mildew in wheat (*Puccinia triticina, Blumeria graminis f.sp. tritici*). *Field Crops Research, 121*(2), 124–131. https://doi.org/[Insert DOI or link if available]

Camargo, A., & Smith, J. S. (2009). Image pattern classification for the identification of disease-causing agents in plants. *Computers and Electronics in Agriculture, 66*(2), 121–125. https://doi.org/10.1016/j.compag.2009.01.003

Cambaza, E., Koseki, S., & Kawamura, S. (2018). *Fusarium graminearum* colors and deoxynivalenol synthesis at different water activity levels. *Foods, 8*(1), Article 7. https://doi.org/10.3390/foods8010007

Carter, G. A., & Knapp, A. K. (2001). Leaf optical properties in higher plants: Linking spectral characteristics to stress and chlorophyll concentration. *American Journal of Botany, 88*(4), 677–684. https://doi.org/10.2307/2657068

Chaerle, L., Hagenbeek, D., De Bruyne, E., Valcke, R., & Van Der Straeten, D. (2004). Thermal and chlorophyll-fluorescence imaging distinguish plant-pathogen interactions at an early stage. *Plant and Cell Physiology, 45*(7), 887–896. https://doi.org/10.1093/pcp/pch097

Chen, Y., Liu, H., & Wang, X. (2021). Hierarchical clustering for analyzing hyperspectral images of wheat leaves infected with *Blumeria graminis*. *Remote Sensing, 13*(9), 1740. https://doi.org/10.3390/rs13091740

Chen, Y., Wang, X., & Zhang, S. (2021). Classification of rice blast disease using drone-captured RGB images with decision trees. *Remote Sensing Applications:*

Society and Environment, 22, Article 100215. https://doi.org/10.1016/j.rsase.2021.100215

De Castro, A. M., Andrade, F. A., & Lima, A. (2021). Scab in apple (*Venturia inaequalis*). *Journal of Plant Pathology, 103*(1), 87–96. https://doi.org/[Insert DOI or link if available]

Dhingra, G., Kumar, V., & Joshi, H. D. (2017). Study of digital image processing techniques for leaf disease detection and classification. *Multimedia Tools and Applications, 77*, 19951–20000. https://doi.org/10.1007/s11042-017-4577-0

Dugan, F. M., Li, X., & Rajasekaran, K. (2021). Detection and identification of *Pythium* species using isozyme patterns. *Plant Disease, 105*(5), 1234–1242. https://doi.org/10.1094/PDIS-05-20-1007-RE

Feng, J., Li, H., Shi, J., Yang, W., & Liao, N. (2009). Cucumber disease diagnosis using multispectral images. In *PIAGENG 2009: Image Processing and Photonics for Agricultural Engineering* (Vol. 7489, p. 74890E). https://doi.org/[Insert DOI or link if available]

Fenu, G., & Malloci, F. M. (2021). DiaMOS Plant: A Dataset for Diagnosis and Monitoring Plant Disease. *Agronomy, 11*(11), 2107. https://doi.org/10.3390/agronomy11112107

Fisher, M. C., Gurr, S. J., Cuomo, C. A., Blehert, D. S., Jin, H., Stukenbrock, E. H., et al. (2020). Threats posed by the fungal kingdom to humans, wildlife, and agriculture. *mBio, 11*(3), e00449–e00420. https://doi.org/10.1128/mBio.00449-20

García-Vera, Y. E., Polochè-Arango, A., Mendivelso-Fajardo, C. A., & Gutiérrez-Bernal, F. J. (2024). Hyperspectral image analysis and machine learning techniques for crop disease detection and identification: A review. *Sustainability, 16*(14), 6064. https://doi.org/10.3390/su16146064

Gavhale, R. K., Gawande, U., Kamal, & Hajari, O. (2014). Unhealthy region of citrus leaf detection using image processing techniques. In *2014 IEEE International Conference for Convergence of Technology* (pp. 978-1-4799-3759-2/14).

Goluguri, N. V., Devi, K., & Vadaparthi, N. (2020). Image classifiers and deep learning classifiers evolved in detection of *Oryza sativa* diseases: Survey. *Artificial Intelligence Review, 54*(1), 1–30.

Gomez, S. (2014). Infection and spread of *Peronospora sparsa* on *Rosa sp.* (Berk.): A microscopic and thermographic approach (Doctoral dissertation). University of Bonn, Germany.

Gonzalez, R. C., & Woods, R. E. (2020). *Digital Image Processing* (4th ed.). Pearson.

Goss, E. M., Schena, L., & Weiland, J. (2020). Isozyme analysis for identifying *Phytophthora* species: Recent advances. *Phytopathology, 110*(6), 1119–1130.

Guarro, G., Gene, J., & Stchigel, A. M. (1999). Developments in fungal taxonomy. *Clinical Microbiology Reviews, 12*(3), 454–500.

Guo, A., Huang, W., Ye, H., Dong, Y., Ma, H., Ren, Y., & Ruan, C. (2020). Identification of wheat yellow rust using spectral and texture features of hyperspectral images. *Remote Sensing, 12*(9), Article 91419.

Gupta, R., Kumar, S., & Verma, R. (2023). Adaptive K-Means for segmenting hyperspectral images of banana leaves affected by Fusarium wilt. *Journal of Plant Pathology and Microbiology, 14*(2), 105–119.

Hasan, M. M., Uddin, A. F., Akhond, M. R., Uddin, M. J., Hossain, M. A., & Hossain, M. A. (2023). Machine Learning and Image Processing Techniques for Rice Disease Detection: A Critical Analysis. *International Journal of Plant Biology, 14*(4), 1190-1207. https://doi.org/10.3390/ijpb14040087

He, K., Zhang, X., Ren, S., & Sun, J. (2021). Deep residual learning for image recognition. *IEEE Transactions on Pattern Analysis and Machine Intelligence,* 43(10), 3346-3355.

Hesamian, M. H., Jia, W., & He, X. (2019). Deep learning techniques for medical image segmentation: A review. *Journal of Biomedical Informatics*, 92, 103120.

Hillnhütter, C., Mahlein, A.-K., Sikora, R. A., & Oerke, E.-C. (2011). Remote sensing to detect plant stress induced by *Heterodera schachtii* and *Rhizoctonia solani* in sugar beet fields. *Field Crops Research, 122*(1), 70-77. https://doi.org/10.1016/j.fcr.2011.02.007

Hong, F., & Huijie, L. (2014). Plant leaves recognition and classification model based on image features and neural network. IJCSI International Journal of Computer Science, 11(1), 100–104.

Horst, R. (2008). "Classification of plant pathogens," in Westcott's Plant Disease Handbook. Ed. R. Horst (Dordrecht: Springer), 43–79.

Hossain, E., Hossain, M. F., & Rahaman, M. A. (2019). A color and texture-based approach for the detection and classification of plant leaf disease using KNN classifier. 2019 International Conference on Electrical, Computer and Communication Engineering (ECCE), 1–6.

Hossain, M. A., Park, S. H., & Kim, J. S. (2020). Isozyme Electrophoresis for the Identification of Fusarium Species: An Updated Review. *Journal of Plant Pathology*, 102(4), 963-973.

Hussain, F., and Usman, F. (2019). "Fungal biotic stresses in plants and its control strategy," in Abiotic and biotic stress in plants. Ed. A. B. de Oliveira (London: Intech Open).

Islam, M. A., Shuvo, M. N. R., Shamsojjaman, M., Hasan, S., Hossain, M. S., & Khatun, T. (2021). An automated convolutional neural network-based approach for paddy leaf disease detection. *International Journal of Advanced Computer Science and Applications, 12*(4), 280–288.

Jacquemoud, S., & Ustin, S. L. (2001). Leaf optical properties: A state of the art. In *Proceedings of the 8th International Symposium on Physical Measurements & Signatures in Remote Sensing* (pp. 223–232). CNES, Aussois, France.

Jafar, A., Bibi, N., Naqvi, R. A., Sadeghi-Niaraki, A., & Jeong, D. (2024). Revolutionizing agriculture with artificial intelligence: Plant disease detection methods, applications, and their limitations. *Frontiers in Plant Science, 15.*

Jaffery, Z., Zaheeruddin, A., & Singh, L. (2013). Performance analysis of image segmentation methods for the detection of masses in mammograms. *International Journal of Computer Applications, 82*(2), 44–50.

Jain, A., Sarsaiya, S., Wu, Q., Lu, Y., & Shi, J. (2019). A review of plant leaf fungal diseases and its environment speciation. *Bioengineered, 10*(1), 409–424.

Jian, Z., & Wei, Z. (2010). Support vector machine for recognition of cucumber leaf diseases. *2nd International Conference on Advanced Computer Control* (Vol. 5, pp. 264–266). IEEE.

Jing, J., Liu, S., Wang, G., Zhang, W., & Sun, C. (2022). Recent advances on image edge detection: A comprehensive review. *Neurocomputing, 503,* 259–271. https://doi.org/10.1016/j.neucom.2022.06.083

Jones, H. G. (2004). Application of thermal imaging and infrared sensing in plant physiology and ecophysiology. *Advances in Botanical Research, 41,* 107–163

Jones, H. G., Stoll, M., Santos, T., de Sousa, C., Chaves, M. M., & Grant, O. M. (2002). Use of infrared thermography for monitoring stomatal closure in the field: Application to grapevine. *Journal of Experimental Botany, 53*(378), 2249–2260. https://doi.org/10.1093/jxb/erf083

Kalaji, H. M., et al. (2017). Chlorophyll fluorescence as a tool for nutrient status identification in field crops. *Scientific Reports, 7,* 14956.

Kamilaris, A., & Prenafeta-Boldú, F. X. (2018). Deep learning in agriculture: A survey. *Computers and Electronics in Agriculture, 147,* 70–90.

Knauer, U., Matros, A., Petrovic, T., Zanker, T., Scott, E. S., & Seiffert, U. (2017). Improved classification accuracy of powdery mildew infection levels of wine grapes by spatial-spectral analysis of hyperspectral images. *Plant Methods, 13*(1), 47.

Kool, J., & Evenhuis, A. (2023). Early detection of *Phytophthora infestans* in potato plants using hyperspectral imaging, local comparison, and a convolutional neural network. *Smart Agricultural Technology, 6,* 100333.

Kuska, M. T., Behmann, J., Großkinsky, D. K., Roitsch, T., & Mahlein, A. K. (2018). Screening of barley resistance against powdery mildew by simultaneous high-throughput enzyme activity signature profiling and multispectral imaging. *Frontiers in Plant Science, 9,* 1074. https://doi.org/10.3389/fpls.2018.01074

Kutty, S. B., Abdullah, N. E., Hashim, H., Rahim, A. A. A., Kusim, A. S., Yaakub, T. N. T., Yunus, P. N. A. M., & Rahman, M. F. A. (2013). Classification of watermelon leaf diseases using neural network analysis. *2013 IEEE Business Engineering and Industrial Applications Colloquium* (BEIAC), 459–464

LeCun, Y., Bengio, Y., & Hinton, G. (2015). Deep learning. *Nature, 521,* 436–444. https://doi.org/10.1038/nature14539

Lenthe, J.-H., Oerke, E.-C., & Dehne, H.-W. (2007). Digital infrared thermography for monitoring canopy health of wheat. *Precision Agriculture, 8*(1), 15–26. https://doi.org/10.1007/s11119-006-9025-6

Li, X., et al. (2021). Detection of early rust symptoms in barley using chlorophyll fluorescence. *Field Crops Research, 265,* 108113.

Li, X., et al. (2022). Integrating infrared thermography and hyperspectral imaging for early detection of plant diseases. *Sensors, 22*(3), 1124.

Lin, F., Guo, S., Tan, C., Zhou, X. and Zhang, D. (2020). Identification of Rice Sheath Blight through Spectral Responses Using Hyperspectral Images. *Sensors* (Basel), 20(21): 6243

Liu, J., Zhang, L., & Zhao, X. (2022). Gaussian Mixture Models for detecting powdery mildew in grapevines using spectral analysis. *Journal of Agricultural and Food Chemistry*, 70(3), 456-467.

Lu, M., Hu, J., & Chen, Q. (2020). Powdery mildew in tomato (*Oidium neolycopersici*). *Plant Science,* 291(6), 110-119.

Luchi N, Ioos R, Santini A. (2020) Fast and reliable molecular methods to detect fungal pathogens in woody plants. *Appl Microbiol Biotechnol.*104(6):2453-2468.

Mahlein, A. K. (2016). Plant disease detection by imaging sensors - Parallels and specific demands for precision agriculture and plant phenotyping. Plant Disease, 100(2), 241–251.

Mahlein, A.-K., Steiner, U., Hillnhütter, C., Dehne, H.-W., & Oerke, E.-C. (2012). Hyperspectral imaging for small-scale analysis of symptoms caused by different sugar beet diseases. *Plant Methods, 8*(3). https://doi.org/10.1186/1746-4811-8-3

Marciulynas, A., Marciulyniene, D., Lynikiene, J., Gedminas, A., Vaiciukyne, M., and Menkis, A. (2020). Fungi and Oomycetes in the Irrigation Water of Forest Nurseries. Forests 11, 459.

Markets and Markets. (2019). Agriculture Technology Market by Offering (Hardware, Software, Services), Application (Precision Farming, Livestock Monitoring, Smart Greenhouse, Fish Farming, Forestry), and Geography - Global Forecast to 2027. Retrieved from https://www.marketsandmarkets.com/.

Mattupalli, C., Moffet, C. A., Shah, K. N. and Young, C. A. (2018). Supervised classification of RGB aerial imagery to evaluate the impact of a root rot disease. *Remote Sensing*, 10(6):

Mehl, P. M., Chen, Y.-R., Kim, S. M., & Chan, D. E. (2004). Development of hyperspectral imaging technique for the detection of apple surface defects and contaminations. *Journal of Food Engineering, 61*(1), 67-81. https://doi.org/10.1016/S0260-8774(03)00188-2

Meunkaewjinda, A., Kumsawat, P., & Attakitmongcol, K. (2008). Grape leaf disease detection from color imagery using a hybrid intelligent system. *5th International Conference on Electrical Engineering/Electronics, Computer, Telecommunications and Information Technology (ECTI-CON 2008)*, Vol. 1, 513–516. Krabi, Thailand.

Micales, J.A., Bonde, M.R., & Peterson, G.L. (1986). The use of isozyme analysis in fungal taxonomy and genetics. *Mycotaxon, 27*, 405-449.

Miller, J. D., Zhang, H., & McDonald, B. A. (2021). Isozyme Variation in Fusarium solani: A Comprehensive Analysis. Mycological Progress, 20(1), 75-85.

Mishra, A., et al. (2016). Chlorophyll fluorescence emission as a marker of leaf senescence during plant stress. *Journal of Experimental Botany,* 67(19), 5611-5623.

Mitra, A., Das, S., & Roy, K. (2021). Adaptive K-Means clustering for unsupervised plant disease segmentation. *IEEE Access*, 9, 118476-118485.

Mitra, S., Patra, S., & Chakraborty, S. (2021). Advances in unsupervised clustering algorithms for plant disease detection. *Computers and Electronics in Agriculture,* 183, 106134.

Mohamed, M. A., Abdullah, A., & Younis, Y. (2018). Advanced Isozyme Techniques for Identification of *Fusarium oxysporum*. *Fungal Biology,* 122(9), 751-759.

Murchie, E. H., & Niyogi, K. K. (2018). Manipulation of photoprotection to improve plant photosynthesis. *Plant Physiology,* 176(2), 1410-1423.

Nagasubramanian, K., Jones, S., Singh, A.K., Sarkar, S., Singh, A. and Ganapathysubramanian, B. (2019). Plant disease identification using explainable 3D deep learning on hyperspectral images. *Plant Methods*. 21, 15:98.

Narayanasamy, P. (2011). "Diagnosis of fungal diseases of plants," in Microbial plant pathogens-detection and disease diagnosis (Dordrecht: Springer), 273 284.

Neumann, M., Hallau, L., Klatt, B., Kersting, K., & Bauckhage, C. (2014). Erosion band features for cell phone image-based plant disease classification. In *Proceedings of the 22nd International Conference on Pattern Recognition (ICPR)* (pp. 3315-3320). Stockholm, Sweden.

Nutter Jr, F.W., Guan, J., 2001. Disease losses, EPP, pp. 340–351.

Oerke, E. C., Herzog, K. and Toepfer, R. (2016). Hyperspectral phenotyping of the reaction of grapevine genotypes to *Plasmopara viticola*. *J. Exp. Bot.,* 67:5529-43.

Oerke, E.C., et al. (2016). Thermographic assessment of plant diseases: A review. Journal of Plant Physiology, 207, 72-82.

Oerke, E.-C., Fröhling, P., & Steiner, U. (2011). Thermographic assessment of scab disease on apple leaves. *Precision Agriculture, 12*(5), 699-715.

Oerke, E.-C., Mahlein, A.-K., & Steiner, U. (2014). Proximal sensing of plant diseases. In M. L. Gullino & P. J. M. Bonants (Eds.), *Detection and Diagnostic of Plant Pathogens, Plant Pathology in the 21st Century* (pp. 55-68). Springer Science and Business Media.

Oerke, E.-C., Steiner, U., Dehne, H.-W., & Lindenthal, M. (2006). Thermal imaging of cucumber leaves affected by downy mildew and environmental conditions. *Journal of Experimental Botany, 57*(9), 2121-2132. https://doi.org/10.1093/jxb/erj170

Olesen, M. H., Carstensen, J. M. and Boelt, B. (2011). Multispectral imaging as a potential tool for seed health testing of spinach (*Spinacia oleracea* L.). *Seed Science and Technology*, 39(1): 140-150.

Omran, E.S.E. (2017). Early sensing of peanut leaf spot using spectroscopy and thermal imaging. *Archives of Agronomy and Soil Science*, 63(7): 883-896.

Patel, M., Prasad, G., & Gupta, S. (2022). Detection and classification of Alternaria leaf spot in cotton using Random Forest and hyperspectral imaging. *Journal of Agricultural Science*, 10(4), 56-72.

Pérez-Roncal, & Lopez, Ainara & Lopez-Molina, & Jaren, Carmen & Urrestarazu, Jorge & Santesteban, Luis & Arazuri, Silvia. (2020). Hyperspectral Imaging to Assess the Presence of Powdery Mildew (Erysiphe necator) in cv. Carignan Noir Grapevine Bunches. Agronomy. 10. 88. 10.3390/agronomy10010088.

Pethybridge, S. J., & Nelson, S. C. (2015). Leaf doctor: A new portable application for quantifying plant disease severity. *Plant Disease, 99*(10), 1310-1316.

Phadikar, S., & Sil, J. (2008). Rice disease identification using pattern recognition techniques. *2008 11th International Conference on Computer and Information Technology (ICCIT),* 420–423. Bangladesh.

Picon, A., Pomares, A., & Jiménez, A. (2021). Deployment of trained models on drones and smartphones for real-time identification of fungal plant diseases. *Journal of Precision Agriculture,* 12(2), 202-215.

Quinlan, J. R. (1986). Induction of decision trees. Machine Learning, 1(1), 81–106. https://doi.org/10.1007/BF00116251

Rahimi, M., Esfahanian, N., & Bahmani, Z. (2023). Self-Organizing Maps for visualizing and interpreting complex plant disease data. *Expert Systems with Applications*, 210, 118343.

Rahimi, M., Khorsand, M., & Ramezanian, A. (2023). Detection of Fusarium head blight in wheat using multispectral imaging and Support Vector Machine. *Field Crops Research*, 289, 108721.

Rahimi, M., Li, Y., & Chen, Z. (2023). Classification of barley leaves infected with *Rhynchosporium commune* using Self-Organizing Maps. *Agricultural Systems,* 193, 103232.

Rashmi, B., Jain, R., & Sharma, S. (2020). Support Vector Machine with kernel tricks for complex disease detection tasks. *Journal of Machine Learning Research,* 21(1), 1-25.

Rathore, N. P. S., & Prasad, L. (2021). A comprehensive review of deep learning models for plant disease identification and prediction. *International Journal of Engineering Systems Modelling and Simulation*, 12(3), 165–179.

Ravi, P. S., et al. (2017). Hyperspectral remote sensing of plant water content. *IEEE Transactions on Geoscience and Remote Sensing,* 55(8), 4643-4652.

Revathi, P., & Hemlatha, M. (2012). Homogeneous segmentation-based edge detection techniques for proficient identification of the cotton leaf spot diseases. International Journal of Computer Applications, 47(2), 0975–8887.

Romero, M., Martínez, J., & López, F. (2023). Septoria blotch in wheat (*Zymoseptoria tritici*). European Journal of Plant Pathology, 165(2), 345-356.

Rossini, M., Alonso, L., Cogliati, S., Damm, A., Guanter, L., Julitta, T., Meroni, M., Moreno, J., Panigada, C., Pinto, F., Rascher, U., Schickling, A., Schüttemeyer, D., Zemek, F., & Colombo, R. (2014). Measuring sun-induced chlorophyll fluorescence: An evaluation and synthesis of existing field data. In *5th International Workshop on Remote Sensing of Vegetation Fluorescence* (pp. 1-5). Paris.

Rumpf, T., Mahlein, A.-K., Steiner, U., Oerke, E.C., Dehne, H.W., & Plümer, L. (2010). Early detection and classification of plant diseases with Support Vector Machines based on hyperspectral reflectance. *Computers and Electronics in Agriculture, 74*(1), 91-99. https://doi.org/10.1016/j.compag.2010.06.009

Sancho-Adamson, M., Trillas, M. I., Bort, J., Fernandez-Gallego, J. A. and Romanyà, J. (2019). Use of RGB vegetation indexes in assessing early effects of Verticillium wilt of olive in asymptomatic plants in high and low fertility scenarios. Remote Sensing, 11(6): 607.

Sankaran, S., Mishra, A., Ehsani, R., & Davis, C. (2010). A review of advanced techniques for detecting plant diseases. *Computers and Electronics in Agriculture, 72,* 1-13.

Schaad, N. W., Wang, J. F., & Leite, B. (2019). Revised Methods for the Identification of Phytophthora infestans Using Isozyme Electrophoresis. *Plant Disease,* 103(9), 2487-2495.

Schor, N., Berman, S., Dombrovsky, A., Elad, Y., Ignat, T. and Bechar, A. (2017). Development of a robotic detection system for greenhouse pepper plant diseases. *Precis. Agric.,* 18:394-409.

Sharma, L. K., et al. (2021). Hyperspectral imaging for nitrogen estimation in cereal crops: A review. *Journal of Precision Agriculture,* 22(3), 604-623.

Shen, W., et al. (2020). Application of chlorophyll fluorescence imaging in plant disease detection. *Sensors,* 20(22), 6615.

Singh, D., et al. (2020). Precision agriculture using infrared thermography and hyperspectral imaging: A review. *Precision Agriculture*, 21(4), 837-857.

Singh, V.; Misra, A.K. Detection of plant leaf diseases using image segmentation and soft computing techniques. Inf. Process. Agric. 2017, 4, 41–49.

Sinha, A., & Shekhawat, R. (2020). Review of Image Processing Approaches for detecting Plant diseases. IET Image Processing. 14. 10.1049/iet-ipr.2018.6210.

Song, K., Liu, Z., Su, H., & Guo, C. (2011). A research of maize disease image recognition based on BP networks. Third International Conference on Measuring Technology and Mechatronics Automation, 246–249. https://doi.org/10.1109/ICMTMA.2011.66

Steiner, U., Bürling, K., & Oerke, E.-C. (2008). Sensorik für einen präzisierten Pflanzenschutz. *Gesunde Pflanzen, 60,* 131-141. https://doi.org/10.1007/s10343-008-0194-2

Sugiura, R., Tsuda, S., Tamiya, S., Itoh, A., Nishiwaki, K., Murakami, N., Shibuya, Y., Hirafuji, M. and Nuske, S. (2016). Field phenotyping system for the assessment of potato late blight resistance using RGB imagery from an unmanned aerial vehicle. *Biosystems Engineering*, 148: 1-10.

Thind, B. S. (2015). "Diagnosis and management of bacterial plant diseases," in Recent advances in the diagnosis and management of plant diseases. Ed. L. P. Awasthi (India: Springer), 101–117. doi: 10.1007/978-81-322-2571-3_10.

Tör, M., and Woods-Tör, A. (2017). "Fungal and oomycete diseases," in Encyclopedia of applied plant sciences. Eds. B. Thomas, B. G. Murray and D. J. Murphy (New York, USA: Academic Press), 77–82.

Ugarte, J. F., Bayona, O. A., Criollo, R. B., Cevallos-Cevallos, J., Mariduena-Zavala, M., Ochoa, D. D. and Vicente, J. V. (2020). Early detection of black Sigatoka in banana leaves using hyperspectral images. *Applications in plant sciences,* 8(8).

Van den Bosch, F., Hendriks, J., & Murphy, K. (2017). Isozyme Patterns of Phytophthora Species: A Review of Recent Studies. *Plant Pathology Journal*, 16(2), 83-94.

Veys, C., Chatziavgerinos, F., Al Suwaidi, A., Hibbert, J., Hansen, M., Bernotas, G., Smith, M., Yin, H., Rolfe, S. and Grieve, B., (2019). Multispectral imaging for pre symptomatic analysis of light leaf spot in oilseed rape. *Plant methods,* 15(1): 1-12.

Vishnoi, Vibhor & Kumar, Krishan & Kumar, Brajesh. (2022). A comprehensive study of feature extraction techniques for plant leaf disease detection. Multimedia Tools and Applications. 80. 10.1007/s11042-021-11375-0.

Wang, X., Li, J., & Zhang, X. (2023). Classification of grapevine diseases using pre-trained ResNet model. Computers and Electronics in Agriculture, 207, 107759.

Wani, J. A., Sharma, S., Muzamil, M., Ahmed, S., Sharma, S., & Singh, S. (2022). Machine learning and deep learning-based computational techniques in automatic agricultural diseases detection: Methodologies, applications, and challenges. Computational Methods in Engineering, 29(3), 641–677.

Wenneker, M., and Thomma, B. P. H. J. (2020). Latent postharvest pathogens of pome fruit and their management: from single measures to a systems intervention approach. *Eur. J. Plant Pathol.* 156, 663–681.

West, J. S., Bravo, C., Oberti, R., Lemaire, D., Moshou, D., & McCartney, H. A. (2003). The potential of optical canopy measurement for targeted control of field crop diseases. *Annual Review of Phytopathology, 41,* 593-614.

Wijekoon, C. P., Goodwin, P. H., & Hsiang, T. (2008). Quantifying fungal infection of plant leaves by digital image analysis using Scion Image software. *Journal of Microbiological Methods, 74,* 94-101.

Xie, X., Chen, Z., & Liang, J. (2022). Transfer learning with Convolutional Neural Networks for detecting Botrytis cinerea in tomatoes. *Computers and Electronics in Agriculture*, 192, 106361.

Xu, M., Yoon, S., Jeong, Y., & Park, D. S. (2022). Transfer learning for versatile plant disease recognition with limited data. Frontiers in Plant Science, 13, 1010981.

Yadav, M., & Verma, T. (2016). Hybrid approach of neural network and genetic algorithm to recognize black mold disease in tomato. International Journal for Research in Applied Science & Engineering Technology (IJRASET), 4(5).

Yang, N., Yuan, M., Wang, P., Zhang, R., Sun, J. and Mao, H. (2019). Tea diseases detection based on fast infrared thermal image processing technology. *Journal of the Science of Food and Agriculture,* 99(7): 3459-3466.

Yang, W., Zhang, Q., & Chen, X. (2022). Powdery mildew in cucumber (*Podosphaera xanthii*). *Remote Sensing*, 14(5), 987-995.

Youwen, T., Tianlai, L., & Yan, N. (2008). The recognition of cucumber disease based on image processing and support vector machine. In Proceedings of the 2008 Congress on Image and Signal Processing (Vol. 2, pp. 262–267). Sanya, People's Republic of China.

Yu, H., Song, J., Chen, C., Heidari, A. A., Liu, J., Chen, H., & Mafarja, M. (2022). Image segmentation of leaf spot diseases on maize using multi-stage Cauchy-enabled grey wolf algorithm. Engineering Applications of Artificial Intelligence, 109, 104653.

Zarco Tejada, P. J., Camino, C., Beck, P. S. A., Calderon, R., Hornero, A., Hernández-Clemente, R., Kattenborn, T., Montes Borrego, M., Susca, L., Morelli, M., et al. (2018). Previsual symptoms of *Xylella fastidiosa* infection revealed in spectral plant trait alterations. Nature Plants, 4, 432–439.

Zhan, L., Wang, Z., & He, M. (2022). Fuzzy C-Means clustering for early-stage plant disease detection using hyperspectral images. Information Sciences, 587, 65-78.

Zhan, Y., Xu, J., & Chen, Q. (2022). Fuzzy C-Means clustering for early-stage detection of *Magnaporthe oryzae* in rice crops. Plant Disease, 106(5), 940–951.

Zhang, D., Wang, Q., Lin, F., Yin, X., Gu, C., & Qiao, H. (2020). Development and evaluation of a new spectral disease index to detect wheat Fusarium head blight using hyperspectral imaging. Sensors, 20(8), 2260.

Zhang, W., Zhu, Q., Huang, M., Guo, Y., & Qin, J. (2019). Detection and classification of potato defects using multispectral imaging system based on single-shot method. Food Analytical Methods, 12, 2920–2929.

Zhang, Y., Chen, X., & Wu, J. (2021). Enhancement of support vector machines performance using ensemble methods. Pattern Recognition Letters, 141, 157-163.

Zhao, J., Fang, Y., Chu, G., Yan, H., Hu, L., & Huang, L. (2020). Identification of leaf-scale wheat powdery mildew (*Blumeria graminis* f. sp. *tritici*) combining hyperspectral imaging and an SVM classifier. Plants, 9(8), 936.

Zhao, X., Lu, Y., & Wu, L. (2019). Application of isozyme electrophoresis in identifying Rhizoctonia solani variants. Fungal Genetics and Biology, 125, 40-50.

Zhou, R., Kaneko, S. I., Tanaka, F., Kayamori, M., & Shimizu, M. (2014). Disease detection of Cercospora leaf spot in Sugar beet by robust template matching. Computers and Electronics in Agriculture, 108, 58-70.

Zhu, X., et al. (2020). Deep learning-based detection of wheat leaf rust using hyperspectral imagery. Computers and Electronics in Agriculture, 175, 105543.

Zhu, Y., Liu, S., & Zheng, C. (2022). Using convolutional neural networks for detection of Botrytis cinerea in tomato plants. International Journal of Plant Pathology, 15(2), 189-202. https://doi.org/10.3923/ijpp.2022.189.202

Chapter 9

Biosensors and Lab-On-Chip Technology for Detection and Diagnosis of Fungal Pathogen and Mycotoxins

***Krishnaveni Kurapati*[1], *Dara Prasanth Kumar*[2], *Lohitha Bommagani*[3]**

[1, 2,3] *Department of Plant Pathology, IARI-IIHR, Bengaluru, Karnataka, India*
[1,3] *Division of Crop Protection, ICAR- CTCRI, Thiruvanathapuram, Kerala, India*
Corresponding author-krishnavenirao218@gmail.com

Abstract

Biosensors are reliable analytical instruments with a broad range of detecting applications, such as drug discovery, food safety, medical diagnostics, security, defense and more. Biosensors based on chips have become powerful tools for developing small-scale sensing systems for medical diagnosis and identification at the location enabling the development of isolated reaction sites for multiplex analysis, speedy detection, reduced sample and reagent costs. These on-chip sensors have good specificity, sensitivity, reproducibility and throughput and can detect the target analyte with small amounts of samples. Promising solutions with nanomaterial integration and microfluidic systems are provided by the introduction of micro-nano biosensor systems. By converting interactions between biological substances (such as biomolecules, cells, and pathogens) and a biological identification element into electrical signals, an electrochemical biosensor can be used as a tool for biological substance detection and quantification. Nanozyme-based microfluidic biosensors have provided a promising strategy for the rapid and on-site determination of mycotoxins. Paper-based lab-on-a-chip (pLOC) instruments have become a viable method for the quick and affordable detection and reduced power consumption. Significant progress in the field could be made by using on-chip detection to close the distance between on-site and sophisticated laboratory detection.

Key words: Biosensors, Food safety, Lab-on-chip sensors, Micro-nano biosensor systems, Mycotoxins.

Introduction

The role of food and agriculture in achieving sustainable development goals needs to increase. Consequently, it is imperative to stop the impact of infections on the safety and quality of food. One of the main causes of a decline in economic food production has been identified is pathogens. For farmers in rural locations, standard pathogen detection methods are costly and time-consuming. This chapter examines the biosensors that have been created to identify biological threats in the food and agriculture sectors. For illiterate farmers in rural communities, nanotechnology is helping to provide a fast, dependable, more sensitive, accessible, user-friendly, and easily adaptable technology.

In recent years, the increasing recognition of the health risks posed by fungal pathogens and mycotoxins has necessitated the development of advanced diagnostic tools. Mycotoxins and fungal diseases are serious hazards to agriculture, human health, and the environment. Pathogen-related food losses can result in reduced quantity and quality of yield, decreased value, and financial losses for producers (Buja *et al.,* 2021). Fungal infections, particularly those caused by opportunistic pathogens, represent a significant public health challenge, especially for immunocompromised individuals.

Toxic compounds released by certain fungi in food, food products, and feed are referred to as mycotoxin. Aflatoxin (AFs), deoxynivalenol, zearalenone, ochratoxin, fumonisin, and patulin are among the most well-known mycotoxins. Of these, certain substances—like AFs B1—have been shown to be potent carcinogens, while others may also cause cancer. Mycotoxicoses is a risk if such food and feed are consumed contaminated with mycotoxin (Rai *et al.*, 2015).

Mycotoxins may be found in the finished product when materials contaminated with fungi, such as the juice of various fruits, grapes, and cereals, are used to make beverages like beer and wines. Furthermore, because these dangerous poisons were first found in animal feed for cattle, they can potentially find their way into human food through tainted milk, eggs, and meat. In terms of mycotoxin production in food, *Aspergillus, Fusarium* and *Penicillium* are the three main fungal genera that stand out. A unique set of mycotoxins is highlighted, taking into account the possible harm to human health and the economic consequences. These include fumonisins, ergopeptine alkaloids, aflatoxins, ochratoxin A (OTA), zearalenone (ZEN), and trichothecenes such T-2 and HT-2 toxins (Oliveira *et al.,* 2019). Mycotoxins, toxic metabolites produced by certain fungi, further exacerbate this issue, posing threats to both human health and food safety. Traditional methods for the detection of these agents, while effective, are often time-consuming, labour-intensive, and require specialized laboratory facilities. There exist several established methods for identifying mycotoxins; nonetheless, due to their limitations, it is imperative to develop novel and expedient ways. In this context, biosensors and lab-on-chip technology emerge as innovative solutions, offering rapid, sensitive, and cost-effective alternatives for the detection and diagnosis of fungal pathogens and mycotoxins. On the whole, the past and most recent biosensors that could ensure the quick delivery of vision 2030 which aims to end hunger and poverty are discussed.

Even though these sensors don't directly interact with biological systems, the word "biosensor" is frequently used to describe any sensor device that measures the concentration of chemicals and other biologically relevant properties. Biosensors are analytical tools that use a physicochemical detector in conjunction with a biological component to detect and measure particular analytes. They utilize various biorecognition elements, such as enzymes, antibodies, or nucleic acids, to selectively bind to target molecules, translating biological interactions into measurable signals like electrical, optical, or thermal responses. In the context of fungal pathogens and mycotoxins, biosensors can detect specific species or toxins with high sensitivity and specificity.

For instance, DNA-based biosensors can utilize polymerase chain reaction (PCR)-linked techniques to identify fungal pathogens at the genetic level. Enzyme-linked immunosorbent assays (ELISA) can be adapted into biosensor formats, enabling the rapid quantification of mycotoxins through antigen-antibody interactions. Moreover, electrochemical biosensors, which can detect changes in electric currents, have shown promise in monitoring mycotoxin levels in various substrates, including food and feed, allowing for real-time assessment of contamination levels.

A biosensor is an analytical tool that generates a measurable signal in direct proportion to the concentration of an analyte, such as a pathogen, a cell component of the pathogen, or a toxin molecule. Biosensors represent the culmination of a rapidly expanding field that combines computing, engineering, and fundamental sciences to address the pressing needs in a variety of fields where their use is necessary.

Elements Of A Biosensor

Biological: Polysaccharides, Microorganisms, Nucleic Acids, Tissue, Enzyme, Antibody

Physical: Transducers, A/D Converter, Amplifier, Display

Analytical tools known as biosensors translate a biological reaction into an electrical signal. Usually, they are composed of three parts:

The three important parts of a biosensor are the transducer, chemical interface, and biospecific capture entity. Biosensors employ a range of biospecific capture entities, including antibodies, enzymes, DNA, and whole cells, to facilitate the biodetection of the target analyte. The binding mechanism used for the biosensor chemical interface may be covalent, entrapment, absorption, encapsulation, or cross-linking. Furthermore, the transducer that is employed for the transmission and measurement of signals may be based on an acoustic, optical, or electrochemical signal.

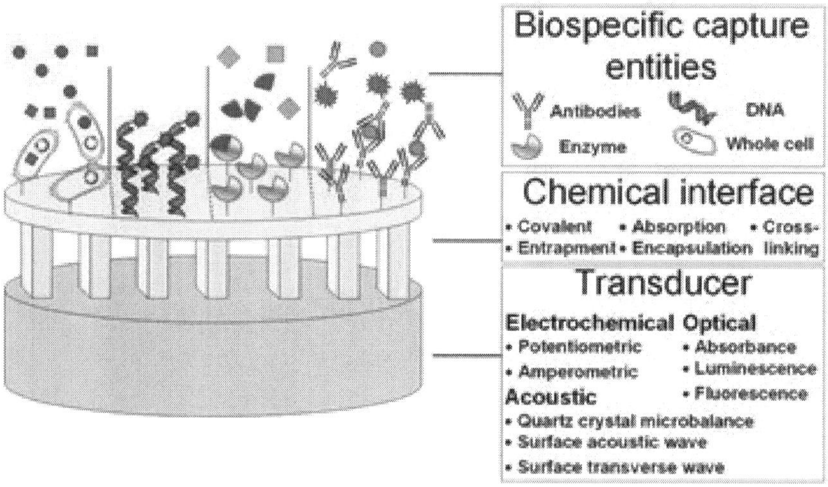

Figure 1. Structure of Biosensor (Harpaz *et al.,* 2017)

The schematic diagram of the biosensor is shown below and consists of three main elements. "Transducer" refers to the detector element that modifies the signal resulting from the analyte's interaction with the biological material in a way that is user-friendly; "sensor" refers to a biological element that is sensitive to changes in biological material, such as tissue, microorganisms, organelles, cell receptors, enzymes, antibodies, nucleic acids, etc. The third component consists of the related electronics, which comprise a CPU, a display unit, and a signal conditioning circuit (amplifier).

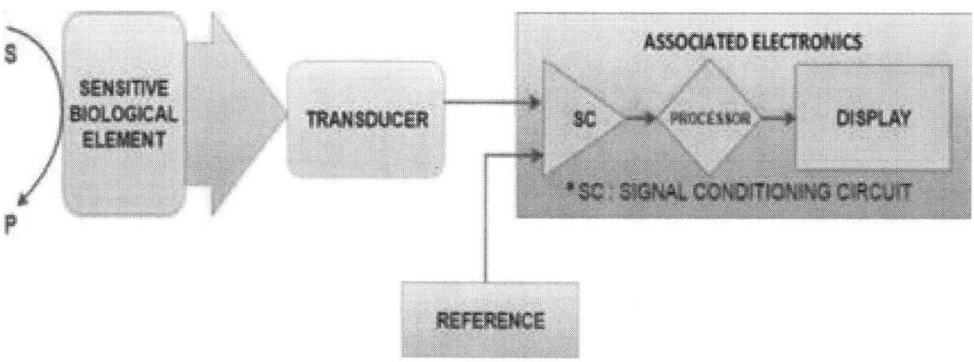

Figure 2. Schematic diagram of biosensor (Malhotra *et al.,* 2017)

Principle of a Biosensor: Conventional techniques (physical or membrane entrapment, non-covalent or covalent binding) are used to immobilize the desired biological material, which is typically a particular enzyme. The transducer and this immobilized biological material are in close proximity to one another. The biological material and the analyte combine to form a bound analyte, which in turn generates the measurable electrical response. The analyte may occasionally be transformed into a product that involves the emission of heat, gas (oxygen), electrons, or hydrogen ions (Malhotra et al., 2017).

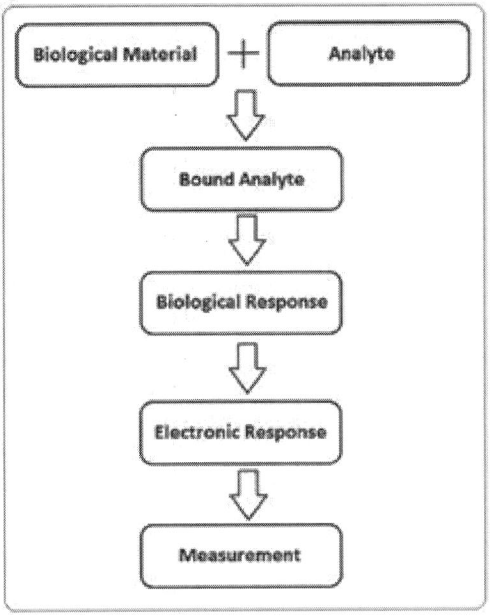

Figure 3. Measurement flow for a biosensor (Malhotra *et al.*, 2017)

The transducer can convert the product-linked changes into detectable and amplified electrical signals. Biosensors are extremely sensitive diagnostic tools that identify biological analytes, such as mycotoxins and pathogenic fungus, by using biological components. Because of these instruments' ease of use and potential for quick detection in clinical, food safety, and environmental applications, mycology has become quite interested in them. However, the development of ultrasensitive devices brought about by nanotechnology has also benefited mycotoxin analysis through advancements in the use of nanomaterials in sensor development (Tothill, 2011). An overview of the various biosensor types used to identify and diagnose mycotoxins and fungal infections is provided below:

Types of Biosensors

Enzymatic Biosensors: These biosensors work by employing enzymes that, when combined with the target analyte (such as mycotoxins), produce a signal that can be measured. Enzymes that catalyze the degradation of particular mycotoxins, for instance, can result in quantifiable changes in electrical or optical properties.

Affinity-based Biosensors: Affinity biosensors employ antibodies, aptamers, or additional particular binding molecules that selectively bind to fungal pathogens or mycotoxins. Common techniques include:

- ☆ **ELISA (Enzyme-Linked Immunosorbent Assay):** A method for detecting proteins (like mycotoxins) through antigen-antibody interactions.
- ☆ **Lateral Flow Assays:** Simple devices that provide qualitative results, often used for quick field tests.

Microbial Sensors: Living microorganisms can be integrated into biosensors to detect fungal pathogens. When these microbes encounter the target pathogen, they produce detectable signals (e.g., fluorescence or electrochemical response).

Genosensors: These biosensors are made to identify particular nucleic acid sequences linked to fungi by using nucleic acid (DNA or RNA) hybridization techniques. Pathogen DNA and RNA amplification can also be accomplished by integrating techniques like PCR (Polymerase Chain Reaction).

Fungi Biosensors can be classified into five different types based on detection.

1. Optical biosensors, 2. Electrochemical Biosensors, 3. Mass-Sensitive Biosensors, 4. Nanoparticles based Biosensors and 5. Lab-on-Chip Biosensors

Optical biosensors: These use light-based detection methods and are adaptable to changes in fluorescence, absorbance, or refractive index caused by fungal presence.

Electrochemical Biosensors: They rely on electrochemical reactions between fungal biomolecules and electrodes, providing rapid and specific detection. Detection of mycotoxins using electrochemical biosensors are prominent owing to their sensitivity, selectivity, low cost, simplicity, and in some cases miniaturization, portability and integration in automated devices (Vidal *et al.,* 2013).

Mass-Sensitive Biosensors: They detect changes in mass due to fungal binding, often using quartz crystal microbalances (QCMs) or surface acoustic wave (SAW) devices.

Nanoparticles based biosensors: Nanoparticles or nanomaterials enhance sensitivity and selectivity in fungal detection and synthesizing new nanomaterials to build analytical platforms with enhanced sensitivity and specificity. These materials include gold-or silver-based nanoparticles (NPs) that have a wide range of uses, including signal amplification, target recognition, and immobilization of biomolecules. Other nano-sized materials that have a multitude of uses, including the bioassays include quantum dots, molecular beacons, DNA dendrimers, carbon nanotubes, liposomes, nanowires, and nucleic acids (Hussain *et al.,* 2020).

Lab-on-Chip Biosensors: Lab-on-Chip Biosensors miniaturized devices integrate multiple functions (sample preparation, detection, and analysis) on a single chip, enabling efficient fungal detection.

Biosensors come in an incredible array of varieties, but they can be broadly classified into two categories: biosensors where the recognition element is immobilized onto a solid support (a well, membrane and fibre etc.) and biosensors where a toxin or toxin-protein conjugate is immobilized. The terms "direct" and "indirect" assays, or less precisely, antibody-immobilized or antigen-immobilized, are frequently used to describe the two formats. Additionally, there are immunosensors for which there is no immobilization of reagents. For mycotoxins, multiplexed assays using the latter approach have not been extensively documented. In multiplexed biosensors for mycotoxins, the antigen-immobilized format is by far the most often utilized of the two immobilization methods (Maragos, 2016).

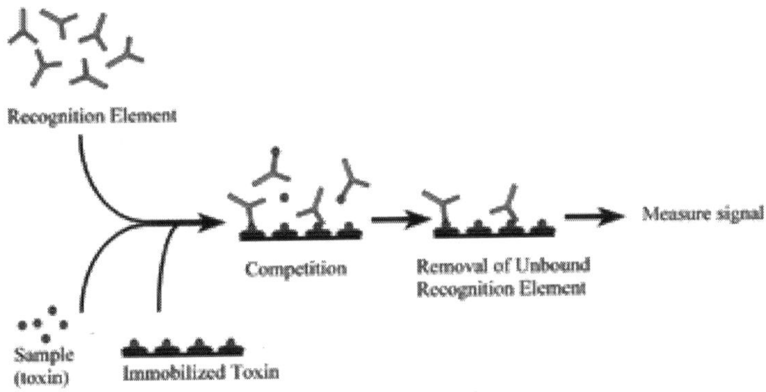

Figure 4. Immunoassay using the antigen-immobilized format (Maragos, 2016)

In 2009, Mendes et al. created a biosensor that can identify *Phakopsora pachyrhizi*, a pathogenic fungus that has been linked to soybean rust.

From the cost point of view, many biosensors must be developed based on label-free methods. Label-free detection methods also reduce the time and steps involved, including the expensive labelling protocols, which facilitate real-time detection. Among different label-free methods, electrochemical-based detection such as numerous benefits have been demonstrated by electrochemical impedance spectroscopy (EIS), including simplicity, scalability, portability, ease to use, sensitivity and low power consumption. Lab on-chip-based biosensors have become powerful emergent tools for developing miniaturized sensing platforms for on-site detection and diagnostics. These lab on-chip-based sensors have the capacity of accurately detecting the analyte of interest using small volumes of samples with excellent specificity and sensitivity, reproducibility, and throughput. Thus, enabling (a) rapid detection, (b) reduced costs due to lower sample and reagents and (c) creating isolated reaction sites for multiplex analysis, and (d) reduced power consumption.

Biosensors represent an innovative approach for the timely detection and diagnosis of fungal pathogens and mycotoxins. Their adaptability to different environments and their ability to provide rapid results make them invaluable tools in clinical, food safety, and environmental contexts. Continued research and development in biosensor technology are likely to further enhance their efficacy and application range.

Applications of Biosensors

Pathogen Detection: Lab-on-a-chip biosensors can detect fungal pathogens such as viruses and bacteria. They enable rapid, point-of-care testing, critical for early intervention and improved survival rates.

Disease Diagnosis: These biosensors can detect antibodies, DNA, RNA, exosomes, and cells associated with fungal infections. Early diagnosis enhances treatment outcomes.

In research areas, they are used in Micro/Nano Biosensing, Microfluidics and Nanofluidic Biosensors, Micro and Nanofabrication.

Micro/Nano Biosensing: Advances in micro/nanoscale biosensors contribute to fungal detection

Microfluidics and Nanofluidic Biosensors: These technologies allow precise manipulation of small volumes of samples and reagents, improving sensitivity and efficiency.

Micro and Nanofabrication: Techniques for creating miniaturized devices are essential for lab-on-a-chip biosensors.

Food Safety: Monitoring mycotoxins, like aflatoxins and ochratoxins in food products, can prevent foodborne diseases and protect public health. Biosensors offer rapid screening methods to certify the safety of food goods for consumption.

Environmental Monitoring: Biosensors can be used to assess fungal presence in water, soil, and air, which is crucial for ecological studies and biodegradation processes.

Lab-on-Chip Technology: A Breakthrough in Diagnostics

Lab-on-a-chip technology plays a important place in pathogen detection and disease diagnosis. It offers more affordable, more sensitive biosensors, and the potential for mass production. The term "lab-on-a-chip" coined in 1979, refers to the imparted microfluidic chip's ability of biosensors to scale down fluid flow, heat, and mass transfer to the micrometre or nanometer scales. Lab-on-chip (LoC) technology integrates multiple laboratory functions on a single microchip, enabling the miniaturization and automation of complex chemical and biological analyses. This technology has a number of benefits including reduced sample and reagent volumes, shorter analysis times and the ability to perform multiple tests simultaneously (multiplexing). In the context of fungal pathogens and mycotoxins, LoC devices can facilitate rapid screening and diagnosis, especially in resource-limited settings where access to conventional laboratory facilities may be restricted.

This allows many analyses to be performed remotely outside the laboratory (at the point of need) Therefore, on-chip detection could be used as the bridging gap between detection in advanced laboratories and detection on-site, providing substantial advancements in the field of point-of-care (POC) diagnostic.

Paper-based lab-on-a-chip (pLOC) devices have emerged as a promising approach for rapid and low-cost detection of agri-food contaminants. Notably, they are cheap, flexible, and available in different forms and compositions.

Different biosensing components can be incorporated into LoC devices to improve their diagnostic capabilities. For instance, fungal pathogens can be separated from complicated biological samples utilizing microfluidic devices, and then their presence can be detected by integrated biosensors. The direct connection between sample preparation, response, and standard diagnostic techniques—which might take days to produce results because of the requirement for culture growth and extensive analysis. In contrast, biosensors can produce results in minutes to hours, facilitating rapid clinical decisions and timely interventions.

Components of Lab-on-Chip Devices

Using a single microfluidic chip, lab-on-a-chip (LOC) technology combines several analytical processes, such as sample preparation, detection, and analysis. Rapid, accurate, and portable diagnostics are made possible by this downsizing.

- **Microfluidics:** Allows precise control and manipulation of small volumes of liquids, essential for sample handling and reaction processes.
- **Sensing Mechanisms**: Employs various techniques such as electrochemical, optical, or mass-sensitive methods to detect fungal pathogens or mycotoxins.
- **Integration Features:** Combines sample preparation, reaction chambers, and detection systems on a single chip, enhancing efficiency and reducing contamination risk.
- **Point-of-care diagnostics**: LoC-based biosensors have the potential to enable quick and decentralized testing in a variety of contexts, such as food processing factories, healthcare institutions, and agricultural settings.

Detection of Fungal Pathogens

Fungal pathogens can lead to significant health issues, especially in immunocompromised individuals, and can also affect crops, leading to economic losses. Lab on chip (LoC) technology can facilitate:

- **Molecular Detection:** Utilizing techniques like PCR (Polymerase Chain Reaction) for the amplification of fungal DNA or RNA, allowing for species-specific detection.
- **Immunoassays**: Employing antibodies linked with detection systems to identify fungal antigens with high specificity.
- **Cellular Analysis**: Direct detection using living cells on a chip which can offer perceptions into the pathogenicity and resistance of fungal strains.

Detection of mycotoxins

Toxic substances known as mycotoxins are produced by fungi, commonly found in contaminated food supplies. Current challenges in mycotoxin detection, such as high sensitivity and specificity requirements, can be addressed by

- **Chemical Sensors:** Utilizing nanoparticle-based sensors that can offer high sensitivity and rapid response times.
- **Bioassays**: Integrating living cells that respond to mycotoxin exposure, providing real-time data on toxin levels.
- **Multi-target Detection**: Allowing simultaneous detection of multiple mycotoxins using a single chip, which is crucial for comprehensive food safety assessments.

A fluorometric biosensor was created by Carlson and Campbell (2009) to identify and measure aflatoxins. These poisons are frequently present in a range of agricultural goods and are generated by a family of fungi. The mechanism by Carlson and Campbell (2009) relies on fluorescence for a quantitative measurement and immunoaffinity for specificity.

The beneficial characteristics of biosensors have made it necessary to utilize them in the bio-monitoring of biological risks, which are frequently observed in the food and agriculture sectors. In addition to pathogen resistance, the persistent use of pesticides to combat diseases has caused the chemicals to bioaccumulate and become more magnified, posing a health risk and contaminating the environment. As a result, there is a huge growth in market demand for biosensors. Food handlers and agro-allied sectors should have access to biosensors so they can monitor and identify the presence of infections in their food and agricultural goods (Oluwaseun *et al.*, 2018).

Lab-on-Chip technology represents a promising frontier in the detection and diagnosis of fungal pathogens and mycotoxins. Its ability to provide fast, accurate, and cost-effective solutions aligns well with the increasing need for efficient diagnostics in healthcare and food safety. Ongoing research, coupled with advancements in microfabrication and analytical techniques, will likely lead to broader applications and improved performance of LoC devices in identifying and managing fungal threats.

Advantages Over Traditional Methods

The integration of biosensors and lab-on-chip technology into the detection and diagnosis of fungal pathogens and mycotoxins offers a number of significant advantages over traditional methods.

Speed and Efficiency: Traditional diagnostic methods can take days to yield results due to the need for culture growth and extensive analysis. In contrast, biosensors can produce results in minutes to hours, facilitating rapid clinical decisions and timely interventions.

Sensitivity and Specificity: Biosensors can identify small concentrations of mycotoxins or fungal DNA with the accuracy of molecular techniques, which increases the possibility of early detection. This characteristic is essential for controlling outbreaks and guaranteeing food safety.

Portability and Accessibility: Numerous biosensors and LOC devices are made to be portable, making point-of-care testing possible in a variety of venues, such as clinics, hospitals, and isolated areas. This kind of accessibility can greatly improve disease surveillance in areas with limited laboratory resources.

Cost-Effectiveness: By reducing the need for expensive equipment and extensive labor, biosensors and LOC technology can lower the overall costs of diagnostics. The ability to perform multiple tests on a single platform can further enhance economic efficiency.

Conclusion

Biosensors and lab-on-chip technology represent a transformative advancement in the detection and diagnosis of fungal pathogens and mycotoxins. These innovative tools have the potential to enhance public health responses, streamline clinical workflows, and improve food safety monitoring. As research continues to address existing challenges and expand the capabilities of these technologies, their integration into routine diagnostic practice will be invaluable in combating the threats posed by fungal infections and mycotoxin contaminants. With ongoing developments, the promise of a safer, more efficient, and more accessible diagnostic landscape is on the horizon.

Biosensors are being developed to achieve high levels of accuracy for the detection of mycotoxins because of their potential, benefits over other analytical methods, and real-time detection capabilities.

Challenges and Future Directions

Despite the promising capabilities of biosensors and lab-on-chip technology, several challenges remain. The development of universally applicable biosensing platforms for a wide range of fungal pathogens and mycotoxins is still a work in progress. Furthermore, ensuring the stability and durability of the biosensing elements in real-world conditions is a key concern.

Future research should focus on enhancing the specificity of biosensors through innovative biorecognition elements and developing new materials to improve the performance of LoC devices. Collaborations between multidisciplinary teams, including microbiologists, engineers, and clinicians, will be essential in driving these advancements. Additionally, there is an urgent need to standardize procedures for biosensor and LOC device fabrication, validation, and application. Regulatory frameworks will be crucial in ensuring the safety and efficacy of these novel diagnostic tools. Continued innovation is needed to enhance the sensitivity, specificity, and versatility of LoC technologies for fungal pathogen and mycotoxin detection.

References

Buja, I., Sabella, E., Monteduro, A. G., Chiriacò, M. S., De Bellis, L., Luvisi, A., & Maruccio, G. (2021). Advances in plant disease detection and monitoring: From traditional assays to in-field diagnostics. *Sensors*, **21**(6): 2129.

Carlson, H. J., & Campbell, R. E. (2009). Genetically encoded FRET-based biosensors for multiparameter fluorescence imaging. *Current opinion in biotechnology*, **20**(1): 19-27.

Harpaz, D., Eltzov, E., Seet, R. C., Marks, R. S., & Tok, A. I. (2017). Point-of-care-testing in acute stroke management: an unmet need ripe for technological harvest. *Biosensors*, **7**(3): 30.

Hussain, K., Malavia, D., M. Johnson, E., Littlechild, J., Winlove, C. P., Vollmer, F., & Gow, N. A. (2020). Biosensors and diagnostics for fungal detection. *Journal of Fungi*, **6**(4): 349.

Malhotra, S., Verma, A., Tyagi, N., & Kumar, V. (2017). Biosensors: principle, types and applications. *Int. J. Adv. Res. Innovative Ideas Edu*cation. **3**(2): 3639-3644.

Maragos, C. M. (2016). Multiplexed biosensors for mycotoxins. *Journal of AOAC International*, **99**(4): 849-860.

Mendes, R. K., Ferreira, D. C. M., Carvalhal, R. F., Peroni, L. A., Stach-Machado, D. R., & Kubota, L. T. (2009). Development of an electrochemical immunosensor for *Phakopsora pachyrhizi* detection in the early diagnosis of soybean rust. *Journal of the Brazilian Chemical Society*, **20:** 795-801.

Oliveira, I. S., da Silva Junior, A. G., de Andrade, C. A. S., & Oliveira, M. D. L. (2019). Biosensors for early detection of fungi spoilage and toxigenic and mycotoxins in food. *Current Opinion in Food Science*, **29**: 64-79.

Oluwaseun, A. C., Phazang, P., & Sarin, N. B. (2018). Biosensors: a fast-growing technology for pathogen detection in agriculture and food sector. *Biosensing Technologies For The Detection of Pathogens-A Prospective Way for Rapid Analysis, 1st edition. INTECH*, 37-52.

Rai, M., Jogee, P. S., & Ingle, A. P. (2015). Emerging nanotechnology for detection of mycotoxins in food and feed. *International Journal of Food Sciences and Nutrition*, **66**(4): 363-370.

Tothill, I. J. W. M. J. (2011). Biosensors and nanomaterials and their application for mycotoxin determination. *World Mycotoxin Journal*, **4**(4):361-374.

Vidal, J. C., Bonel, L., Ezquerra, A., Hernández, S., Bertolín, J. R., Cubel, C., & Castillo, J. R. (2013). Electrochemical affinity biosensors for detection of mycotoxins: A review. *Biosensors and Bioelectronics*, **49**:146-158.

Chapter 10

Immunological Techniques for Fungal Pathogen and Mycotoxin Detection and Diagnosis

Vinod Chouhan

Division of Plant Pathology,
ICAR-Indian Agricultural Research Institute, New Delhi-110012
Email: Vinod.chouhan1994@gmail.com

Abstract

Immunological techniques play a crucial role in the detection and diagnosis of plant fungal diseases and mycotoxins, providing rapid, sensitive, and specific methods to protect crops and ensure food safety. Fungal pathogens can significantly impact plant health, leading to reduced yield and quality, while mycotoxins produced by certain fungi pose serious health risks to humans and animals. Traditional diagnostic methods, such as visual inspection and culture-based techniques, are often time-consuming and lack specificity. In contrast, immunological techniques, including enzyme-linked immunosorbent assays (ELISA), lateral flow immunoassays (LFIAs), and immunofluorescence assays, offer more efficient alternatives. ELISA, for example, is widely used due to its high sensitivity and ability to quantify specific antigens or antibodies in plant tissues, facilitating early disease detection. LFIAs provide rapid, on-site testing capabilities, making them invaluable for field diagnostics. Immunofluorescence assays enable the visualization of fungal pathogens within plant tissues, providing detailed insights into pathogen-host interactions. Recent advancements have improved the specificity and sensitivity of these assays through the development of monoclonal antibodies and recombinant antibody technologies, allowing for the detection of low-abundance fungal antigens and mycotoxins. In conclusion, immunological techniques are indispensable tools in the detection and diagnosis of plant fungal diseases and mycotoxins, offering rapid, accurate, and versatile solutions that support sustainable agriculture and food security. Future advancements in immunodiagnostic technologies will likely further improve their efficacy, enabling better disease monitoring and control in agricultural systems.

Keywords: Immunological, Mycotoxins, Detection, Diagnosis

Introduction

Toxigenic fungi, including those from the genera Aspergillus, Fusarium, Penicillium, and Alternaria, generate mycotoxins that serve as harmful contaminants (Xie et al., 2022). Mycotoxins can be found at various stages in the food crop supply chain, including planting, harvest, post-harvest, crop storage, the production line, and in the finished goods. A variety of crops are susceptible to contamination by mycotoxins, which encompasses cereal grains such as wheat, maize, barley, rye, and oats, as well as rice, nuts, dried fruits, spices, and feedstuffs (Haque et al., 2020). Mycotoxins, generated by various fungal diseases, are anticipated to rise in food, animal feed, and agricultural products worldwide due to climate change (Liu and Van der Fels-Klerx, 2021). Carcinogenicity, hepatotoxicity, nephrotoxicity, neurotoxicity, immunotoxicity, and oestrogenic effects represent just a fraction of the detrimental impact mycotoxins have on both human and animal health. Moreover, they undermine international trade and the global economy. Moreover, 400 mycotoxin metabolites have been identified (Ji et al., 2016). Aflatoxins (AFs), ochratoxin A (OTA), fumonisins (FMs), zearalenone (ZEA), deoxynivalenol (DON), T-2 toxin (T-2), and HT-2 toxin (HT-2) are among the most detrimental mycotoxins identified in food and animal feed. Due to the stricter regulations, levels of mycotoxin contamination in animal feed are generally higher than those found in human food. The European Union (EU) has established that the highest allowable concentration of aflatoxin B1 (AFB1) in maize intended for food use is 5 μg/kg. The highest permissible concentration of AFB1 for various feed components, such as soybean, wheat, barley, and maize, is set at 20 μg/kg (Pereira et al., 2019). Li et al. (2014) indicates that the regulatory threshold for ZEA is set at 60 μg/kg in food corn and 500 μg/kg in feed corn, as established by NHFPC and CFDA. It is concerning that the prevalence of contamination with various mycotoxins has been increasing globally. One reason for the coexistence of mycotoxins is that certain fungal strains are capable of producing multiple types of toxins. Additionally, samples may experience contamination from various types of fungi simultaneously. Finally, the presence of mycotoxins can arise when a product is composed of multiple contaminated ingredients (Alassane-Kpembi et al., 2017). Pinotti et al. (2016) demonstrated the presence of mycotoxins in food, feed ingredients, and complete feedstuffs globally, with varying levels ranging from 30 to 100 percent. Out of the 7049 feed samples analysed, various mycotoxins were identified in more than 40% of the samples from the US, EU, and Asia. Rodrigues and Naehrer (2012) discovered that 82% of the feed samples tested positive for two or more mycotoxins in Asia, highlighting a significant level of co-contamination. Reports indicate that various samples, such as feed components, maize, wheat, barley, and silage, have been found to be contaminated with mycotoxins and their metabolites. The samples were gathered from diverse regions across the globe, encompassing Europe, America, and Australia. Pereira et al. (2019) discovered that fusarium toxins were consistently present in the samples. The simultaneous presence of multiple mycotoxins, referred to as co-contamination, poses a growing concern due to its combined toxic effects, which are more severe than those caused by individual mycotoxins on human and animal health (Pinotti et al., 2016; Pereira et al., 2019). The mutagenesis activity of AFB1 was notably enhanced in the presence of DON and T-2 toxin (Alassane-Kpembi et al., 2017). Consequently, prominent regulatory authorities such as the FDA, the EU, and the Codex Alimentarius Commission have established stringent thresholds for permissible mycotoxin levels in consumable products (Xing et al., 2020b). Despite ongoing efforts to mitigate their impact through public awareness initiatives, legislative measures, and advancements in production techniques, mycotoxins continue to pose a significant challenge. Consequently, there is an urgent requirement for swift analytical techniques to monitor different mycotoxins in natural environments. Effective management and prevention of mycotoxin contamination in feed and food supplies is crucial. Detection of mycotoxins can be achieved through two main methods: conventional techniques and expedited strategies (Anfossi et al., 2016). Conventional techniques such as UHPLC-MS/MS and high-performance liquid chromatography (HPLC) have proven effective in delivering quantitative analyses of mycotoxins. Nonetheless, these approaches are constrained to examining a limited range of mycotoxins within a controlled laboratory setting, primarily because they depend on costly apparatus, intricate sample preparation, labour-intensive processes, and vulnerability to matrix interference. As a result, several efficient techniques requiring minimal sample preparation have been established. Most rapid methods for testing mycotoxins rely on immunoassays. The enzyme-linked immunosorbent test, commonly referred to as ELISA, is among the most recognised methods in immunoassay techniques. This method offers numerous benefits compared to alternatives, such as cost-effectiveness, high sensitivity, user-friendliness, rapid detection, and environmental safety (Jia et al., 2021). Some of the significant drawbacks of ELISA include the reliance on costly and time-intensive detection equipment, protracted procedures, and the need for specialised experts (Sakamoto et al., 2018). Another approach involves employing alternative, more rapid techniques that adhere to the immunoassay principle, akin to ELISA, to address these challenges. This article aims to explore rapid techniques for the simultaneous detection of multiple mycotoxins through the application of the immunoassay concept. Immunological techniques have significantly enhanced the capacity to identify and diagnose mycotoxins and fungal infections when compared to older methods. The methods employed allow for the detection of toxin contaminations and fungal infections with remarkable sensitivity and specificity, leveraging the unique characteristics of antigen-antibody interactions. This comprehensive review will examine a variety of immunological techniques essential in mycology and toxicology, including ELISA, lateral flow assays, radioimmunoassay, Surface Plasmon Resonance (SPR), among others.

Current methodologies for the examination of multiple mycotoxins utilising immunoassays

Eliminating mycotoxins from the agricultural and food industries at the same time is not possible. A straightforward, accessible, economical, and accurate rapid multiplex detection method has been established to ensure that mycotoxins are intercepted before they can enter the food chain. Immunoassays are often utilised for the simultaneous detection of various mycotoxins because of their specificity, sensitivity, and accuracy. The results of these tests are contingent upon the unique interaction between antibodies and their specific target antigens. There are two main categories of rapid immuno-based techniques for the multiplex analysis of mycotoxins: label and label-free sensors. Chauhan et al. (2016) indicate that these sensors can be classified into two main categories: non-competitive and competitive, based on their use of direct or indirect methods. Mycotoxins are small molecules with a single binding site, leading to the discovery that employing a competitive format allows for the rapid development of effective immuno-based methods (Wang et al., 2018). In the direct competitive mode, mycotoxins present in samples engage in competition with labelled mycotoxins for binding to fixed mycotoxin-specific antibodies. In instances where the mycotoxin concentration in the sample is elevated, the immobilised antibody encounters increased difficulty in binding to the labelled mycotoxin, leading to a diminished signal strength. Consequently, the relationship between the signal and mycotoxin ratio is inversely related to the concentration of mycotoxins present in the sample. The sample's mycotoxins engage in competition with immobilised labelled mycotoxins or their conjugates for a finite quantity of specific antibodies in the indirect competitive format. The latest rapid immuno-based multiplex techniques for mycotoxins significantly integrate ideas from competing formats. Enzyme-Linked Immunosorbent Assay (ELISA) is a widely utilised analytical method that allows for the detection and quantification of proteins, antibodies, and hormones in various samples. This technique is essential in both clinical and research settings, providing valuable insights into biological processes and disease mechanisms. A widely used immunological technique for the detection and quantification of specific antigens or antibodies in various samples is the Enzyme-Linked Immunosorbent Assay, commonly referred to as ELISA. This approach utilises the precise nature of antigen-antibody interactions along with the sensitivity inherent in enzyme-substrate reactions to attain measurable results. Due to its remarkable sensitivity, specificity, and adaptability, the enzyme-linked immunosorbent assay (ELISA) has emerged as a crucial instrument in diagnostic microbiology, especially for identifying mycotoxins and fungal infections.

Fundamentals of ELISA

The process of ELISA encompasses several essential stages:

1. A solid-phase surface, usually a 96-well microplate, is prepared by applying a coating of either an antigen for the detection of antibodies or an antibody for the detection of antigens. The coating undergoes an incubation period, facilitating the adherence of the antigen or antibody to the surface.
2. Blocking: The treatment of the plate with a blocking buffer is essential to inhibit non-specific binding. This step addresses any vacant sites on the plate, minimising background interference and guaranteeing that only targeted interactions are identified.

3. The sample that includes the target analyte, such as fungal antigens or mycotoxins, is introduced into the wells. Upon the presence of the target, it associates with the fixed antibody or antigen.
4. Following the binding process, an enzyme-linked secondary antibody is introduced for detection. This secondary antibody is designed to specifically interact with the target analyte or the primary antibody in the context of an indirect ELISA.
5. The addition of a substrate for the enzyme initiates a reaction, resulting in the generation of a detectable signal. Typically, this leads to a measurable colour change that can be analysed with a spectrophotometer.
6. Examination: The strength of the colour generated correlates directly with the quantity of the target analyte found in the sample. The optical density (OD) values are analysed in relation to a standard curve for the purpose of quantifying the analyte.

Utilisation of ELISA for the Detection of Fungal Pathogens

1. Identification of fungal antigens is a critical process, with ELISA being widely employed to detect these antigens in clinical specimens like serum or urine, as well as in environmental samples. This software demonstrates notable efficacy in the diagnosis of systemic fungal infections.
2. Identification of fungal biomarkers involves recognising specific indicators produced by certain fungal infections, which can be detected through ELISA methods. These biomarkers are capable of detecting the presence of the infection and, in specific situations, may also suggest its pathogenicity.

Benefits of ELISA

1. ELISA demonstrates remarkable sensitivity and specificity in detecting trace amounts of antigens or mycotoxins, thanks to the use of targeted antibodies.
2. ELISA demonstrates versatility by detecting various fungal diseases and mycotoxins, making it advantageous for applications in clinical settings, environmental monitoring, and food safety assessments.
3. ELISA enables accurate measurement of analyte concentrations.
4. High Throughput: ELISA is ideal for large-scale screening as it allows for the simultaneous analysis of multiple samples using microplates.

Constraints of ELISA

1. Cross-reactivity occurs when similar antigens or mycotoxins lead to interference, which can impact the specificity of assays.
2. Creating antibodies for particular fungal infections or mycotoxins presents significant challenges, requiring extensive investigation and verification.

3. The intricacies of sample preparation are critical; inadequate preparation may result in erroneous outcomes.

Western Blotting

Western blotting serves as a widely utilised technique for the detection and analysis of specific proteins within a complex mixture. This method integrates electrophoretic protein separation with targeted immunodetection, allowing for the identification and quantification of proteins according to their molecular weight and antigenic properties. Western blotting serves to identify fungal antigens, analyse protein expression profiles, and detect mycotoxins through interactions with specific antibodies.

Fundamentals of Western Blotting

Isolation of Proteins: The Western blotting technique begins with the separation of proteins based on size using SDS-PAGE (sodium dodecyl sulfate-polyacrylamide gel electrophoresis). Proteins undergo denaturation and are encapsulated in SDS, imparting a negative charge proportional to their size. In electrophoresis, proteins move through the gel matrix, where smaller proteins migrate at a faster rate compared to their larger counterparts.

Membrane Transfer: After electrophoresis, proteins are transferred from the gel to a nitrocellulose or polyvinylidene fluoride (PVDF) membrane. The transfer process is typically achieved through electroblotting, where an electric current drives the proteins from the gel to the membrane, preserving their size and position.

Blockage: To prevent non-specific antibody binding, the membrane is blocked with a protein-rich solution, like bovine serum albumin (BSA) or non-fat dry milk. This step ensures that antibodies specifically bind to the intended target proteins, avoiding interactions with other areas of the membrane. Antibody Incubation: The membrane undergoes treatment with a primary antibody that specifically attaches to the target protein. Following the washing step to remove unbound antibodies, a secondary antibody linked to an enzyme (such as horseradish peroxidase, HRP) or a fluorescent dye is added. The secondary antibody binds to the primary antibody, enabling detection. Recognition: The identification of the target protein is achieved through the utilisation of an enzyme substrate or a fluorescent reagent. The interaction between the substrate and the enzyme in enzyme-based detection results in the generation of a chemiluminescent or colorimetric signal, which can subsequently be visualised and measured. The detection process involves direct imaging of the fluorescent signal using a fluorescence scanner.

Utilisation in the identification of fungal diseases

Identification of Fungal Antigens: Western blotting is utilised to identify specific fungal antigens in clinical specimens such as blood, cerebrospinal fluid, and tissue biopsies. This approach proves to be efficient in confirming the existence of fungal infections and assessing immune responses. Protein Expression Profiling: Western blotting enables the analysis of protein expression levels in fungal infections. This is advantageous for exploring pathogenic mechanisms, pinpointing virulence factors, and understanding the interactions between fungi and their hosts. Diagnostic Confirmation: Western blotting is utilised to confirm fungal infections by detecting specific fungal proteins. This is especially beneficial when other diagnostic approaches yield unsatisfactory results.

Utilisation in the identification of mycotoxins

Identification of Mycotoxin-Protein Conjugates: Western blotting serves as a method to detect mycotoxins that are conjugated to proteins within food matrices or biological specimens. This application enables the identification of mycotoxins and clarifies their interactions with proteins. Protein Alterations: Western blotting serves as a tool to detect changes in protein expression or modifications caused by exposure to mycotoxins. This tool enhances understanding of the molecular mechanisms that contribute to mycotoxin toxicity.

Validation of Mycotoxin Detection Methods: Western blotting confirms the presence of specific mycotoxin-protein interactions, thereby supporting alternative mycotoxin detection techniques. The validation process is crucial for guaranteeing the accuracy and reliability of detection tests.

Benefits of Western Blotting

1. Enhanced specificity and sensitivity are hallmarks of Western blotting, attributed to the utilisation of specific antibodies and the ability to detect minimal quantities of target proteins.
2. Protein Size Determination: This method enables the evaluation of protein sizes, thereby aiding in the identification of specific proteins and their isoforms.
3. Quantitative Analysis: Western blotting provides quantitative data on protein levels, enabling the assessment of changes in protein expression or the effects of mycotoxins.

Constraints of Western Blotting

1. The Western blotting process is characterised by its multiple steps, which include protein separation, transfer, and detection, resulting in a labour-intensive and time-consuming procedure.
2. Demands Specific Tools: Western blotting demands specific tools, including electrophoresis apparatus, transfer systems, and imaging devices, which can be expensive.
3. The possibility of non-specific binding presents a challenge, as it may result in false positives or increased background noise. Meticulous adjustment of antibody concentrations and blocking conditions is essential to mitigate these challenges.

Lateral Flow Immunoassay (LFIA)

The convenient, rapid, portable, and highly specific nature of LFIA, also known as the immunochromatographic assay or strip test, has been utilised to detect various mycotoxins (Chen et al., 2016). The Lateral Flow Immunoassay (LFIA) consists of several components: a sample pad, a conjugate pad, a membrane, a test line, a control line, an adsorbent pad, and a support pad. A variety of lateral flow immunoassays (LFIA) designed for mycotoxin detection utilise a competitive assay format. In this setup, the specific antibody competes with the mycotoxin present in the sample for binding to the mycotoxin-conjugate that is immobilised on the test line. The presence of mycotoxin in the sample hinders the anti-mycotoxin antibody's ability to bind to the mycotoxin-conjugate fixed on the test line. This leads to a decrease or complete lack of signal on the test line. If no mycotoxin is present, the anti-mycotoxin antibody will attach to the immobilised mycotoxin-conjugate on the test line, leading to a visible test line. The presence of the control line ensures that the test procedure has been conducted correctly. Although the LFIA concept is straightforward, applying this technique to analyse a wide range of controlled mycotoxins presents certain challenges. A number of efficient multiplex lateral flow immunoassays (LFIAs) have been created to concurrently identify two to five mycotoxins, focussing on either quality or quantity. Numerous variations of multiplex lateral flow immunoassays (LFIAs) have been documented and can be categorised into the following categories: (1) formats - line versus spots, and (2) reporters - nanoparticles (NPs) (including colour nanoparticles (colloidal gold nanoparticles: AuNPs) and luminescence nanoparticles (quantum dots: QDs, quantum dot nanobeads: QBs). The standard line format is commonly utilised in most multiplex lateral flow immunoassays for mycotoxins. Applying the LFIA concept previously discussed, the line format is remarkably simple. However, this format had constraints in detecting only a few mycotoxin targets at the same time and required a significant amount of reagents and sample volumes for analysis. On the other hand, the spot or microarray format of LFIA has shown the ability to overcome the limitations of the line format by creating an array of target mycotoxins through the production of small concentrations of mycotoxins on the membrane. This array represents the contaminants that need to be identified. Nonetheless, the detection of mycotoxins in trace amounts on the LFIA membrane could pose difficulties regarding visibility and would necessitate the utilisation of an expensive, specialised detector. A novel microarray lateral flow strip test has been developed, demonstrating effectiveness in the simultaneous and quantitative detection of five mycotoxins (Charlermroj et al., 2021). The experiment employed a highly fluorescent marker, showcasing a significant and notable wavelength difference between the excitation and emission of light under ultraviolet light exposure. This facilitated the seamless acquisition of the signals using an economical digital camera. The method successfully identified AFB1, DON, FB1, T-2, and ZEA at detection limits of 1.3, 0.5, 0.4, 0.4, and 0.9 ng/mL, respectively. This threshold is below the constraints set forth by the European Commission Regulation. The test demonstrated acceptable recoveries ranging from 70.7% to 119.5% for intra-assay and from 80.4% to 124.8%

for inter-assay. Although the test demonstrates remarkable sensitivity, specificity, and rapidity in the simultaneous analysis of five mycotoxins, it is essential that it undergo a thorough validation process with sample matrices. To attain a high degree of sensitivity in assays, various nanoparticles (NPs) have been utilised as signal labels, such as coloured NPs, luminous NPs, magnetic NPs (MnNPs), and additional compounds (Wu et al., 2020). Many commercial lateral flow immunoassay (LFIA) instruments designed for mycotoxin detection often utilise coloured nanoparticles, especially gold nanoparticles (AuNPs). This preference can be explained by the various benefits that AuNPs offer, including their affordability, chemical stability, and straightforward production process (Hou et al., 2020). Nonetheless, these coloured nanoparticles (NPs) present several challenges, including the potential for false positives or negatives, insufficient sensitivity, reduced signal intensity, and interference from background sources (Xing et al., 2020a). As a result, to enhance the sensitivity of lateral flow immunoassays (LFIA), advancements have been made with luminescent nanoparticles, such as quantum dots (QDs) and quantum bars (QBs), along with various fluorescent compounds. The reporters demonstrate a wide range of light absorption, significant differences between the wavelengths of absorbed and emitted light, strong fluorescence, low interference from background signals, and durability against dimming due to light exposure. A recent advancement in multiplex LFIA sensor technology involves detecting distinct analytes using a range of coloured bands (Duan et al., 2019; Wu et al., 2020). To minimise the chances of misinterpretation in multiplex mycotoxin analysis, the narrow strip test area and closely spaced test lines were enhanced with multicolour labels. This was particularly important in cases where a single colour label was used to identify a subtle positive signal. A new multicolour immunochromatographic test strip has been developed, utilising four different coloured AuNPs (red, purple, blue, and black). This test strip enables the simultaneous detection of four mycotoxins in maize samples. The detection is easily observable without the need for specialised equipment, and it can be accomplished in just 20 minutes. The detection limits (LOD) for the four mycotoxins, FB1, ZEA, OTA, and AFB1, were found to be 3.27, 0.70, 0.10, and 0.06 ng/mL, respectively. The average recoveries of four mycotoxin detections in maize samples varied between 82.36% and 116.23%, exhibiting a coefficient of variation of less than 12.98%. Duan et al. (2019) created a semi-quantitative lateral flow immunoassay (LFIA) for the detection of mycotoxins (ZEA, OTA, and FB1) in maize samples. The assay results can be accessed in just 10 minutes. The optical limits of detection (LODs) for ZEA, OTA, and FB1 were determined to be 10, 5, and 20 ng/mL, respectively. This method shows promise for monitoring various mycotoxins, as it can detect them visibly and provides benefits in precision, accuracy, reproducibility, selectivity, and repeatability.

Fluorescence Polarisation Immunoassay (FPIA)

FPIA represents a competitive format characterised by its homogeneity and mix-and-read capabilities. This method for sample preparation and analysis eliminates the need for any separation or rinsing stages (Raysyan et al., 2020). The interaction between a mycotoxin and a mycotoxin-fluorophore conjugate (tracer) in their competition to bind to an anti-mycotoxin antibody forms the foundation for detecting a mycotoxin through FPIA. Fluorescence polarisation was utilised to measure the interaction between the anti-mycotoxin antibody and the tracer. This involved observing the orientation of fluorescence emission in both horizontal and vertical directions, which is affected by the size and rotation rate of the fluorophore. When mycotoxin is not present, the anti-mycotoxin antibody is able to attach to the tracer, leading to a rise in polarization The presence of free mycotoxin reduces the availability of the antibody for the trace, leading to a decrease in polarisation. The FPIA principle has seen growing application for the simultaneous detection of various mycotoxins in the field. This is achieved using a multiplexed wavelength label strategy. A dual-wavelength Fabry-Perot interferometer (FPI) was utilised to identify total aflatoxins (AFs) and the Zearalenones (ZEAs) family in maize. The FPI was marked with unique fluorescent tracers. According to the findings of Zhang et al. (2018a), this method demonstrated enhanced sensitivity and accurate quantification when compared to High-Performance Liquid Chromatography-Tandem Mass Spectrometry (HPLC-MS/MS). The complete process, encompassing sample preparation through to detection, was accomplished in under 30 minutes. The detection limits for total aflatoxins and zearalenone in the maize matrix were determined to be 4.98 μg/kg and 11.03 μg/kg, respectively. The recovery rates for AFs ranged from 78.6% to 103.6%, whereas the recoveries for ZEAs varied from 86.1% to 102.9%. Lippolis et al. (2019) introduced a novel FPIA platform aimed at detecting the simultaneous presence of T-2, HT-2, T-2-glucoside, and HT-2-glucoside in wheat. This was achieved through the concurrent use of a monosubstituted fluorescein-labelled HT-2 toxin tracer alongside an anti-HT-2 antibody. The method showcased remarkable sensitivity, achieving an inhibitory concentration (IC50) of 2.0 ng/mL, and displayed a clear and effective performance. The quantification limit (LOQ) was determined to be 15 μg/kg, whereas the detection limit (LOD) was set at 10 μg/kg. However, it could not distinguish between T-2 and HT-2. Furthermore, the fluorescence polarization of tracers may be affected by the variety of food commodities. To ensure the precision of multiplex FPIA, it is essential to implement a validation procedure for each matrix.

Chemiluminescence immunoassay (CLIA)

CLIA presents a viable method for mycotoxin detection, offering the significant benefit of requiring only minimal optical equipment without the need for an external light source. The CLIA principle involves the direct measurement of the intensity of chemiluminescence (CL) resulting from a chemical reaction. To improve the chemiluminescence (CL) signal, it is often essential to utilise a catalyst, including enzymes, transition metal ions, or noble metal nanoparticles, due to the reaction's variable luminosity and the low intensity of emitted photons. The luminol-hydrogen peroxidechemiluminescence (CL) system was utilised to simultaneously detect multiple mycotoxins, with a horseradish peroxidase (HRP) enzyme acting as a catalyst (Zhang et al., 2018b). Multiplex CLIA has effectively created a technique for the concurrent identification of various mycotoxins using a minimal sample volume. Employing a 10 μL volume of multiplex CLIA, Xu et al. (2016) effectively established a technique for assessing AFB1, OTA, and FB1 in cereal samples. This method involved the use of silica-hydrogel photonic crystal microsphere suspension arrays, which were created through the self-assembly of silica nanoparticles and hydrogel techniques. This approach showcases an extensive dynamic linear detection range of 0.0001 to 1 ng/mL for AFB1 and OTA, and 0.001 to 10 ng/mL for FB1. Furthermore, it exhibited an exceptionally high sensitivity, with detection limits (LODs) of 0.0017, 0.0021, and 0.0004 ng/mL for AFB1, FB1, and OTA, respectively. The background signal was minimal, and there were no signals from cross reactions, making it appropriate for the simultaneous detection of multiple mycotoxins. Additionally, a monolith was utilised to create a novel chemiluminescence immune-affinity 96-spot array that demonstrates remarkable sensitivity, featuring customisable functionality, straightforward fabrication, and cost-effectiveness. This array aimed to concurrently identify Fusarium mycotoxins present in maize. A 10 μL sample with a limited volume was utilised to identify the contaminated mycotoxins. The limits of detection (LODs) established for four Fusarium mycotoxins are as follows: 0.0036 ng/mL for DON, 0.0048 ng/mL for ZEA, 0.0039 ng/mL for T-2, and 0.0017 ng/mL for FB1. The detection range for these mycotoxins' spans from 0.01 to 0.1 ng/mL. A dual signal amplification strategy has been employed to enhance the signal for the CLIA method. This methodology involves the application of two techniques: enzyme-mediated tyramine signal amplification and multi-HRP-wrapped AuNPs. The implementation of this strategy has led to a notable enhancement in the sensitivity of CL detection, resulting in a 50-57-fold increase. This improvement allows for the concurrent detection of multiple mycotoxins (Zong et al., 2021). Citrinin (CIT), AFB1, and OTA were identified at concentrations lower than picomolar (pM). The CLIA methodology combines the benefits of the electrochemical approach, including its simplicity and stability, with the strengths of CL features, which offer high sensitivity and a broad dynamic range. Nonetheless, these approaches might prove inadequate for performing on-site testing involving a large volume of samples, given that they require an intricate preparatory process and an expensive CL detector.

Surface Plasmon Resonance (SPR)

The SPR immunosensor functions as a plasmonic optical device, leveraging refractive index variations on a metal or dielectric sensor surface to identify the binding interactions between antigens and antibodies, all without the need for labelling. Mahmoudpour et al. (2019) and Jia et al. (2021) have demonstrated that the concentration of analytes correlates directly with changes in the refractive surface plasmon resonance (SPR) signal. The SPR biosensor has demonstrated its effectiveness as an analytical instrument in detecting mycotoxin contamination in agricultural products and feed ingredients, playing a crucial role in ensuring food safety. This approach offers numerous advantages, including the ability to reuse the sensor chip, enhanced sensitivity, reduced sample consumption, accurate and real-time detection, as well as quick and straightforward operation (Mahmoudpour et al., 2019). The SPR immunosensor is commonly employed to examine the measurable change in the refractive signal at the surface, which correlates directly with the mass of the target molecules (Karczmarczyk et al., 2016). The generation of a refractive index signal (SPR signal) proved inadequate for targets with lower molecular weights, like mycotoxins, because of their insufficient mass, leading to restricted sensitivity in detection capabilities. To address this limitation, a sandwich or indirect competitive inhibition format was utilised for the study of mycotoxins (Karczmarczyk et al., 2016). Karczmarczyk et al. (2016) and Mahmoudpour et al. (2019) have shown that modifying sensor chips with nanoparticles, including gold nanoparticles (AuNPs), can improve sensitivity and boost signal amplification. The swift detection capabilities of an imaging surface plasmon resonance (iSPR) immunosensor facilitate the concurrent identification of various mycotoxins with a single sensor device, highlighting its potential as an effective method. A multiplexed iSPR approach was created for the detection of Fusarium mycotoxins in wheat. This approach utilises a competitive inhibition assay that incorporates secondary antibodies (Ab_2) linked to AuNPs. The platform significantly enhanced the SPR signal, achieving improvements of around 25 times for DON, 90 times for ZEA, and 12 times for T-2. The analysis was completed in 17.5 minutes, encompassing the regeneration of the sensor. The detection limits established were 15 μg/kg for deoxynivalenol (DON), 24 μg/kg for zearalenone (ZEA), and 12 μg/kg for T-2 toxin. The sensor chip demonstrates the capability to function effectively for more than 46 cycles, exhibiting a signal loss of less than 12%, which is not significant. To enable effective field deployment, a prototype portable nanostructured iSPR apparatus was utilised to develop a double 3-plex competitive inhibition assay aimed at detecting six mycotoxins in barley. The detection limits for DON, ZEA, T-2, OTA, FB1, and AFB1 were found to be 26, 6, 0.6, 3, 2, and 0.6 μg/kg, respectively. The sensor circuits could undergo at least 60 regeneration cycles to enable effective detection. Although this technique can identify DON, T-2, ZEA, and FB1 within the regulatory limits set by the European Union, it is crucial to enhance the sensitivity to detect OTA and AFB1. Li et al. (2018) and Wei et al. (2019) employed a self-assembled monolayer featuring a hydrazone linkage to create a multiplexed SPR immunosensor. The sensor provided accurate measurements of four mycotoxin levels in maize and wheat samples. The electrical signal produced from the interaction of mycotoxins with their specific antibodies allowed for the determination of the quantitative detection limits for AFB1, OTA, ZEA, and DON, which were found to be 0.59, 1.27, 7.07, and 3.26 ng/mL, respectively. The use of reusable SPR sensor processors in compact or portable surface plasmon resonance (SPR) devices seems to be the emerging direction for on-site detection of various mycotoxins.

Immunosensor utilising Surface Enhanced Raman Scattering (SERS)

An advanced form of Raman scattering, a SERS immunosensor involves the adsorption of molecules onto metal surfaces. The SERS immunosensor utilises a distinctive spectrum that originates from the molecular vibration of the analyte resulting from an antigen-antibody interaction. The SERS immunosensor presents a promising analytical approach for the concurrent detection of multiple mycotoxins in food, distinguished by its remarkable sensitivity, swift response time, and ease of use (Li et al., 2018). Nonetheless, the multiplex detection of mycotoxins using SERS faces significant challenges regarding precision, accuracy, and selectivity. The challenges arise from the weak SERS signal and the interference caused by Raman active compounds found in food matrices. Currently, the main approaches for improving the Raman signal in fingerprinting identification and SERS labelling detection are electromagnetic and chemical augmentation techniques (Wu et al., 2021). The primary focus of this review will be on SERS tagging detection, a method that identifies target mycotoxins through the interaction between the mycotoxin and SERS-active substrates, eliminating the necessity for immunological identification. The core principle of SERS labelling detection involves the attachment of Raman reporters, which are highly effective molecules for Raman scattering, to recognition components. This process enhances the stability of the SERS signal and produces distinctive Raman signals (Wu et al., 2021). The mycotoxin in question competes with SERS tags or SERS nanoprobe to form a connection with the mycotoxin-specific antibody. The intensity of spectral features is inversely related to mycotoxin content, allowing for the determination of mycotoxin levels in a sample through the analysis of SERS signals. The combination of SERS immunosensor and microarray concept has enabled the simultaneous detection of various mycotoxins, including AFB1, ZEA, and OTA (Li et al., 2018). Gold nanoparticles (AuNPs) were labelled with 5,5-dithiobis(succinimidyl-2-nitrobenzoate) (DSNB) for use as a Raman reporter. During this period, the gold surface of a microarray was chemically bonded to AFB1-bovine serum albumin (BSA), ZEA-BSA, and OTA-BSA conjugates, serving as specific capture sites. The detection limits for AFB1 in food products, ZEA, and OTA were found to be 0.53 to 0.57 μg/kg, 0.26 to 0.29 μg/kg, and 0.061 to 0.066 μg/kg, respectively. The coefficients of variation were below 15%, and the average recoveries spanned from 83.8% to 108.1%. The SERS tagging platform shows promise as a tool for multiplex detection of mycotoxins. Nonetheless, the non-uniformity of the plasmonic matrix platform and the variation in patches pose challenges that result in the uneven immobilisation of antibodies on the surface, as well as the heterogeneous binding of antigens and externally captured surface Raman tags (Jia et al., 2021). To address the problem of non-uniformity, it is essential to gather several final Raman intensity values. The technology of SERS-based mapping has been developed for the quantitative identification of various mycotoxins. This approach involves measuring the signals from every pixel location on a plasmonic substrate. The OTA, AFB1, and FB were examined using the development of SERS mapping on three-dimensional plasmonic nanopillar arrays. Wang et al. (2018) reported that the limits of detection (LODs) for OTA, AFB1, and FB were 0.00509 ng/mL, 0.00511 ng/mL, and 0.00607 ng/mL, respectively. The SERS-based mapping technique is considered a promising and extremely sensitive analytical tool for quantifying various mycotoxins found in food products. To effectively construct a multiplex SERS immunosensor, it is essential to meticulously design the SERS tags, guarantee the homogeneity of the SERS substrate, and choose appropriate Raman reporters.

Electrochemical immunosensor

An electrochemical immunosensor is frequently utilised for the quantitative analysis of samples through the measurement of electrical signals with the aid of an electrochemical transducer (Yin et al., 2019). The signal in the mycotoxin EC immunosensor is produced when the mycotoxin attaches to an anti-mycotoxin antibody that is immobilised on the electrode's surface. An electrochemical immunosensor, recognised for its exceptional sensitivity, rapid detection capabilities, cost-effectiveness, and dependability, has been employed to assess various mycotoxins in a range of food products (Lu et al., 2016; Jia et al., 2021). To date, various types of electrodes have been examined, alongside methods for immobilising specific recognition elements, strategies for amplifying the electrochemical signal, and the advancement of innovative biometric recognition techniques. The objective of these studies is to improve the sensitivity of electrochemical immunosensors for the concurrent detection of multiple mycotoxins. The working electrode serves as an essential element in the development of a multiplex electrochemical immunosensor. An SPCE immunosensing platform utilising AuNPs and PPy-ErGO was developed to improve electrical conductivity, biocompatibility, and sensitivity for the detection of FB1 and DON in maize (Lu et al., 2016). The detection limit for FB1 was identified as 4.2 ng/mL, whereas the detection limit for DON was noted to be 8.6 ng/mL. The analysis revealed that these values exhibited minimal interference from the sample matrix. A dual-channel three-electrode electrochemical sensor array was employed to improve the system's performance, facilitating the simultaneous detection of FB1 and DON (Lu and Gunasekaran, 2019). The system was developed on a glass slide that was coated with transparent indium tin oxide (ITO) through the application of the photolithographic technique. The combination was achieved with a polydimethylsiloxane microfluidic channel that functions via capillary action. Upon the binding of the mycotoxin to the modified electrode surface, which has been immobilised with an anti-mycotoxin antibody, the electrochemical signals (EC signals) will be recorded and analysed in relation to the control signal. This will enable the accurate assessment of mycotoxin contamination levels. The detection limits established for FB1 in this system were 0.097 ng/mL, whereas the detection limit for DON was determined to be 0.035 ng/mL. The sensor exhibited consistent performance throughout a two-week storage period. To develop a portable electrochemical immunosensor capable of detecting multiple mycotoxins in diverse food samples, it is essential to ensure the stability of the antigen or antibody on the electrode surface, the conduction of the electrochemical signal, and the creation of a portable multiplex electrochemical sensing device, as these elements will be critical for its effectiveness.

Nanoarray Immunoassay

The Nanoarray represents a cutting-edge technology platform that is based on the foundational concepts of microarray (Matson, 2021). Nanoarrays offer advantages such as compact size, exceptional sensitivity, and straightforward sample preparation requirements. Consequently, they can function as a different method for examining a range of biomolecules, such as DNA, proteins, bacteria, and mycotoxins. The platform has been successfully employed for various purposes, such as identifying plant diseases (Charlermroj et al., 2014). A mycotoxin nanoarray was developed to enable the simultaneous detection of various mycotoxins, such as ZEA, T-2, and FB1. Mycotoxin conjugates were carefully placed into the wells of a microtitre plate in minimal volumes, creating a grid of multiple spots (4 × 3 array) within each well. The IC50 values were analysed to determine the sensitivity levels for ZEA, T2, and FB1. The sensitivity levels recorded in phosphate buffer saline were 70.1, 2.8, and 90.9 ng/mL. In methanol, the levels were 172.4, 3.2, and 129.3 ng/mL. For wheat, the sensitivity levels were 197.4, 0.7, and 216.7 ng/mL, while in maize, they were 43.6, 0.5, and 25.9 ng/mL, respectively. This advanced multiplex nanoarray offers numerous benefits, such as the capacity for semi-quantitative analysis, minimal antibody consumption, and high-throughput functionality. However, a commercially available detector that works with a nanoarray tends to be expensive and is not portable for on-site detection.

Immunoaffinity column clean-up

The immunoaffinity column (IAC) is extensively utilised for sample clean-up in the analysis of mycotoxins. The IAC features an anti-mycotoxin antibody that is fixed onto a solid support like agarose gel in phosphate buffer, all housed within a compact plastic cartridge. The sample extract is introduced to an immunoaffinity column that contains antibodies specific to a particular mycotoxin. The mycotoxin interacts with the antibody, and water is subsequently flowed through the column to eliminate any impurities. Subsequently, by introducing a solvent like methanol into the column, the bound mycotoxin is detached from the antibody and consequently eluted from the column. The mycotoxin present in the methanol elute undergoes further development through the addition of a chemical substance, which either enhances fluorescence or makes the mycotoxin fluorescent prior to measurement in a fluorometer. The methanol solution can also be utilised for HPLC analysis prior to the addition of a fluorescent enhancing chemical. The clean-up process involving IAC allows for the concentration of the mycotoxin within the column, which enhances the sensitivity of the fluorometric assay or reduces its limit of detection. Nonetheless, IACs exhibit a restricted loading capacity, and the procedures for sample clean-up are more intricate when compared to alternative rapid methods for mycotoxins.

Column clean-up using solid-phase extraction (SPE)

The utilisation of solid phase extraction columns for purification processes is both efficient and cost-effective. The predominant packing materials utilised in SPE columns or cartridges include silica gel, C-18 bonded to silica gel, florisil, and ion exchange resins. Standard SPE columns hold the analytes on the adsorbent, allowing the non-mycotoxin materials to be eluted first, followed by the elution of the mycotoxins. A one-step SPE cleanup column has been created for the efficient purification of mycotoxins for use in a fluorometric method. The packing material of the one-step SPE column consists of a porous frit positioned at the top of the column packing, contained within a funnel-shaped, robust plastic tube that features plastic caps at both ends. A sample extract is introduced into the sample reservoir, and a rubber syringe plunger, or an equivalent device, is employed to propel the sample extract through the one-step SPE column. The purified extract obtained from the lower end of the tube contains the mycotoxin, which can be promptly derivatised and analysed using a fluorometer. The procedure is efficient and straightforward, as the elution of the purified mycotoxin from the column involves a single pass using the extract solvent as the eluting solvent. The extracts from the clean-up column are also applicable for multi-toxin analysis using alternative methods. The column exhibits an extended shelf-life due to the absence of biological reagents, allowing for storage at room temperature. In contrast to immunoaffinity columns, one-step SPE columns do not have the capability to concentrate the analyte during the clean-up process, and the recovery rates may fluctuate based on the complexity of the food samples.

Fluorescent polarization

Fluorescence polarisation immunoassay (FP) relies on the competitive interaction between mycotoxin and a mycotoxin-fluorescein tracer for binding to a mycotoxin-specific antibody. The polarisation serves as an indicator of the orientation of fluorescence emission from both horizontal and vertical directions, yet it does not directly quantify fluorophore concentration. The orientation of the fluorescence observed is connected to the rotation rate of the fluorophore in solution, which is, in turn, associated with the size of the fluorophore in that solution. Smaller molecules exhibit increased rates of rotation and reduced polarisation compared to their larger counterparts. The FP employs the engagement of a mycotoxin-specific antibody with a mycotoxin–fluorophore conjugate (tracer) to efficiently reduce the rotational speed of the tracer. The attachment of the antibody to the tracer enhances polarisation; however, when free mycotoxin is present, a reduced amount of antibody binds to the tracer, leading to decreased polarisation. The polarisation value exhibits an inverse relationship with mycotoxin concentration. Fluorescence polarisation immunoassay presents two significant distinctions from ELISA: it does not rely on an enzyme reaction for detection, and there is no need to separate the bound and free compounds. Consequently, FP assays eliminate the need for a wash step and the waiting period for enzyme reactions to develop colour.

A study demonstrated a rapid extraction method lasting 3 minutes, followed by a 2-minute detection step using fluorescence polarization, achieving a recovery rate of 71.2% for spiked deoxynivalenol in wheat. The approach is straightforward and can be easily transported for field use. Nonetheless, like ELISA methods, matrix effects could be present in FP, necessitating a comprehensive validation study for the method's application across various commodities. Furthermore, the FP does not offer a high throughput approach. Samples must be analysed one at a time, rather than in a batch process.

Optical fiber-based immunosensor

A wave that diminishes over time is produced at the boundary where an optical fibre meets a material with a lower refractive index, such as a liquid or cladding. In this region, fluorescent molecules have the capability to absorb energy from the evanescent wave and subsequently emit fluorescence. A segment of the fluorescence will be redirected into the fibre and is detectable. Immobilising antibodies on the surface of an optical fibre significantly reduces fluorescent interference from the bulk solution. The signal produced in the assay is indicative of the toxin concentration, though it fluctuates based on the specific assay format employed. Investigations into the detection of fumonisin B1 utilising a fiber-optic immunosensor revealed that the sensor could identify fumonisin B1 within a quantitation range of 10–1000 ng/mL, with a limit of detection established at 10 ng/mL [32]. The method offers several advantages: (1) high specificity; (2) ease of miniaturisation; (3) real-time monitoring; and (4) adaptability for remote sensing. Nonetheless, the approach might exhibit constraints in its sensitivity. Utilising immunoaffinity column clean-up can lead to an enhancement in sensitivity. Furthermore, solvents have the potential to influence the precision of the method, as they can alter the refractive index of a medium.

Conclusion

In conclusion, immunological techniques serve as essential instruments for the identification and diagnosis of fungal pathogens and mycotoxins. Their exceptional sensitivity and specificity, coupled with quick turnaround times, render them suitable for various applications, including clinical diagnostics, food safety, and environmental monitoring. Despite existing challenges, continuous progress in antibody technology, assay design, and the integration with various diagnostic platforms is consistently improving the capabilities and applications of these techniques. In agricultural and food safety contexts, detecting minute quantities of mycotoxins is essential for preventing contamination and safeguarding public health. Mycotoxins, generated by fungi such as Aspergillus, Penicillium, and Fusarium species, present considerable dangers to both human and animal health owing to their toxic, carcinogenic, and immunosuppressive characteristics. While immunological techniques offer numerous benefits, they also come with certain limitations. A significant challenge is the possibility of cross-reactivity, where antibodies might attach to non-target antigens, resulting in false positives. The advancement of more targeted antibodies and the implementation of blocking agents or alternative detection techniques can assist in addressing this issue; however, it continues to pose a considerable challenge. A further constraint is the necessity for thoroughly characterised antibodies, which can incur significant costs and demand considerable time for production. In the case of certain emerging or less common fungal pathogens and mycotoxins, the absence of commercially available antibodies may require the creation of custom reagents. Immunological techniques extend their utility to various fields beyond just clinical diagnostics and food safety. In agriculture, these techniques are employed to assess crop health and identify fungal pathogens that may jeopardise yield and quality. In the realm of food safety. Immunological techniques play a crucial role in verifying that food products are devoid of harmful concentrations of mycotoxins, thereby aiding in the prevention of widespread foodborne disease outbreaks and minimising the economic repercussions tied to food recalls and disposal.

References

Alassane-Kpembi, I., Schatzmayr, G., Taranu, I., Marin, D., Puel, O., Oswald, I.P., 2017. Mycotoxins co-contamination: Methodological aspects and biological relevance of combined toxicity studies. Crit. Rev. Food Sci. Nutr. 57, 3489–3507. https://doi.org/ 10.1080/10408398.2016.1140632.

Anfossi, L., Giovannoli, C., Baggiani, C., 2016. Mycotoxin detection. Curr. Opin. Biotechnol. 37, 120–126. https://doi.org/10.1016/j.copbio.2015.11.005.

Charlermroj, R., Himananto, O., Seepiban, C., Kumpoosiri, M., Warin, N., Gajanandana, O., Elliott, C.T., Karoonuthaisiri, N., 2014. Antibody array in a multiwell plate format for the sensitive and multiplexed detection of important plant pathogens. Anal. Chem. 86, 7049–7056. https://doi.org/10.1021/ac501424k.

Charlermroj, R., Phuengwas, S., Makornwattana, M., Sooksimuang, T., Sahasithiwat, S., Panchan, W., Sukbangnop, W., Elliott, C.T., Karoonuthaisiri, N., 2021. Development of a microarray lateral flow strip test using a luminescent organic compound for multiplex detection of five mycotoxins. Talanta 233, 122540. https://doi.org/ 10.1016/j.talanta.2021.122540

Chauhan, R., Singh, J., Sachdev, T., Basu, T., Malhotra, B.D., 2016. Recent advances in mycotoxins detection. Biosens. Bioelectron. 81, 532–545. https://doi.org/10.1016/j. bios.2016.03.004.

Chen, Y., Chen, Q., Han, M., Zhou, J., Gong, L., Niu, Y., Zhang, Y., He, L., Zhang, L., 2016. Development and optimization of a multiplex lateral flow immunoassay for the simultaneous determination of three mycotoxins in corn, rice and peanut. Food Chem. 213, 478–484. https://doi.org/10.1016/j.foodchem.2016.06.116.

Duan, H., Li, Y., Shao, Y., Huang, X., Xiong, Y., 2019. Multicolor quantum dot nanobeads for simultaneous multiplex immunochromatographic detection of mycotoxins in maize. Sens. Actuators B Chem. 291, 411–417. https://doi.org/10.1016/j. snb.2019.04.101.

Haque, M.A., Wang, Y., Shen, Z., Li, X., Saleemi, M.K., He, C., 2020. Mycotoxin contamination and control strategy in human, domestic animal and poultry: a review. Microb. Pathog. 142, 104095 https://doi.org/10.1016/j.micpath.2020.104095.

Hou, S., Ma, J., Cheng, Y., Wang, H., Sun, J., Yan, Y., 2020. One-step rapid detection of fumonisin B1, dexyonivalenol and zearalenone in grains. Food Control 117, 107107. https://doi.org/10.1016/j.foodcont.2020.107107

Ji, C., Fan, Y., Zhao, L., 2016. Review on biological degradation of mycotoxins. Anim. Nutr. 2, 127–133. https://doi.org/10.1016/j.aninu.2016.07.003.

Jia, M., Liao, X., Fang, L., Jia, B., Liu, M., Li, D., Zhou, L., Kong, W., 2021. Recent advances on immunosensors for mycotoxins in foods and other commodities. TrAC - Trends Anal. Chem. 136, 116193 https://doi.org/10.1016/j.trac.2021.116193.

Karczmarczyk, A., Reiner-Rozman, C., Hageneder, S., Dubiak-Szepietowska, M., Dost´alek, J., Feller, K.H., 2016. Fast and sensitive detection of ochratoxin A in red wine by nanoparticle-enhanced SPR. Anal. Chim. Acta 937, 143–150. https://doi. org/10.1016/j.aca.2016.07.034.

Li, X., Zhao, L., Fan, Y., Jia, Y., Sun, L., Ma, S., Ji, C., Ma, Q., Zhang, J., 2014. Occurrence of mycotoxins in feed ingredients and complete feeds obtained from the Beijing region of China. J. Anim. Sci. Biotechnol. 5, 37. https://doi.org/10.1186/2049-1891-5-37.

Li, Y., Chen, Q., Xu, X., Jin, Y., Wang, Y., Zhang, L., Yang, W., He, L., Feng, X., Chen, Y., 2018. Microarray surface enhanced Raman scattering based immunosensor for multiplexing detection of mycotoxin in foodstuff. Sens. Actuators B Chem. 266. https://doi.org/10.1016/j.snb.2018.03.040.

Lippolis, V., Porricelli, A.C.R., Mancini, E., Ciasca, B., Lattanzio, V.M.T., De Girolamo, A., Maragos, C.M., McCormick, S., Li, P., Logrieco, A.F., Pascale, M., 2019. Fluorescence polarization immunoassay for the determination of T-2 and HT-2 toxins and their glucosides in wheat. Toxins 11. https://doi.org/10.3390/toxins11070380.

Liu, C., Van der Fels-Klerx, H.J., 2021. Quantitative modeling of climate change impacts on mycotoxins in cereals: a review. Toxins 13. https://doi.org/10.3390/toxins13040276.

Lu, L., Gunasekaran, S., 2019. Dual-channel ITO-microfluidic electrochemical immunosensor for simultaneous detection of two mycotoxins. Talanta 194, 709–716. https://doi.org/10.1016/j.talanta.2018.10.091.

Lu, L., Seenivasan, R., Wang, Y.-C., Yu, J.-H., Gunasekaran, S., 2016. An electrochemical immunosensor for rapid and sensitive detection of mycotoxins fumonisin B1 and deoxynivalenol. Electrochim. Acta 213, 89–97. https://doi.org/10.1016/j.electacta.2016.07.096.

Mahmoudpour, M., Ezzati Nazhad Dolatabadi, J., Torbati, M., Pirpour Tazehkand, A., Homayouni-Rad, A., de la Guardia, M., 2019. Nanomaterials and new biorecognition molecules based surface plasmon resonance biosensors for mycotoxin detection. Biosens. Bioelectron. 143, 111603 https://doi.org/10.1016/j.bios.2019.111603

Matson, R.S., 2021. Well-based antibody arrays. Methods Mol. Biol. 2237, 11–38. https://doi.org/10.1007/978-1-0716-1064-0_2.

Pereira, C.S., Cunha, S.C., Fernandes, J.O., 2019. Prevalent mycotoxins in animal feed: occurrence and analytical methods. Toxins 11, 290. https://doi.org/10.3390/toxins11050290.

Pinotti, L., Ottoboni, M., Giromini, C., Dell'Orto, V., Cheli, F., 2016. Mycotoxin contamination in the EU feed supply chain: a focus on cereal byproducts. Toxins 8. https://doi.org/10.3390/toxins8020045.

Raysyan, A., Moerer, R., Coesfeld, B., Eremin, S., Schneider, R., 2020. Fluorescence polarization immunoassay for the determination of diclofenac in wastewater. Anal. Bioanal. Chem. 413. https://doi.org/10.1007/s00216-020-03058-w.

Rodrigues, I., Naehrer, K., 2012. A three-year survey on the worldwide occurrence of mycotoxins in feedstuffs and feed. Toxins 4, 663–675. https://doi.org/10.3390/toxins4090663.

Sakamoto, S., Putalun, W., Vimolmangkang, S., Phoolcharoen, W., Shoyama, Y., Tanaka, H., Morimoto, S., 2018. Enzyme-linked immunosorbent assay for the quantitative/qualitative analysis of plant secondary metabolites. J. Nat. Med. 72, 32–42. https://doi.org/10.1007/s11418-017-1144-z.

Wang, X., Park, S.-G., Ko, J., Xiao, X., Giannini, V., Maier, S.A., Kim, D.-H., Choo, J., 2018. Sensitive and reproducible immunoassay of multiple mycotoxins using surface-enhanced Raman scattering mapping on 3D plasmonic nanopillar arrays. Small 14, 1801623. https://doi.org/10.1002/smll.201801623

Wei, T., Ren, P., Huang, L., Ouyang, Z., Wang, Z., Kong, X., Li, T., Yin, Y., Wu, Y., He, Q., 2019. Simultaneous detection of aflatoxin B1, ochratoxin A, zearalenone and deoxynivalenol in corn and wheat using surface plasmon resonance. Food Chem. 300, 125176 https://doi.org/10.1016/j.foodchem.2019.125176.

Wu, Y., Zhou, Y., Huang, H., Chen, X., Leng, Y., Lai, W., Huang, X., Xiong, Y., 2020. Engineered gold nanoparticles as multicolor labels for simultaneous multi-mycotoxin detection on the immunochromatographic test strip nanosensor. Sens. Actuators B Chem. 316, 128107 https://doi.org/10.1016/j.snb.2020.128107

Wu, Z., Pu, H., Sun, D.-W., 2021. Fingerprinting and tagging detection of mycotoxins in agri-food products by surface-enhanced Raman spectroscopy: principles and recent applications. Trends Food Sci. Technol. 110, 393–404. https://doi.org/10.1016/j. tifs.2021.02.013.

Xie, H., Wang, X., van der Hooft, J.J.J., Medema, M.H., Chen, Z.-Y., Yue, X., Zhang, Q., Li, P., 2022. Fungi population metabolomics and molecular network study reveal novel biomarkers for early detection of aflatoxigenic *Aspergillus* species. J. Hazard. Mater. 424, 127173 https://doi.org/10.1016/j.jhazmat.2021.127173

Xing, C., Dong, X., Xu, T., Yuan, J., Yan, W., Sui, X., Zhao, X., 2020a. Analysis of multiple mycotoxins-contaminated wheat by a smart analysis platform. Anal. Biochem. 610, 113928 https://doi.org/10.1016/j.ab.2020.113928.

Xing, K.-Y., Shan, S., Liu, D.-F., Lai, W.-H., 2020b. Recent advances of lateral flow immunoassay for mycotoxins detection. TrAC Trends Anal. Chem. 133, 116087 https://doi.org/10.1016/j.trac.2020.116087.

Xu, J., Li, W., Liu, R., Yang, Y., Lin, Q., Xu, J., Shen, P., Zheng, Q., Zhang, Y., Han, Z., Li, J., Zheng, T., 2016. Ultrasensitive low-background multiplex mycotoxin chemiluminescence immunoassay by silica-hydrogel photonic crystal microsphere suspension arrays in cereal samples. Sens. Actuators B Chem. 232, 577–584. https:// doi.org/10.1016/j.snb.2016.03.123.

Yin, X., Hou, T., Huang, B., Yang, L., Li, F., 2019. Aptamer recognition-trigged label-free homogeneous electrochemical strategy for an ultrasensitive cancer-derived exosome assay. Chem. Commun. 55, 13705–13708. https://doi.org/10.1039/C9CC07253E.

Zhang, X., Tang, Q., Mi, T., Zhao, S., Wen, K., Guo, L., Mi, J., Zhang, S., Shi, W., Shen, J., Ke, Y., Wang, Z., 2018a. Dual-wavelength fluorescence polarization immunoassay to increase information content per screen: applications for simultaneous detection of total aflatoxins and family zearalenones in maize. Food Control 87, 100–108.

Zhang, Z., Lai, J., Wu, K., Huang, X., Guo, S., Zhang, L., Liu, J., 2018b. Peroxidasecatalyzed chemiluminescence system and its application in immunoassay. Talanta 180, 260–270. https://doi.org/10.1016/j.talanta.2017.12.024.

Zong, C., Jiang, F., Wang, X., Li, P., Xu, L., Yang, H., 2021. Imaging sensor array coupled with dual-signal amplification strategy for ultrasensitive chemiluminescence immunoassay of multiple mycotoxins. Biosens. Bioelectron. 177, 112998 https:// doi.org/10.1016/j.bios.2021.112998.

Chapter 11

Chromatographic Techniques for Mycotoxin Detection and Quantification

Waghamare Minal Bhujangrao[1*], Dipayan Hazra[2], Shelke Prajakta Vijay[3], Jammugani Vinod Kumar[4], Vaishnavi Gajananrao Wuike[5]

[1,2]Department of Plant Pathology, Uttar Banga Krishi Vishwavidyalaya, Pundibari, Cooch Behar, West Bengal- 736165, India
[3]Department of Plant Pathology, Dr. Panjabrao Deshmukh Krishi Vidyapeeth, Akola, Maharashtra- 444104, India
[4,5]Department of Biochemistry, Uttar Banga Krishi Vishwavidyalaya, Pundibari, Cooch Behar, West Bengal- 736165, India
**Corresponding author: minalwaghamare2020@gmail.com*

Abstract

Pathogenic filamentous fungus produces mycotoxins which are low-molecular-weight, secondary metabolites, it contaminates feeds and meals. Because of its inclusion in widely used grains, it is an unavoidable hazard that has a variety of acute and chronic effects on the health of people and animals. Quantification of mycotoxins in foodstuffs is particularly challenging since only a small number of toxins are known to be present within food samples. However, in recent years, there has been development in mycotoxin analysis investigations. Chromatographic methods provide excellent sensitivity and accuracy, as well as the capacity to identify many mycotoxins simultaneously, independent of their chemistry or biological activity. Advanced chromatography methods, like high-performance liquid chromatography, have made sample cleaning easier. Thin layer chromatography (TLC) is regarded the best analytical method for preliminary qualitative detection of mycotoxin, and it is widely employed in the food business. However, liquid chromatographic methods, such as High-Performance Liquid Chromatography (HPLC) and Liquid Chromatography-mass Spectrometry (LC-MS), are the most effective analytical techniques established to date for quantifying mycotoxin present in food items. These wide range of analytical techniques are continually being improved, as the most important confirmation tool for assessing numerous mycotoxins due to its high specificity, sensitivity, accuracy, and reliability.

Keywords: Chromatographic analytical techniques, Mycotoxins, Food safety

Introduction

Mycotoxins are low molecular weight secondary metabolites generated by filamentous fungus that can harm people and animals at low doses. They have a huge influence on economies and worldwide commerce. Mycotoxigenic fungi are those capable of generating mycotoxins (Umereweneza *et al.,* 2018). Mycotoxins are an important topic in phytopathology, the study of plant diseases, since they can have substantial consequences for plant health and production in agriculture. Mycotoxin-producing fungus contaminate food and feed, posing socioeconomic and health risks (Abd-Elghany and Sallam, 2015). Mycotoxin contamination in animal feed can be caused by fungal infections in crops or through the consumption of moldy grains and forage. Fungi can develop mycotoxins on plants before harvesting (pre-harvest toxins) or after harvest, during crop storage and transportation (postharvest toxins). Environmental stressors, including high temperatures, dampness, and insect damage, can lead to mold development and mycotoxin contamination in plants (Medina *et al.,* 2015). Improper harvesting, drying, handling, packing, and transportation can increase mycotoxin production (Bhat *et al.,* 2010).

Mycotoxins are predicted to contaminate 60-80% of the world's agricultural supply, with 20% exceeding the legally necessary food safety criteria imposed by the European Union (Eskola *et al.,* 2020). According to Mesterhazy *et al.* (2020), mycotoxin contamination causes around 1.3 billion metric tons of food waste every year, accounting for one-third of worldwide production of food. Around 400 mycotoxins have been identified with only 30 recognized to be toxic to people and animals (Vila-Donat *et al.*, 2018). Pathogenic fungi create mycotoxins, which constitute a health concern when they infect cereal crops, fruits, and vegetables. Around 25% of worldwide grain products are contaminated with mycotoxins (Streit *et al.,* 2013).

Mycotoxin contamination is a global risk to food crops, including cereals (maize, wheat, rice, and barley), legumes, fruits and vegetables, peanuts, walnuts, almonds, Pistachios coffee beans, seeds of cotton, and herbs and spices (pepper, paprika and ginger), contaminating food and feed, with socioeconomic and health consequences (Montanari *et al.,* 2016). Agricultural commodities such as grain, barley, wheat, and peanuts are vulnerable to mycotoxin contamination, particularly under circumstances such as insect damage, drought stress, and inadequate storage (Frisvad *et al.*, 2019)

Mycotoxin Toxicity

Certain fungus produces mycotoxins such as *Aspergillus, Fusarium,* and *Penicillium* that can cause significant infections in individuals as well as animals.

Mycotoxins can potentially cause cancer, allergies, and organ damage (Alshannaq and Yu, 2017). The intensity of their effects is determined by the level of exposure as well as their mutagenic and teratogenic properties. Consuming some fungal-contaminated foods, for example, might have long-term health consequences such as liver cancer, immunological suppression, abdominal discomfort endocrine disruption stunted development and genotoxicity. Long-term exposure to several mycotoxins may have synergistic health effects and efficient mycotoxin detection is critical for ensuring food safety and avoiding health concerns (Amuzie and Pestka, 2010). Among mycotoxins, AFs are particularly problematic since they are very toxic and classified as Group 1 human carcinogens by the International Agency for Research on Cancer (IARC).

Mycotoxin Sample preparation, Extraction method and Purification process

Sample preparation

Sample preparation is all about finding solvents that are safe for the environment and getting rapid results. The extraction process and cleanup are the two most important phases prior to the mycotoxin analysis. Using the proper solvents, the contaminated food and feed samples are extracted in an effort to eliminate mycotoxins from the sample (Somsubsin *et al.,* 2018). The success of the extraction is greatly influenced by the solvent selection and extraction technique. A good extraction solvent is one that has the best efficacy in removing only the mycotoxins from the sample. It lessens the matrix impact, is affordable, and safe for the user (Zhang *et al.,* 2018)

Extraction solution

The two most commonly utilised extraction solvents in mycotoxin analysis are mixtures of methanol and water, as well as acetonitrile and water in varying ratios. The presence of pigments, essential oils, and fatty acids in the samples makes extraction challenging and necessitates the use of several extraction solvents, such as a formic acid–ethyl acetate mixture. Furthermore, mycotoxins have been extracted using a variety of extraction solvents, including 1-octanol and toluene, dichloromethane, acetone, and chloroform (De Saeger, 2011). Whereas polar mycotoxins, such as FBs, are soluble in water, hydrophobic mycotoxins are readily dissolved in all the previously mentioned solvents. A new green solvent called Deep Eutectic Solvent (DES) was recently reported that successfully extracts aflatoxins from rice samples (He *et al.,* 2020). Tetramethyl ammonium chloride and malonic acid, two safe, affordable, and biodegradable components, were combined to create this extractant. Additionally, using this extract adheres to the green chemistry principles by reducing the need for solvents and derivatisation of chemicals

Extraction methods

In mycotoxin analysis, new extraction methods with numerous benefits have been developed. Although the QuEChERS (QuEChERS stands for 'Quick, Easy, Cheap,

Effective, Rugged and Safe) method was originally created for the detection of pesticides, reports of its application to the analysis of mycotoxin have been made, yielding outstanding results. This extraction method is quick, inexpensive, and does not require specialised staff. It uses small amounts of acetonitrile as an organic solvent (Anastassiades *et al.,* 2002). Two immiscible phases from two distinct solvents are utilised in LLE (Liquid-liquid extraction), wherein the solubility of the mycotoxins varies between the two phases. The removal of nontarget chemicals occurs in one phase, while the removal of mycotoxins occurs in the other. This approach is typically popular in liquid samples, while the SLE (Supported liquid extraction), which primarily uses ultrasonic extraction, homogenisation, and shaking, is employed for mycotoxin analysis in agricultural goods with solid structures (Yang *et al.,* 2014). The ASE (Accelerated solvent extraction) extraction method, which uses traditional solvents and temperatures above the boiling point, is quick, easy, and environmentally benign. Because it operates at pressures between 1500 and 2000 psi, ASE is similar with pressurised liquid extraction (PLE). The SFE (Supercritical fluid chromatography) approach makes use of the properties of supercritical fluids, primarily CO_2 under critical circumstances. Few organic solvents using MAE (Microwave-assisted extraction) is a green approach, that takes less time to extract results. This method involves rapidly heating organic solvents using microwave energy in order to seperate mycotoxins in the solvent (Du *et al.,* 2018). Three different solvents are often used with the VALDS-ME: water, a dispersive solvent, and an extraction solvent. Two nontoxic solvents, 1-octanol and toluene, were used as the extractant in a recent work by (Somsubsin *et al.,* 2018). This allowed for the rapid extraction of AFs from rice samples using a vortex and the addition of Na_2SO_4 as a demulsifier. All these extraction techniques, solvents, and their advantages and disadvantages are listed in Table 1.

Cleanup process

The mandatory steps of sample concentration and clean-up are followed before the analysis (Woo *et al.,* 2020). After extraction, the cleanup process aids in further eliminating interference from the matrix, which increases selectivity. SPE (Solid phase extraction) is a widely used, quick method that uses minimal solvent to purify mycotoxins through sorption on a solid absorbent. Rinsing the solid absorbents—which are typically packaged in cartridges—is intended to get rid of impurities and collect the mycotoxin. In addition to the standard strong anionic exchange cartridges and the C-8, C-18 hydrophilic–lipophilic balance cartridges, novel and interesting absorbents have been created for the removal of mycotoxin (Aiko *et al.,* 2016). Recently, high-absorption capacity substitute sorbent materials such as carbon nanomaterials and magnetic carbon nanomaterials have been employed (Huertas-Pérez *et al.,* 2017). Among them, multi-walled carbon nanotubes were utilised concurrently for study of TCs and graphene oxide for studying of AFs. The beneficial materials with high absorption capacity are carbon nanostructures because of their distinct electrical, mechanical, and chemical capabilities. In terms of the disadvantages of the new alternative sorbent materials over the previous ones, their self-synthesis may have limited their applicability, and they need to be tested on a variety of mycotoxin kinds to be deemed suitable. IACs (Immunoaffinity chromatography) are typically used for cleanup in order to obtain higher quantification, precision, and accuracy limits. When the test sample passes through the column, the material that IACs have packed with particular antibodies for certain mycotoxins interacts with the test sample's antigens. Simultaneously, the sample's interfering substances are eliminated through column washing, and the mycotoxin is eluted using a solvent that is often miscible. A popular technique for single- and multi-mycotoxin analyses is the IAC clean-up (Zhang *et al.*, 2018).

Multi-mycotoxin IACs, which analyse AFs, OTA, ZEA, and *Fusarium* toxins simultaneously, have been developed for multi-mycotoxin analysis. While using IACs to analyse mycotoxin yields reliable results, the process is time-consuming and requires the use of numerous harmful organic chemicals (Ahmad *et al.*, 2014). These columns might be unable to fully absorb mycotoxins because of a variety of matrix components that obstruct the antibodies' ability to attach to them, or because the antibodies are inhibited against mycotoxin structures. In addition, IACs are costly and have a limited lifespan (Zhang *et al.*, 2018).

Mycotoxins are also cleaned up using the BSA-based sample clean-up columns. Serum albumins are widely available and inexpensive, particularly those of bovine origin. They can be utilised in the cleanup phase because they have a great affinity for some mycotoxins. Designed and validated a new BSA-based sample clean-up column to be utilised for OTA in wine determination (Leal *et al.*, 2019). The technique was predicated on OTA's adequate binding to BSA following its immobilisation on agarose (Al-Jaal *et al.*, 2019). In comparison to IACs, the method's implementation was deemed satisfactory. One kind of globular protein that forms robust complexes with OTA is serum albumin. One major drawback of utilising BSA as a blocking agent is that it might cause OTA interference due to its binding to the protein, which can either improve or diminish the assay findings. To substitute the antibodies in IACs, synthetic systems such as MIPs, aptamers, and peptides have been produced. Synthetic polymers called MIPs are less expensive than antibodies and have a high level of specialisation in recognising particular analytes (Leal, *et al.*, 2019). Additionally, aptamers are less expensive than antibodies and, once affixed to a column's solid surface, are reusable. *In vitro* selection can be used to generate aptamers, which are then employed for quick on-site detection of food contaminants and are suggested as a substitute for mycotoxin and toxigenic fungal detection. Dual-terminal proximity aptamer probes were introduced by Xia *et al.*, (2019) as a rapid, enzyme-free amplification technique for detecting AFB1 that can be completed in a single tube. One minute was all it took to detect AFB1, making this test among the quickest for AFB1. Following the implementation of this structure, AFB1 was found in samples of bean paste and peanut oil, with recovery rates of 90.3% and 114.8%, respectively (Xia *et al.*, 2019).

Table 1. List of the solvents and extraction techniques for mycotoxins

Extraction Solvents	*Extraction Methods*	*Advantages*	*Limitations*
Organic solvents or mixtures (CH_3CN, MeOH, MeOH/CH_3CN)	QuEChERS (Quick, Easy, Cheap, Effective, Rugged and Safe)	Economical, fast, simple, detection of low ppb levels, better eproducibility and accuracy	Modifications of the original procedure, need of an additional enrichment step
Mixture of organic solvents dispersive solvent and water	VALDS–ME	Use of low-density solvents, simple, fast, effective	Optimization after control a lot of parameters
Mixture of organic Solvents (MeOH/ CH_3CN, CH_3CN/water)	ASE or PLE (Accelerated solvent extraction/ pressurised liquid extraction)	Fully automated, faster extraction compared to the conventional ones, minimal solvents, higher extraction efficiency	Expensive instruments, matrix components excessively coextracted
Mixture of organic Solvents (hexane, cyclohexane) with diluted acids or water	LLE (Liquid-liquid extraction)	Effective, for small-scale Preparations	Time consuming, the sample can be absorbed by the glass equipment depending on the matrix and the determined compounds
Aqueous solution	MAE (Microwave-assisted extraction)	Reduced extraction time and extraction solvent	Only applicable for thermally stable compounds, costly instruments
Mixture of organic Solvents with diluted acids or water	SLE (Supported liquid extraction)	Smaller volumes of solvent	Matrix effects
Supercritical fluids (CO_2), MeOH, ethanol, acetone	SFE (Supercritical fluid chromatography)	Fast technique, small solvent volumes, preconcentration effect, extraction of temperature sensible analytes	Need for specialized and very expensive equipment, not suitable for routine analysis

Source: (Agriopoulou et al., 2020)

Thin-layer chromatography in Analysis and Detection of Mycotoxins

In many countries of the world the establishment of maximum permissible limits requires the application of techniques capable of delivering incorrect and reliable results in the mycotoxin analysis (Andrade *et al.,*2013). In food and feed mycotoxin analysis, chromatography is used as the predominant analytical technique. TLC is regarded as the most traditional chromatographic method and is useful for quickly identifying mycotoxins. Although the method is inexpensive, the measurements cannot be regarded as sensitive and accurate. Furthermore, the physical and chemical structure of each mycotoxin dictates the cleanup process, and sample preparation is essential. Table 2 provides examples of TLC methods for mycotoxin identification (Agriopoulou *et al.,* 2020).

Among the chromatographic techniques, TLC is considered the oldest and can be used for rapid screening of mycotoxins. It is a low-cost technique, but the measurements cannot be considered accurate and sensitive (Andrade *et al.,*2013). Moreover, the sample preparation step is necessary, and the physical and chemical structure of each mycotoxin determines the clean-up step. Examples of TLC techniques for mycotoxin detection are presented in Table 2.

Table 2. Examples of Alfatoxin mycotoxin detection using thin-layer chromatography (TLC)

Mycotoxins	*Sample*	*Origin*	*References*
AFB1, AFB2, AFG1, AFG2, CIT	Herbs and herbal products	India	Bunny *et al.,* 2024
AFB1, AFB2, AFG1, AFG2	Brazil nuts	Brazil	Rai *et al.,* 2024
AFB1, AFB2, AFG1, AFG2	Herbal medicines	Nigeria	Bunny *et al.,* 2024
AFB1, AFB2, AFG1, AFG2, CIT, OTs, PAT, T-2, ZEA, ST, DAS	Almonds, cashew nuts, chestnuts, hazelnuts, pistachio nuts, walnuts	Saudi Arabia	Kim *et al.,* 2024
AFB1, AFB2, AFG1, AFG2	Corn-based food Products	Brazil	Antunovic *et al.,* 2022
AFB1, AFB2, AFG1, AFG2, OTA	Medicinal plants	Pakistan	Bunny *et al.,* 2024

High-Performance Thin-Layer Chromatography (HPTLC)

High performance Thin Layer Chromatography (HPTLC) is a highly effective method for the detection and quantification of mycotoxins, a group of toxic secondary metabolites produced by fungi (Abdel-Gawad *et al.,* 1993). A further development in TLC is high-performance thin-layer chromatography (HPTLC). Reduction of layer thickness (down to 100 microns) and particle size (2 to 10 microns) of the stationary phase leads to an improved separation within a shorter time. Therefore, qualitative and quantitative analysis of aflatoxins has been developed using multidimensional–HPTLC–fluorescence excitation, such as applied in peanut butter samples (Lin *et al.,* 1998).

HPTLC offers several advantages over other analytical techniques, including high sensitivity, versatility, cost-effective and high throughput. Minimum migration distance of the separated components with high mobile phase velocity and the reconcentration effect of the migrated spot obtained in the continuous multiple development process are responsible for the great sensitivity and repeatability of the HPTLC system. HPTLC can detect mycotoxins at very low levels, making it suitable for monitoring food and feed samples for contamination. HPTLC can be used to analyze a wide range of mycotoxins, including aflatoxins, ochratoxins, trichothecenes, and zearalenone (Attimarad *et al.,* 2011).

HPTLC is a relatively simple technique that does not require complex instrumentation or extensive training. Compared to other analytical methods, HPTLC can be a more cost-effective option, especially for small-scale laboratories. With automated HPTLC systems, many samples can be analyzed in a short period of time (Lee *et al.*,1998). HPTLC offers a great sensitivity for the detection of less amounts, can provide good resolution of complex mycotoxin combinations, and can scan the plate in-situ at greater spectral parameters useful for verifying the identification and integrity of the spot can be obtained at more than one wavelength. Using a two-dimensional separation method, the HPTLC-CCD system can successfully confirm the presence of aflatoxins (Felicio *et al.,* 2011).

HPTLC procedure for mycotoxin detection includes sample preparation, derivatization, sample application, development, visualization and quantification. The mycotoxin-containing sample is extracted using appropriate solvents. In some cases, derivatization may be necessary to enhance the detectability or selectivity of certain mycotoxins. The extracted and derivatized sample is applied to a pre-coated HPTLC plate. The plate is developed using a suitable solvent system, which causes the mycotoxins to separate based on their polarity. The separated mycotoxins are visualized using appropriate detection methods, such as UV light, iodine vapor, or specific reagents (Nicol, 1998).

HPTLC Detection Methods for Mycotoxins

- UV Detection: Many mycotoxins absorb UV light and can be directly visualized under a UV lamp.
- Post-Chromatographic Derivatization: Chemical reactions can be performed on the plate to enhance the sensitivity or selectivity of certain mycotoxins.

- Densitometry: Densitometers can be used to quantitatively measure the intensity of the mycotoxin spots (Chilaka *et al.*, 2012).

High-Performance Liquid Chromatography (HPLC):

High-Performance Liquid Chromatography (HPLC) is a widely used analytical technique for the detection and quantification of mycotoxins in food, feed, and agricultural products. The high sensitivity and selectivity of HPLC over a wide range of concentrations makes this method more advantageous in mycotoxin analysis. In the analysis of mycotoxins by liquid chromatography, UV, fluorescence or MS detectors are preferred. It can also be combined with different detection systems such as MS/MS to increase sensitivity (Keskin and Eyupoglu, 2023). Mycotoxins are toxic secondary metabolites produced by certain fungi, which can contaminate crops and pose significant health risks to humans and animals. HPLC is favored in mycotoxin analysis due to its sensitivity, specificity, and ability to handle complex matrices.

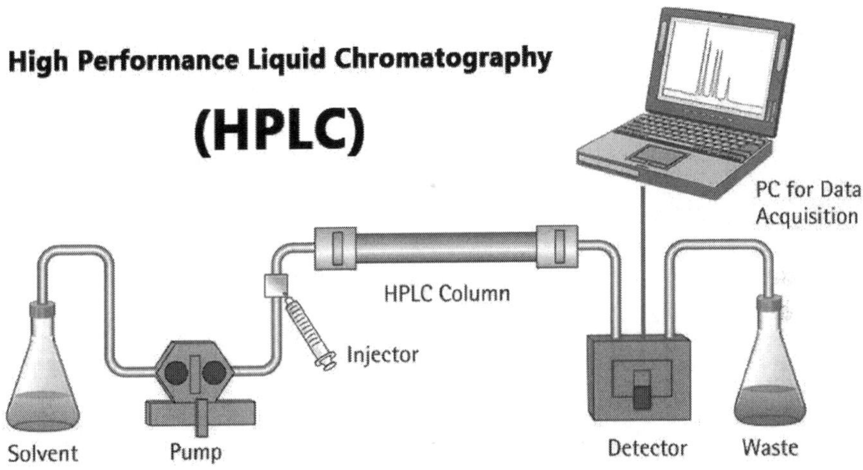

Figure 1: Schematic representation of HPLC unit

(Source: https://microbenotes.com/high-performance-liquid-chromatography-hplc/)

Principle of HPLC

The distribution of the analyte (sample) between a stationary phase (column packing material) and a mobile phase (eluent) is the foundation of the HPLC separation principle. The analyte's chemical structure determines whether the molecules are retarded during the stationary phase.

A sample's "on-column" duration is determined by the intermolecular interactions that occur between its molecules and the packing material. As a result, various sample components elute at various periods. As a result, the sample ingredients are separated. The analytes are identified by a detection unit (such as a UV detector) once they have exited the column. A data management system (computer software) converts and records the signals, which are then displayed in a chromatogram. The mobile phase may then travel via other detector units, a fraction collecting unit, or the trash after passing the detector unit. A solvent reservoir, a pump, an injection valve, a column, a detector unit, and a data processing unit are the typical components of an HPLC system (Fig. 1). The pump moves the solvent, or eluent, through the system at a steady pace and high pressure. A steady and pulseless flow from the pump is essential to minimize the drift and noise of the detector sine wave. The injection valve supplies the analyte (sample) to the eluent.

Steps in HPLC:

1. **Sample Introduction**: The sample mixture is injected into the HPLC system.
2. **Mobile Phase**: The sample is moved through the system by a liquid solvent, or a combination of solvents, known as the mobile phase. High pressure is applied to the mobile phase as it passes through a column.
3. **Stationary Phase**: The stationary phase is located inside the column and is usually a tightly packed column full of tiny particles coated in a substance that interacts with the sample components.
4. **Separation**: Various components interact with the stationary phase in different ways as the sample moves through the column. This causes the components to move at different speeds and, as a result, separate from one another.
5. **Detection**: Once separated, the components exit the column and pass through a detector, which records their presence, typically based on their absorption of light at a specific wavelength.
6. **Data Analysis**: A chromatogram, or graph showing the separated components over time, is created by a computer using signals from the detector. This graph enables identification and quantification.

Types of HPLC

- **Reverse Phase HPLC (RP-HPLC)**: The stationary phase is non-polar, and the mobile phase is polar. It's the most common type.
- **Normal Phase HPLC (NP-HPLC)**: The stationary phase is polar, and the mobile phase is non-polar.
- **Size-Exclusion HPLC**: Separates molecules based on size.
- **Ion-Exchange HPLC**: Separates ions and polar molecules based on their affinity to the ion exchanger.
- **Chiral HPLC**: Separates enantiomers molecules that are mirror images of each other.

Applications of HPLC in Mycotoxins Detecton (Keskin and Eyupoglu, 2023; Agriopoulou *et al.*, 2020):

- **Aflatoxins**: Produced by *Aspergillus* species, commonly found in nuts, grains, and dairy products.
- **Ochratoxin A**: Found in grains, coffee, and dried fruits, this toxin is produced by *Aspergillus* and *Penicillium* species.
- **Fumonisins**: Typically found in maize, these are produced by *Fusarium* species.
- **Zearalenone**: Also produced by *Fusarium* species, found in maize and wheat.
- **Deoxynivalenol (DON)**: Another toxin from *Fusarium* species, commonly found in grains.

Liquid Chromatography-Mass Spectrometry (LC-MS)

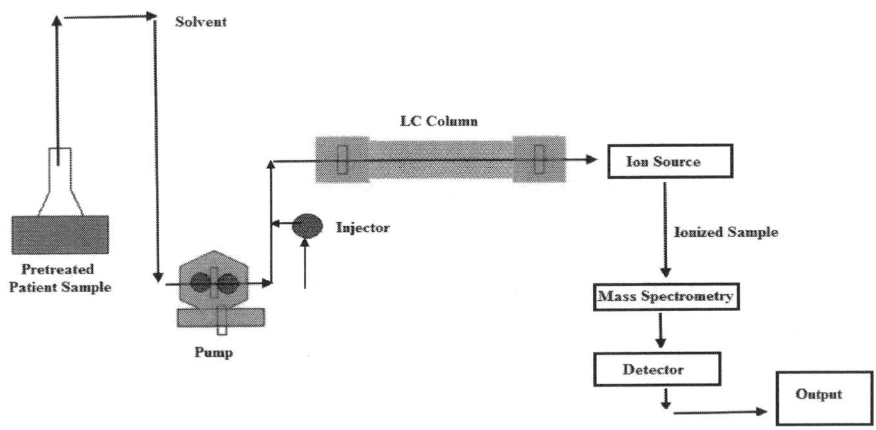

Figure 2: Schematic representation of LCMS.

(Source: https://b2capi.thyrocare.com/Liquid-Chromatography.html)

LC-MS is a highly effective and versatile technique used for the detection and quantification of mycotoxins in food, feed, and agricultural products. It combines the separation power of liquid chromatography (LC) with the detection and structural elucidation capabilities of mass spectrometry (MS), making it particularly valuable for analyzing complex matrices and detecting multiple mycotoxins simultaneously (Ezekwesili-Ofili *et al.*, 2014).

Principle of LC-MS

An ionization source interface is combined HPLC and MS in a typical LC-MS system (Figure 2). Following LC separation of the sample, the separated sample species are sprayed into an atmospheric pressure ion source to undergo gas phase conversion into ions. The ions are then sorted by mass to charge ratio using the mass analyzer, and the detector counts the ions that emerge from the mass analyzer and may also magnify the signal that is produced by each ion. This leads to the creation of the mass spectrum, which is a plot of the ion signal as a function of the mass-to-charge ratio and is used to ascertain the masses of particles and molecules as well as the elemental orisotopic nature of a sample. It also clarifies the structures of chemical of molecules (Kumar and Vijayan, 2014).

Liquid Chromatography (LC) Separation

- Similar to HPLC, the sample is injected into the LC system.
- The sample is separated into its components as it passes through a column containing a stationary phase, with a mobile phase (solvent) carrying it through.
- The separation process depends on the interaction of the sample components with the stationary and mobile phases, based on factors like polarity, size, and charge.

Ionization

- After separation, the eluted components enter the mass spectrometer.
- The compounds are ionized to generate charged particles (ions), typically using techniques like Electrospray Ionization (ESI) or Atmospheric Pressure Chemical Ionization (APCI).
- ESI is commonly used for polar and large biomolecules, while APCI is better suited for less polar, smaller molecules.

Mass Spectrometry (MS) Detection

- The ionized compounds are then introduced into the mass analyzer, where they are separated based on their mass-to-charge ratio (m/z).
- The mass analyzer can be of different types, such as Quadrupole, Time-of-Flight (TOF), Ion Trap, or Orbitrap, each offering different advantages in terms of resolution and sensitivity.
- The separated ions are detected, producing a mass spectrum—a plot of the ion signal as a function of the m/z ratio.

Data Analysis

- The mass spectra provide information about the molecular weight and structure of the compounds.
- By analyzing the fragmentation patterns (how the molecule breaks apart), the structure and identity of the molecules can be inferred.
- Coupled with the retention time from the LC, LC-MS provides a robust method for identifying and quantifying compounds.

Advantages of LC-MS in Mycotoxin Detection

- **High Sensitivity and Specificity**: LC-MS can detect mycotoxins at very low levels, even in complex matrices, and provides specific identification based on molecular weight and fragmentation patterns.
- **Multi-Residue Analysis**: LC-MS can simultaneously detect and quantify multiple mycotoxins in a single run, making it highly efficient for screening large numbers of samples.
- **Minimal Sample Preparation**: LC-MS often requires less sample clean-up compared to traditional methods, as the mass spectrometer can distinguish between the target analyte and matrix interferences.
- **Structural Elucidation**: MS provides detailed structural information through fragmentation, which is valuable for identifying unknown mycotoxins or confirming the presence of known ones.

Applications of LC-MS

- **Food Safety Monitoring**: LC-MS is used to monitor mycotoxin levels in food products to ensure compliance with safety regulations.
- **Agricultural Quality Control**: Analyzing crops and feed for mycotoxins to prevent contamination and ensure product quality.
- **Research and Development**: Investigating the occurrence, distribution, and metabolism of mycotoxins in various environmental and biological samples.

Tandem Mass Spectrometry (LC-MS/MS)

Tandem Mass Spectrometry, or LC-MS/MS, is an advanced extension of Liquid Chromatography-Mass Spectrometry (LC-MS) that involves two stages of mass spectrometry. This technique allows for even more precise identification and quantification of compounds, especially in complex mixtures, by providing additional structural information through fragmentation patterns.

The primary steps in MS/MS are the identification of the precursor ion, its fragmentation (often using collision-induced dissociation, or CID), and the determination of the m/z ratio of the resulting product ions. Tandem in space and tandem in time represent the two essentially distinct approaches to MS/MS. Separate, independent mass analyzers are located in physically distinct parts of tandem-in-space equipment. A hybrid mass spectrometer is a device that integrates many types of analyzers. Quadrupole-time of flight (QqToF), triple/tandem quadrupole (QqQ), and Orbitrap hybrid instruments are a few examples of tandem mass spectrometers. Ion-trapping mass spectrometers, such as 3-D quadrupole ion traps (QIT), linear ion traps (LIT), and Fourier transform ion cyclotron resonance (FT-ICR) devices, are commonly used as tandem-in-time equipment. The three phases of MS are carried out at separate points during the experiment, but they all take place in the same physical trapping volume (Hird *et al.,* 2014).

Liquid Chromatography (LC) Separation

- Just like in standard LC-MS, the sample is first separated into its components using liquid chromatography.
- The separated components elute from the column and are then ionized as they enter the mass spectrometer.

First Mass Spectrometry Stage (MS1)

- In the first mass analyzer, the ionized molecules are separated based on their mass-to-charge ratio (m/z).
- A specific ion (called the precursor ion) corresponding to a particular m/z value is selected for further analysis.

Collision-Induced Dissociation (CID)

- The selected precursor ion is then fragmented in a collision cell by colliding it with an inert gas (like nitrogen or argon). This process is called Collision-Induced Dissociation (CID).
- The collision causes the precursor ion to break apart into smaller fragments (product ions).

Second Mass Spectrometry Stage (MS2)

- The resulting product ions are then analyzed in the second mass spectrometer (MS2) based on their m/z ratios.
- The MS2 stage provides a detailed fragmentation pattern that is characteristic of the structure of the precursor ion.

Data Analysis

- The data from MS2 is used to deduce the structure of the precursor ion by analyzing the fragmentation pattern.
- By combining the retention time from the LC, the precursor ion information from MS1, and the fragmentation data from MS2, LC-MS/MS provides highly specific identification and quantification of compounds.
- This process can be repeated for multiple precursor ions, allowing for the simultaneous analysis of several compounds in a single run.

Applications of LC-MS/MS

- **Pharmacokinetics and Drug Metabolism**: Monitoring how drugs are absorbed, distributed, metabolized, and excreted in the body.
- **Proteomics**: Identifying and quantifying proteins, peptides, and their post-translational modifications.
- **Clinical Diagnostics**: Measuring biomarkers, therapeutic drugs, and detecting metabolic disorders.

- **Environmental Analysis**: Detecting trace levels of pollutants, pesticides, and other contaminants.
- **Food and Beverage Testing**: Identifying contaminants, residues, and adulterants with high specificity.
- **Forensic Toxicology**: Detecting drugs, poisons, and other toxic substances in biological samples.

Advantages of LC-MS/MS

- **High Sensitivity and Specificity**: The two stages of mass spectrometry provide enhanced specificity, reducing the chances of interference from other compounds.
- **Detailed Structural Information**: Fragmentation patterns allow for precise structural elucidation of complex molecules.
- **Quantification**: Provides accurate quantification of compounds even at very low concentrations.
- **Versatility**: Capable of analyzing a wide range of molecules, from small metabolites to large biomolecules.

Gas chromatography

Gas chromatography is an analytical technique for separating, identifying, and quantifying components in a mixture. It works by vaporizing the sample and passing it through a column with a stationary phase. The mixture's components interact with the stationary phase in diverse degrees, causing them to circulate at different rates through the column.

Martin and Synge (1952) Introduced chromatography principles and described gas-liquid chromatography, for which they received the Nobel Prize in Chemistry for their contributions to the field, laying the foundation for modern chromatographic methods (Poole, 2021). Gas chromatography is sometimes referred to as vapor-phase chromatography (VPC) or gas-liquid partition chromatography (GLPC). James and Martin invented the first practical gas chromatograph (GC) equipment in 1952. This equipment was the first to demonstrate the practical application of gas chromatography, a technique used to separate and analyze volatile chemicals (James and Martin, 1952). From the 1950s to the 1960s, GC became widely used in laboratories, with substantial advances in detectors, columns, and data processing, making it a standard in chemical analysis for separating complex mixtures. Also, it is a Standard analytical approach that underpins research, development, and quality control in many industries, including petrochemical manufacturing, as well as environmental, food contaminant, drug residue, and forensic analysis (Bartle and Myers, 2002).

Principles of gas chromatography

Gas chromatography (GC) separates volatile, thermally stable analytes using partition coefficients between a stationary liquid phase and a mobile gas phase.

More volatile substances elute first as they pass through a column heated to 50-300°C to promote volatilization. The stationary phase is often a high-boiling liquid, and appropriate gas flow improves efficiency. GC has great resolution, sensitivity, and speed, making it ideal for studying complex mixtures, particularly when combined with mass spectrometry (Hites, 1997).

The major components of a GC system

According to Poole, (2021), following are major components of GC system:

- Mobile Phase (Sample Injector): An inert gas (often helium or nitrogen) that transports the vaporized sample along the column
- Stationary phase (Carrier Gas): A liquid or solid substance within the column that interacts with the sample components
- Column: A tube, typically formed of glass or metal, through which separation happens. There are two types of columns Packed column and Open tubular column

  Detector: A device at the column's end that detects and records the quantities of separated components as they getaway (Bartle and Myers, 2002).

GC Detectors –

TCD (thermal conductivity detector)

FID (Flame Ionization Detector)

MSD (Mass Selective Detector)

ECD (Electron Capture Detector)

NPD (Nitrogen Phosphorus Detector)

FTIRD (Fourier transformation infrared detector)

AED (Atomic emission detector)

Types of gas chromatography

Gas chromatography is classified into many different types, each with its uses and analyses

- Gas-Liquid Chromatography (GLC)
- Gas-Solid Chromatography (GSC)
- Capillary Gas Chromatography (CapGC)
- High-Resolution Gas Chromatography (HRGC)
- Multidimensional Gas Chromatography (MDGC)

Gas-liquid chromatography (GLC): It is the most known type of gas chromatography. GLC's stationary phase is a liquid that coats the column's interior walls or solid support material. As the sample moves through the column, components with lower boiling temperatures or less affinity for the stationary phase move quicker, causing separation (James and Martin, 1956) Applications:

- ☆ Analyzing environmental toxins
- ☆ Quantifying essential oils in perfumes
- ☆ Analyzing food Flavors and aromas

Gas-Solid Chromatography (GSC): It uses a solid stationary phase, generally a solid adsorbent such as silica gel, alumina, or activated charcoal. The separation method is based on chemical adsorption on a solid surface, which is very useful for separating gases and low molecular weight molecules (Greene and Past, 1958).

Applications

- ☆ The separation of permanent gases such as oxygen, nitrogen, and carbon dioxide
- ☆ Hydrocarbon analysis in petrochemicals

Capillary Gas Chromatography (CapGC): Capillary gas chromatography (CGC) combined with mass spectrometry (MS) is one of the most powerful analytical technologies available today. Gohlke originally detailed the separation of volatile organic compounds (VOCs) by GCMS 60 years ago (Gohlke, 1959). He used packed column GC with on-line time-of-flight (TOF) mass spectrometry. The column used is significantly smaller and longer than those used in standard packed columns. The stationary phase is immediately coated onto the inner wall of the capillary tube, resulting in increased efficiency and better component separation.

Applications

- ☆ Analyze complex mixtures, such as essential oils
- ☆ Detection of relatively small quantities of pollutants in medicines
- ☆ Environmental analysis of volatile organic compounds (VOCs)

High-resolution gas Chromatography (HRGC): HRGC provides higher separation efficiency, resolution, and sensitivity. The main goal is to separate the chemicals in a mixture. It uses capillary columns but with improved detection mechanisms and more exact temperature control (David *et al.,*1999) Applications:

- ☆ Comprehensive study of complex mixtures in biochemical research
- ☆ Analyzing steroids and hormones in biological samples
- ☆ Pesticide residue analysis in agricultural products

Multidimensional Gas Chromatography (MDGC): It involves the utilization of two or more columns with distinct stationary phases linked together in sequence. The material is initially separated on one column before being selectively moved to another column for further separation. This method improves the capacity to resolve complex mixes of comparable qualities (Marriott *et al.*, 2021).

Applications

- Comprehensive analysis of pet agrochemical products
- Detailed profile of diverse Flavors and taste
- Analysis of isomeric chemicals that are difficult to distinguish using one-dimensional GC

Common applications

- In biochemical research, GC helps to understand metabolic pathways, find biomarkers, and analyze the impact of medicines and other compounds on biological systems
- Use official U.S. Environmental Protection Agency (EPA) procedures to quantify contaminants in drinking and wastewater
- Forensic science analyses bulk drugs for identification, impurity profiling, and drug intelligence purposes
- Quantifying multiresidue methods for pesticides and related compounds in food
- Identifying unknown organic chemicals in hazardous waste dumps
- Synthetic organic chemists can identify reaction products
- Used in the food and beverage sector to evaluate Flavors, aromas, and essential oils
- Analysis of emerging and persistent environmental contaminants (Hites, 1997).

Conclusion and future prospects

Mycotoxins cause contamination of food, and while some permissible levels have already been defined, finding sensitive and effective detection technologies is a high concern. Advances in mycotoxin detection are emphasizing present limits in understanding the effective impact on animal and human health caused by their presence in foods. Specifically, demonstrating the coexistence of numerous toxic substances in the same commodity and identifying novel compounds in the mycotoxin group necessitate fresh and devoted toxicological research. Various extraction and clean-up techniques are used prior to detecting and quantifying mycotoxins in contaminated samples. Because the most significant aspects of sample preparation procedures are reduced analysis time, minimal solvent volumes, and extraction scale, extraction techniques that offer these benefits must be chosen.

Large number of techniques for analysis are constantly being optimized and confirmed, and many novel methods are being developed, thin layer chromatography (TLC), gas chromatography (GC), high-performance liquid chromatography (HPLC), and liquid chromatography coupled with a mass spectrometer (LC/MS) techniques remain the fundamental tools for analyzing several mycotoxins. These techniques, with their high level of sensitivity, precision, and reliability, allows for the investigation of many mycotoxins in the same matrix. Continued study and improvement in these areas, as well as an emphasis on understanding the factors that impact the success of these techniques, are critical for mitigating mycotoxin risks, guaranteeing food safety, and safeguarding consumer health (Caldas *et al.*, 2007).

References

Abdel-Gawad, K.M. and Zohri, A.A. (1993). Fungal flora and mycotoxins of six kinds of nut seeds for human consumption in Saudi Arabia. *Mycopathologia* 124: 55–64.

Abd-Elghany, S. M. and Sallam, K. I. (2015). Rapid determination of total aflatoxins and ochratoxins A in meat products by immuno-affinity fluorimetry. *Food chemistry*, 179: 253-256.

Agriopoulou, S., Stamatelopoulou, E. and Varzakas, T. (2020). Advances in analysis and detection of major mycotoxins in foods. *Foods*, *9*(4): 518.

Ahmad, B. Ashiq, S. Hussain, A. Bashir, S. and Hussain, M. (2014) Evaluation of mycotoxins, mycobiota, and toxigenic fungi in selected medicinal plants of Khyber Pakhtunkhwa, *Pakistan. Fungal Boil.* 118: 776–784.

Aiko, V. and Mehta, A. (2016). Prevalence of toxigenic fungi in common medicinal herbs and spices in India. *3 Biotechlogy*, 6: 159.

Al-Bukhaiti, W. Q., Noman, A., Qasim, A. S. and Al-Farga, A. (2017). Gas chromatography: Principles, advantages and applications in food analysis. *International Journal of Agriculture Innovations and Research*, 6(1):2319-1473.

Al-Jaal, B., Salama, S., Al-Qasmi, N. and Jaganjac, M. (2019). Mycotoxin contamination of food and feed in the Gulf Cooperation Council countries and its detection. *toxicon*, 171: 43-50.

Alshannaq, A. and Yu, J. H. (2017). Occurrence, toxicity, and analysis of major mycotoxins in food. *International journal of environmental research and public health*, 14(6): 632.

Amaike, S. and Keller, N. P. (2011). Aspergillus flavus. *Annual review of phytopathology*, 49(1): 107-133.

Amuzie, C. J. and Pestka, J. J. (2010). Suppression of insulin-like growth factor acid-labile subunit expression—A novel mechanism for deoxynivalenol-induced growth retardation. *Toxicological Sciences*, 113(2): 412-421.

Anastassiades, M., Lehotay, S. J. and Štajnbaher, D. (2002). *Quick, easy, cheap, effective, rugged, and safe (QuEChERS) approach for the determination of pesticide residues.*

Andrade, P., Homem-De-Mello, M., França, J. and Caldas, E.D. (2013). Aflatoxins in food products consumed in Brazil: A preliminary dietary risk assessment. *Food Addit. Contam. Part* 30, 127–136.

Antunovic, M., Tomas, M.K., Saric, A., Lucan, C. M., Babic, J., and Kovac, T. (2022). Afatoxin M1 in milk and dairy products–A mini review. *Croatian Journal of Food Science and Technology* 14(2):235–48.

Attimarad, M., Ahmed, K. M., Aldhubaib, B. E. and Harsha, S. (2011). High-performance thin layer chromatography: A powerful analytical technique in pharmaceutical drug discovery. *Pharmaceutical methods*, 2(2): 71-75.

Bartle, K. D. and Myers, P. (2002). History of gas chromatography. *TrAC Trends in Analytical Chemistry*, 21(9-10):547-557.

Bhat, R., Rai, R.V. and Karim, A.A. (2010). Mycotoxins in Food and Feed: Present Status and Future Concerns, *Comprehensive Reviews in Food Science and Food Safety*. 9: 57-81.

Bunny, S. M., Umar, A., Bhatti, H. S. and Honey, S. F. (2024). Aflatoxin risk in the era of climatic change-a comprehensive review. *CABI Agriculture and Bioscience*, *5*(1), 105.

Caldas, E.D. and Silva, A.C.S. (2007). Mycotoxins in Corn-Based Food Products Consumed in Brazil: An Exposure Assessment for Fumonisins. *Journal of Agricultural and Food Chemistry* 55: 7974–7980.

Chhaya, R. S., O'Brien, J. and Cummins, E. (2022). Feed to fork risk assessment of mycotoxins under climate change influences-recent developments. *Trends in Food Science & Technology*, 126: 126-141.

Chilaka, C. A., De Kock, S., Phoku, J. Z., Mwanza, M., Egbuta, M. A. and Dutton, M. F. (2012). Fungal and mycotoxin contamination of South African commercial maize. *Journal of Food, Agriculture and Environment*, 10(2): 296-303.

D Felicio, J., E Freitas, T., H Rossi, M. and Gonçalez, E. (2011). Mycotoxins detection by chromatography. *Recent Patents on Food, Nutrition & Agriculture*, 3(3): 204-211.

David, F., Gere, D. R., Scanlan, F. and Sandra, P. (1999). Instrumentation and applications of fast high-resolution capillary gas chromatography. *Journal of Chromatography A*, 842(1-2):309-319.

De Saeger, S. (Ed.). (2011). *Determining mycotoxins and mycotoxigenic fungi in food and feed*. Elsevier.

Du, L. J., Chu, C., Warner, E., Wang, Q. Y., Hu, Y. H., Chai, K. J. and Zhang, Q. D. (2018). Rapid microwave-assisted dispersive micro-solid phase extraction of mycotoxins in food using zirconia nanoparticles. *Journal of Chromatography A*, 1561: 1-12.

Eaton, D. L. and Groopman, J. D. (2013). *The toxicology of aflatoxins: human health, veterinary, and agricultural significance*. Elsevier.

Eskola, M., Kos, G., Elliott, C. T., Hajšlová, J., Mayar, S. and Krska, R. (2020). Worldwide contamination of food-crops with mycotoxins: Validity of the widely cited 'FAO estimate'of 25%. *Critical reviews in food science and nutrition*, 60(16): 2773-2789.

Ezekwesili-Ofili, J. Onyemelukwe, N., Agwaga, P. and Orji, I.(2014). The Bioload and Aflatoxin Content of Herbal Medicines from Selected States in Nigeria. *African Journal of Traditional, Complementary and Alternative Medicines,*11:143–147.

Frisvad, J. C., Hubka, V., Ezekiel, C. N., Hong, S. B., Novßkovß, A., Chen, A. J. and Houbraken, J. (2019). Taxonomy of Aspergillus section Flavi and their production of aflatoxins, ochratoxins and other mycotoxins. *Studies in mycology*, 93(1): 1-63.

Gohlke, R. S. (1959). Time-of-flight mass spectrometry and gas-liquid partition chromatography. *Analytical Chemistry*, 31(4):535-541.

Greene, S. A. and Past, H. (1958). The determination of heats of adsorption by gas-solid chromatography. *The Journal of Physical Chemistry*, 62(1):55-58.

Gruber, B., David, F. and Sandra, P. (2020). Capillary gas chromatography-mass spectrometry: Current trends and perspectives. *TrAC Trends in Analytical Chemistry*, 124:115475.

He, T., Zhou, T., Wan, Y. and Tan, T. (2020). A simple strategy based on deep eutectic solvent for determination of aflatoxins in rice samples. *Food Analytical Methods*, 13: 542-550.

Hird, S. J., Lau, B. P. Y., Schuhmacher, R. and Krska, R. (2014). Liquid chromatography-mass spectrometry for the determination of chemical contaminants in food. TrAC Trends in Analytical Chemistry, 59: 59-72.

Hites, R. A. (1997). Gas chromatography mass spectrometry. *Handbook of instrumental techniques for analytical chemistry*, 1:609-625.

Huertas-Pérez, J. F., Arroyo-Manzanares, N., García-Campaña, A. M. and Gámiz-Gracia, L. (2017). Solid phase extraction as sample treatment for the determination of Ochratoxin A in foods: A review. *Critical reviews in food science and nutrition*, 57(16): 3405-3420.

Ismaiel, A. A. and Papenbrock, J. (2015). Mycotoxins: producing fungi and mechanisms of phytotoxicity. *Agriculture*, 5(3): 492-537.

James, A. T. and Martin, A. J. P. (1956). Gas–liquid chromatography: the separation and identification of the methyl esters of saturated and unsaturated acids from formic acid to n-octadecanoic acid. *Biochemical Journal*, 63(1):144.

James, A. T., Martin, A. J. P. and Smith, G. H. (1952). Gas-liquid partition chromatography: the separation and micro-estimation of ammonia and the methylamines. *Biochemical Journal*, 52(2): 238.

Kamle, M., Mahato, D. K., Devi, S., Lee, K. E., Kang, S. G. and Kumar, P. (2019). Fumonisins: Impact on agriculture, food, and human health and their management strategies. *Toxins*, 11(6): 328.

Keskin, E. and Eyupoglu, O. E. (2023). Determination of mycotoxins by HPLC, LC-MS/MS and health risk assessment of the mycotoxins in bee products of Turkey. Food Chemistry, 400: 134086.

Khan, R., Anwar, F. and Ghazali, F. M. (2024). A comprehensive review of mycotoxins: Toxicology, detection, and effective mitigation approaches. *Heliyon*.

Kim, J. H., Chan, K. L., Hart-Cooper, W. M., Ford, D., Orcutt, K., Sarreal, S. B. L. and Orts, W. J. (2024). The Inhibitory Activity of Salicylaldehyde Compounds on Aspergillus Species and Their Effects on Aflatoxin Production and Crop Seed Germination. *Applied Sciences*, 14(21): 10000.

Kowalska, A., Walkiewicz, K., Kozieł, P. and Muc-Wierzgoń, M. (2017). Aflatoxins: characteristics and impact on human health. *Advances in Hygiene and Experimental Medicine*, 71: 315-327.

Kumar, K. J. and Vijayan, V. (2014). An overview of liquid chromatography-mass spectroscopy instrumentation. Pharmaceutical methods, 5: 47.

Leal, T., Abrunhosa, L., Domingues, L., Venâncio, A. and Oliveira, C. (2019). BSA-based sample clean-up columns for ochratoxin A determination in wine: Method development and validation. *Food chemistry*, 300: 125204.

Lee, K. Y., Poole, C. F. and Zlatkis, A. (1980). Simultaneous multi-mycotoxin determination by high performance thin-layer chromatography. *Analytical Chemistry*, 52(6): 837-842.

Lee, M. S. and Kerns, E. H. (1999). LC/MS applications in drug development. Mass spectrometry reviews, 18(3-4): 187-279.

Lin, L., Zhang, J., Wang, P., Wang, Y. and Chen, J. (1998). Thin-layer chromatography of mycotoxins and comparison with other chromatographic methods. *Journal of chromatography A*, 815(1): 3-20.

Marriott, P. J., Chin, S. T. and Nolvachai, Y. (2021). Techniques and application in comprehensive multidimensional gas chromatography–mass spectrometry. *Journal of Chromatography A*, 1636:461788.

Medina, A., Rodríguez, A., Sultan, Y. and Magan, N. (2015). Climate change factors and Aspergillus flavus: effects on gene expression, growth and aflatoxin production. *World Mycotoxin Journal* 8: 171-179.

Mesterházy, Á., Oláh, J. and Popp, J. (2020). Losses in the grain supply chain: Causes and solutions. *Sustainability*, 12(6): 2342.

Montanari, C., Bargossi, E., Gardini, A., Lanciotti, R., Magnani, R., Gardini, F. and Tabanelli, G. (2016). Correlation between volatile profiles of Italian fermented sausages and their size and starter culture. *Food Chemistry*, 192: 736-744.

Nicol, R. W. (1998). Analysis of Fusarium toxins in maize and wheat using thin layer chromatography. *Mycopathologia*, 142(2): 107-113.

Ostry, V., Malir, F., Toman, J. and Grosse, Y. (2017). Mycotoxins as human carcinogens—the IARC Monographs classification. *Mycotoxin research*, 33: 65-73.

Peng, W. X., Marchal, J. L. M. and Van der Poel, A. F. B. (2018). Strategies to prevent and reduce mycotoxins for compound feed manufacturing. *Animal Feed Science and Technology*, 237: 129-153.

Poole, C. (Ed.). (2021). *Gas chromatography*. Elsevier.

Rai, J. P., Narware, J., Kumar, R., Kumar, R., Pandey, P., Prakash, N. and Ghatak, A. (2024). Aflatoxin's Toll on Health: Insights into Human and Animal Impact.

Shephard, G. S. (2008). Impact of mycotoxins on human health in developing countries. *Food Additives and contaminants*, 25(2):146-151.

Snyder, L. R., Kirkland, J. J. and Dolan, J. W. (2011). Introduction to modern liquid chromatography.

Somsubsin, S., Seebunrueng, K., Boonchiangma, S. and Srijaranai, S. (2018). A simple solvent based microextraction for high performance liquid chromatographic analysis of aflatoxins in rice samples. *Talanta*, 176: 172-177.

Steinkellner, H., Binaglia, M., Dall'Asta, C., Gutleb, A. C., Metzler, M., Oswald, I. P. and Alexander, J. (2019). Combined hazard assessment of mycotoxins and their modified forms applying relative potency factors: Zearalenone and T2/HT2 toxin. *Food and Chemical Toxicology*, 131: 110599.

Streit, E., Schwab, C., Sulyok, M., Naehrer, K., Krska, R. and Schatzmayr, G. (2013). Multi-mycotoxin screening reveals the occurrence of 139 different secondary metabolites in feed and feed ingredients. *Toxins*, 5(3): 504-523.

Tapia, M. S., Alzamora, S. M. and Chirife, J. (2020). Effects of water activity (aw) on microbial stability as a hurdle in food preservation. *Water activity in foods: Fundamentals and applications*, 323-355.

Thimmaiah, S. K. and Thimmaiah, S. K. (2004). *Standard methods of biochemical analysis*. Kalyani publishers.

Umereweneza, D., Kamizikunze, T. and Muhizi, T. (2018). Assessment of mycotoxins types in some foodstuff consumed in Rwanda. *Food Control*, 85: 432-436.

Van Ruth, S. M. (2001). Methods for gas chromatography-olfactometry: a review. *Biomolecular Engineering*, 17(4-5):121-128.

Vila-Donat, P., Marín, S., Sanchis, V. and Ramos, A. J. (2018). A review of the mycotoxin adsorbing agents, with an emphasis on their multi-binding capacity, for animal feed decontamination. Food and chemical toxicology, 114: 246-259.

Vin, K., Rivière, G., Leconte, S., Cravedi, J. P., Fremy, J. M., Oswald, I. P. and Sirot, V. (2020). Dietary exposure to mycotoxins in the French infant total diet study. *Food and Chemical Toxicology*, 140: 111301.

Woo, S. Y., Ryu, S. Y., Tian, F., Lee, S. Y., Park, S. B. and Chun, H. S. (2019). Simultaneous determination of twenty mycotoxins in the Korean soybean paste doenjang by LC-MS/MS with immunoaffinity cleanup. *Toxins*, 11(10): 594.

Xia, X., Wang, Y., Yang, H., Dong, Y., Zhang, K., Lu, Y. and He, Q. (2019). Enzyme-free amplified and ultrafast detection of aflatoxin B1 using dual-terminal proximity aptamer probes. *Food chemistry*, 283: 32-38.

Yang, J., Li, J., Jiang, Y., Duan, X., Qu, H., Yang, B. and Sivakumar, D. (2014). Natural occurrence, analysis, and prevention of mycotoxins in fruits and their processed products. *Critical reviews in food science and nutrition*, 54(1): 64-83.

Yang, Y., Li, G., Wu, D., Liu, J., Li, X., Luo, P. and Wu, Y. (2020). Recent advances on toxicity and determination methods of mycotoxins in foodstuffs. *Trends in food science & Technology*, 96:233-252.

Zhang, L., Dou, X. W., Zhang, C., Logrieco, A. F. and Yang, M. H. (2018). A review of current methods for analysis of mycotoxins in herbal medicines. *Toxins*, 10(2): 65.

Zhou, H., George, S., Hay, C., Lee, J., Qian, H., and Sun, X. (2017). Individual and combined effects of aflatoxin B1, deoxynivalenol and zearalenone on HepG2 and RAW 264.7 cell lines. *Food and chemical Toxicology*, 103:18-27.

Chapter 12

Role of Artificial Intelligence and Machine Learning in Fungal Pathogen and Mycotoxin Detection and Management

Waghamare Minal Bhujangrao[1*], Dipayan Hazra[2], Shelke Prajakta Vijay[3] and Ritwik Sahoo[4]

[1,2,4]Department of Plant Pathology, Uttar Banga Krishi Vishwavidyalaya, Pundibari, Cooch Behar, West Bengal- 736165, India

[3]Department of Plant Pathology, Dr. Panjabrao Deshmukh Krishi Vidyapeeth, Akola, Maharashtra- 444104, India

**Corresponding author email: minalwaghamare2020@gmail.com*

Abstract

Climate change, intensive agricultural practices and globalization all contribute to the rapid introduction and evolution of plant diseases. Artificial intelligence and machine learning provide ground-breaking opportunities for understanding the complexities of plant Patho-systems and gaining practical insights for disease control. The ability of artificial intelligence to evaluate massive amounts of agricultural data allows for the discovery of connections that are beyond human cognitive powers, making it a powerful tool for better understanding the nature of plant-disease relationships. Machine learning can detect behavioural patterns that humans cannot. Mycotoxins are harmful secondary metabolites, are one of the most serious safety threats to the nutrition/food chain, affecting human, animal well-being, the economy and worldwide commerce in food products. Despite the presence of numerous solutions for preventing and controlling fungal contamination, mycotoxin contamination in food and agricultural products is unavoidable and unexpected. Traditional lab test techniques for mycotoxin identification are time-consuming and may not always be appropriate for massive screenings. However, due to their accuracy and timeliness, artificial intelligence and machine learning approaches have grown in favour in recent years for application in mycotoxin detection and the food safety business. Artificial intelligence and machine learning can assist agricultural systems throughout the world become more productive and sustainable.

Keywords: Artificial Intelligence, Machine Learning, Mycotoxins, Phytopathology and Disease Control

Introduction

Mycotoxins are toxic compounds generated by fungus that grow on crops and foods such as cereals, nuts, spices, and dairy products (World Health Organization, 2023). Ingestion of certain mycotoxins has been connected to a range of adverse health implications on both people and animals. From short-term toxicity to long-term repercussions like liver cancer, some cases result in death (Marroqu'ın-Cardona *et al.,* 2014; Liu and Wu 2010; Mavrommatis *et al.,* 2021). Mycotoxins are low molecular weight secondary metabolites produced by fungal filaments that can injure animals as well as humans in small quantities. They have a significant impact on economies and international commerce. Mycotoxigenic fungi are those that produce microbial toxins (Umereweneza *et al.,* 2018). Mycotoxins are a major concern in phytopathology, since they may have major effects on plant wellness and agricultural productivity. Mycotoxin-producing fungi contaminate food and feed, endangering societal and health consequences (Abd-Elghany and Sallam, 2015).

Mycotoxins in the global food supply chain pose a threat to both human health and agricultural financial stability (Marrón-Cardona *et al.,* 2014; Alshannaq and Yu, 2017). Mycotoxin contamination has a significant economic impact, spending billions of Euros yearly to identify, enforce, and mitigate the presence of it in food and feed (Wu, 2015). Mycotoxin contamination in feed for animals can result from fungal diseases in crops or by consuming moldy grains and forage (Vila-Donat *et al.,* 2018). Fungi can produbhatce mycotoxins on plants prior to harvesting (pre-harvest toxins) or after harvest, during crop storage and transportation (postharvest toxins). Environmental factors such as extreme temperatures, moisture, and damage from insects can cause mold growth and mycotoxin contamination in plants (Ostry *et al.,* 2017). Inappropriate harvesting, drying, handling, packaging, and transportation can lead to increased mycotoxin generation.

Different analytical techniques in analysis and detection of mycotoxins

- ✰ Chromatographic methods in mycotoxin detection
- ✰ Immunological Methods in mycotoxin detection
- ✰ Biosensors in Mycotoxins Detection
- ✰ Molecular Techniques

Importance of Artificial intelligence and machine learning in detection of mycotoxin and its management

While conventional detection methods such as LC-MS, HPLC, GC-MS, and ELISA produce valid results, they frequently result in big, complicated datasets that need detailed interpretation and analysis. In recent years, there has been an increase in the use of machine learning (ML) technologies for both detecting and predicting the existence of mycotoxins. At its foundation, machine learning uses statistical approaches to construct algorithms that enable computers to learn from data and make decisions based on detected patterns and conclusions, without having to be explicitly programmed for each task (Liakos *et al.*, 2018). ML approaches are widely classified into three categories: supervised learning (SL), unsupervised learning (UL), and reinforcement learning (RL).

In SL, an algorithm is trained on a dataset including inputs as well as outputs. The system learns to correlate the inputs and outputs. After training, the framework may use the learnt connection to predict the outputs for fresh, unknown parameters (Baştanlar and Ozuysal, 2014). In UL, an algorithm is supplied just the input data and discovers structures and patterns in its contents using only the inputs. After training, it can categorize fresh inputs using the patterns it discovered. In RL, an algorithm learns to make judgments by taking activities that lead to a desired outcome. It processes input through incentives or penalties linked with its activities and uses this knowledge to construct a structure for making decisions that attempts to maximize rewards (Alpaydin, 2020).

The main innovations that have played a significant part in altering the agricultural industry across the pandemic are artificial intelligence (AI), machine learning (ML), and the internet of things (IoT). Constant disease attacks endanger agriculture, decreasing the quality and number of agricultural items while also having a considerable negative influence on the national economy. Large-scale agricultural monitoring now involves checking for symptoms of plant diseases (Ismaiel and Papenbrock, 2015). As a result, detecting diseases early can reduce their severity and save more crops. Manual disease detection, on the other hand, is arduous, prone to errors, and requires a thorough grasp of plant pathogens. Moving from one technique of disease control to another is a significant difficulty for farmers (Babu *et al.*,2004). However, automated techniques are more efficient. Traditional plant disease detection and identification procedures rely on specialists' trained eyes. In this paper, we provide an automated technique for identifying and detecting plant diseases using our application.

Plant diseases have a substantial influence on crop quantity and quality. Early detection and diagnosis of these diseases are critical for preventing crop loss and increasing output. Agriculture accounts for around 17% of India's GDP (Agarwal *et al.*, 2019). India leads in essential crops such as potato, tomato, and peppers (Thapa and Subash, 2019). The development and spread of diseases in agricultural areas are influenced by a variety of variables, including environmental conditions and cross-contamination. Various crops are growing in every continent of agricultural farming, and they are available for research (Eskola *et al.*, 2020). Pest attacks generate an annual loss in agricultural output of 30-33 percent. Because of the many diseases that have contributing variables, agricultural practitioners require assistance in transitioning from one infection management method to a different one in order to reduce the burden of these infections. As a result, this scenario has a direct influence on both the quantity and quality of agricultural output in general (Streit *et al.,* 2013).

Despite substantial technical improvements, farmers continue to use conventional methods for identifying diseases in crops. Rather of relying on contemporary specialized technologies, farmers continue to manually and visually examine crops for symptoms of illness (Shaikh *et al.,* 2022). The conventional method of visually assessing and evaluating crops relying only on the farmer's competence provide several obstacles and constraints in research on agriculture. In the worst-case situation, an undiscovered crop infection could lead the entire crop to fail, reducing productivity. Certain agricultural illnesses may cause subtle signs, making it difficult to determine the best course of treatment. In such cases, it might be difficult to determine the best judgment, the natural world, and intervention strategy. Therefore, it becomes essential to conduct advanced and comprehensive research (Tapia *et al.,* 2020).

Computer-aided automated research such as machine learning and deep learning can be useful in aiding precise, quick, and early disease detection. The benefits of using these technologies stem from their capacity to give quick and precise results using machine-learning detection and image processing processes. Using AI techniques in agriculture can save labor costs, minimize time inefficiencies, and improve crop quality and output. The use of proper management methodologies can help to speed up disease control strategies by employing the most recent data on crop disease and health location (Eaton and Groopman, 2013).

In understanding artificial intelligence's role in this sector, it's important to recognize that machine learning is a subset of AI. ML seeks to enable computers to learn from experience. Currently, we encounter several subtypes inside the ML domain, each adapted to a distinct learning situation. Supervised training entails feeding input data into the system and predicting goal outcomes from the data. Radivojevic *et al.* (2020) want to create a system that can anticipate outcomes based on inputs. This entails developing algorithms to categorize leaves into diseases of plants categories using labeled data. In contrast, unsupervised learning employs a different technique. In this situation, the system receives data without clear the input-output specifications.

It seeks to identify hidden patterns or correlations in data. Semi-supervised learning occurs when some data is labeled, as in supervised learning, but some data is without labels, as in unsupervised learning. Distinguishing between classification and regression tasks in machine learning is also important since they create distinct types of data for output. Classification projects seek qualitative outcomes and arrange inputs into categories. One method of categorization is dividing plant leaf diseases into various categories (Shoaib *et al.,* 2022).

Deep learning model (DL) is an AI and ML discipline that has had a considerable influence on the classification of images, object identification, and natural language processing (Sarker, 2021). DL uses neural systems to pick features autonomously, reducing the need for extensive artificial feature engineering. It enhances efficiency and generalizability in tasks like image recognition and recognizing targets by integrating low-level data into abstract, high-level characteristics.

Steps in typical machine learning process

Data collection begins with gathering raw data from various sources or sites.

Data Preparation: The raw data is then processed for analysis. This step usually include cleaning and organizing the data.

Data Splitting: Once prepared, the data may be divided into three sections. These are training data, validation data, and test data

Model Selection: The kind of data determines whether an unsupervised or supervised learning model (or models) is used.

Model Training, Evaluation, and Prediction: This method includes training a model using data for training, optimizing the model's hyperparameters with data from validation, and assessing its overall performance with data from tests.

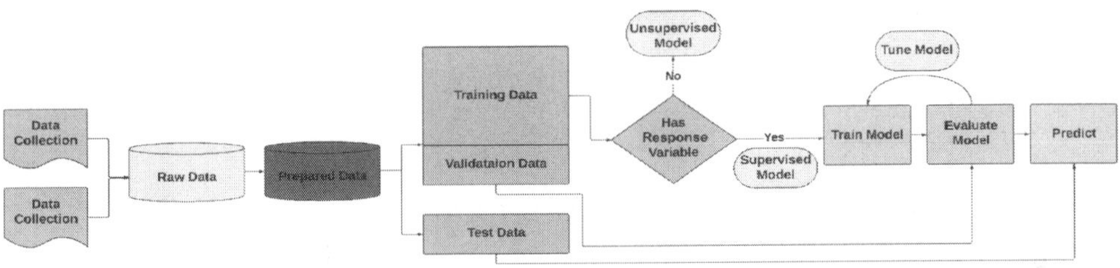

Figure1. Typical machine learning process (Inglis *et al.*, 2024).

Artificial intelligence (AI) automatic categorization of fungal vegetable diseases

Plant pathology classifies plant diseases into biotic and abiotic types. Biotic diseases are caused by fungi, bacteria, insects, and viruses. Non-living causes of abiotic diseases include environmental nutritional inadequacies, chemical imbalances, metal toxic effects, and physical traumas. Crops can also exhibit symptoms of abiotic diseases if subjected to adverse environmental conditions such as hot temperatures, excess precipitation, insufficient light, a shortage of critical nutrients, a highly acidic soil pH, as well as greenhouse gasses. Plant diseases can be difficult to identify with the naked eye, complicating identification and categorization (Liu and Wang, 2021).

It's also crucial to realize that many plant diseases have similar symptoms. It might be challenging to determine which disease of plants is causing harm since they seem so similar. Several difficult to study and detect indicators include abnormal development of leaves, distortion of leaf coloration, delayed growth, diminished and weaker pods, and so on (Frisvad *et al.,* 2020). These visual symptoms, such as diseased leaves, help in the diagnosis. To preserve a healthy ecology, boosting vegetable output and sustaining the agricultural sector's economic sustainability are critical (Mitra, 2021).

The major purpose of this study is to identify the underlying causes of diseases of the leaves. Previous research has repeatedly demonstrated that a plant's leaf health is closely proportional to its defence strength (Qiu *et al.,* 2022). If a plant's leaves remain healthy, its immune system increases and develops better prepared to combat illnesses that may develop in other areas of the plant. Vegetable plant illnesses are common and can be destructive to the foliage (Wani *et al.,* 2022). These infections are highly serious since they may spread quickly and cause a severe damage.

Artificial intelligence automated tomato fungal disease detection

Combine KNN and probabilistic neural networks (PNN) to identify and characterize different diseases of tomato leaves. The collection includes image samples of healthy and infected tomato leaves in the field. The model correctly recognized Verticillium wilt, powdery mildew, leaf miners, Septoria leaf spot, and spider mites. The findings showed that the PNN model outperformed the KNN model in classification accuracy, with a score of 91.88%.

Leaf Mold	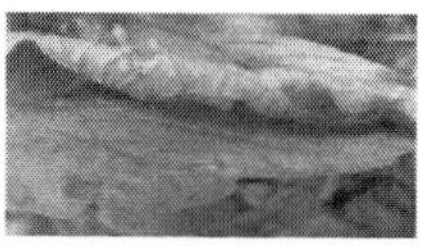	Fungus	The elder leaves have light greenish-to-yellow dots on the top surface	(Zhang et al., 2020)
Tomato Gray Leaf Spot	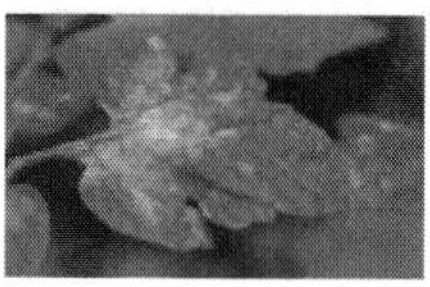	Fungus	Circular, grayish-brown lesions on leaves, which have a yellow halo and severe infections can lead to defoliation	(Liu and Wang, 2020b)

Figure1: tomato crop diseases with their symptoms based on causative agents (Jafar *et al.,* **2024).**

Artificial intelligence automated chilli fungal disease detection

Naik *et al.,* (2022) investigate the efficacy of DL and ML algorithms for classifying chilli leaf fungal diseases. Twelve pre-trained DL networks were used, and the dataset included photos of essential disorders. Without augmentation, VGG19 had the best accuracy (83.54%), however DarkNet53 did remarkably well with augmentation. A distinctive squeeze-and-excitation-based convolutional neural network (SECNN) model beat the others, achieving 98.63% accuracy without augmentation and 99.12% with augmentation, respectively.

Anthracnose		Fungus	Circular, sunken lesions on leaves, and stems, develop black, and spore-bearing structures	(Varur et al., 2023)
Fusarium Wilt		Fungus	Yellowing and wilting of leaves, often starting with lower leaves, and eventually plant death	(Sambrani et al., 2023)

Figure 2: diseases of chilli with their symptoms based on causative agents (Jafar *et al.,* 2024).

Artificial intelligence automated potato fungal disease detection

Patil *et al.* (2017) examined three machine learning algorithms, RF, SVM, and ANN, for detecting potato blight in potato leaf pictures. These approaches were developed and validated using the Plant village dataset provided by the University of Agricultural Sciences India. The collection included photos of leaves with early blight, healthy leaves and late blight. Applying the fuzzy c-mean clustering algorithm to each image assisted in identifying and distinguishing between healthy and unhealthy groups. The simulation demonstrated that the ANN was the most accurate machine learning approach for disease detection. ANN achieves an astounding SVM at 84% and RF at 79% and 92% accuracy.

Rhizoctonia Canker		Fungus	Lesions and spots surround the leaf surface	(Wani et al., 2022)
Potato Wart Disease		Fungus	Warty, tumor-like growths on tubers, which can vary in size and color	(Oppenheim et al., 2019)
Potato Late Blight	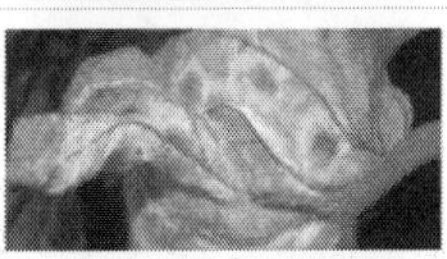	Oomycete	In damp conditions, reddish lesions develop erratically, unusual white fleecy sporulation beneath leaf	(Nur et al., 2023)
Early Blight	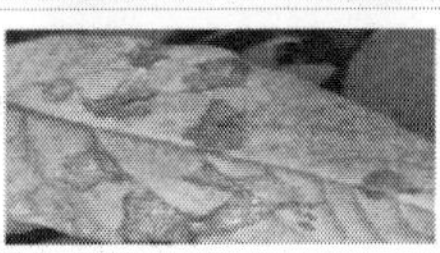	Fungus	Infections with a golden border that develop concentric hoops	(Afzaal et al., 2021)

Figure3: Potato crop diseases with their symptoms based on causative agents (Jafar *et al.,* 2024).

Artificial intelligence automated cucumber fungal disease detection

The suggested cucumber identification approach (Zhang *et al.,* 2017) includes three steps: K-means clustering, shape/color feature extraction, and sparse representation classification. It resolves the constraint of considering all characteristics identically, resulting in efficient calculation and enhanced performance. Several cucumber diseases were recognized, including mildew, bacterial, and powdery mildew. Compared to four other approaches, the SR classifier accurately classifies seven important cucumber diseases, with an overall detection rate of 85.7%.

Powdery Mildew	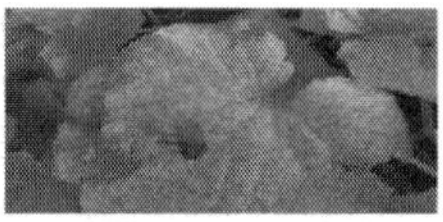	Fungus	White powdery spots on both sides of leaves, yellowing and wilting leaves	(Lin et al., 2019b)
Anthracnose	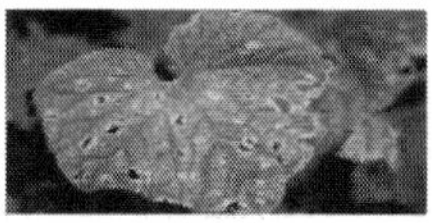	Fungus	Small, water-soaked lesions become sunken and dark, lesions are pink spore masses, leaf distortion, and curling	(Zhang et al., 2021)

Table 4: The cucumber crop fungal diseases with their symptoms based on causative agents (Jafar *et al.*, 2024).

In this study (Zhang *et al.*, 2020), artificial intelligence is utilized to identify and categorize diseases affecting greenhouse plants, specifically those affecting cucumber leaves. The dataset comprised powdery mildew, downy mildew, healthy leaves, and multiple combinations of these diseases. They employed the cutting-edge EfficientNet-B4-Ranger architecture to develop a classification model that achieved 97% accuracy. The design was considered to be the most appropriate alternative for this kind of application.

Application of (AI)artificial intelligence and (ML) machine learning for mycotoxin detection and management

Weather Data

Although agricultural systems have a significant influence on reducing the impact of weather, mycotoxin-producing fungus and the ensuing contamination in food grains are mostly driven by meteorological circumstances (Camardo *et al.*, 2021). It is difficult to forecast the weather in a place where climate variations happen on their own. Because of the many factors that affect climate change, including humidity, wind speed, sea level, and air density, weather is a non-linear system. In agriculture, having a reliable forecasting system can be quite important. Robust weather prediction is achieved by the application of machine learning and data mining approaches that leverage Artificial Neural Networks (ANN) and Support Vector Regression (SVR) (Montanari *et al.*, 2016). For regression challenges, machine learning techniques such as Support Vector Regression (SVR) are employed. It is a type of Support Vector Machine (SVM) that can be used for various tasks such as time series forecasting, because of its ability to predict continuous numerical values. Whether, in order to solve issues, artificial neural networks (ANNs) simulate the human brain mathematically using a sizable collection of artificial neurons. An ANN can translate input data to an output target as a linear or non-linear function. One of the most well-known neural networks is the multilayer perceptron (MLP), which can learn any nonlinear function given enough training data and the right number of neurons (Rasel *et al.*, 2018).

Agronomic Data

Mycotoxin-producing fungus and their contamination in food grains are also driven by different agronomic practices also. Field experiments are used to validate the study of crop-soil-environment relationships in agronomic research. The design of field experiments was greatly enhanced by the use of contemporary statistical methods, which also made it possible to draw important conclusions for the development of sound agronomic practices that will boost crop productivity, improve input efficiency, and protect the environment (Baştanlar, and Ozuysal, 2014). Yet, translating research into actual field settings requires prompt analysis of massive agronomic data sets with enormous spatiotemporal differences. To expand the scope of agronomic study, georeferenced satellite data and new big data analytics technologies should be used. Agronomic field tests could evaluate the imaging and artificial intelligence (AI)-based methods that can process massive data sets and handle unmanned aerial vehicles (UAVs). Internet of Things (IoT), AI (machine and deep learning), big data analytics, and satellite and UAV-based data collection are some data science applications in agronomy. According to recent studies, the accuracy of yield prediction and image categorization can be increased by up to 85% by utilizing AI-based algorithms (Ramesh *et al.*, 2020).

Crop Phenology Data

The incorporation of particular cultivars is a crucial component of spatiotemporal data. Crop varieties can differ in how susceptible they are to mycotoxin contamination and fungus colonization. Monitoring crop phenology is important in precision farming also. We have the ability to accurately track crop phenology at a high spatial resolution track by the Sentinel-1 and Sentinel-2 satellites (Baştanlar, and Ozuysal, 2014). By analyzing the temporal behavior of Sentinel-1 characteristics and Sentinel-2 indices, the data from Sentinel-1, Sentinel-2, and their combination are used to monitor crop phenological stages. Mainly employed are seven machine learning models: naïve Bayes, fuzzy rule-based systems, artificial neural network, logistic regression, decision tree, random forest, and support vector machine. Performance measurements include kappa value, area under the receiver operating characteristic curve, F score, accuracy, precision, specificity, sensitivity, or recall (Yeasin *et al.*, 2022).

Spectral Data

Spectral data are highly recognized for their non-invasiveness and are main often employed types in mycotoxin detection. This kind of data entails measuring the material being studied's absorbance or reflectance of light at different wavelengths. Multispectral and hyperspectral data are two further classifications for spectral data that provide varying degrees of information and detail. The multispectral imaging method is use to differentiate material based on their spectral characteristics since it records information at a small number of distinct wavelength bands. Multispectral imaging concentrates on discrete bands as opposed to hyperspectral imaging, which records continuous spectral information over a broad range of wavelengths (Baştanlar, and Ozuysal, 2014). This simplifies data collecting and processing while maintaining useful information for particular applications. In contrast to multispectral imaging, hyperspectral imaging provides much more detailed information by capturing data over a continuous spectrum of wavelengths. This technique is very useful for accurately identifying mycotoxins and toxic fungal pollutants. Numerous platforms, such as unmanned aerial vehicles (UAVs) and ground-based devices, may use to obtain hyperspectral photographs (Inglis *et al.*, 2024).

Application of AI in Phytopathology

Approaches to phytopathological research and plant disease control are being revolutionized by artificial intelligence (AI). Every year, plant diseases reduce major crop yields by 10% to 16%, necessitating immediate innovation (Steinkellner *et al.,* 2019). With stated accuracy rates of 95% and superior to human visual evaluation, artificial intelligence (AI) demonstrates a propensity for automated disease identification and diagnosis through picture recognition techniques. AI-powered precision agriculture maximizes data-driven, customized crop protection tactics that increase resilience (Baştanlar, and Ozuysal, 2014). AI-powered real-time monitoring uses plant and environmental data to identify pre-symptomatic abnormalities for early warnings. These uses demonstrate AI's ability to reveal hidden disease trends in ever-more-complex agricultural data. AI has the potential to revolutionize precision management and illness prevention in the field. In general, artificial intelligence (AI) is a strategic innovation avenue to improve ecological plant health management (González-Rodríguez *et al.*, 2024).

Application of AI in detection of plant disease and diagnosis

Detecting plant diseases quickly and accurately is essential to increasing agricultural output over the long run. The biggest threat to crop productivity is disease infection, which can result in financial losses. The roots, stems, and leaves of plants are only a few of the sections that can be impacted by viruses, fungus, bacteria, and other infectious organisms. Conventional methods for identifying plant diseases are resource-intensive, time-consuming, and require special knowledge (Baştanlar, and Ozuysal, 2014). Consequently, techniques for the analysis and detection of automated leaf infection diagnostics utilizing artificial intelligence (AI) and Internet of Things (IoT) sensors are taken into consideration. AI is used in a variety of preset procedures to identify and diagnose plant diseases. Image acquisition, preprocessing, segmentation, feature selection, and classification are predefined processes (Jafar *et al.*, 2024).

Advancement in plant disease propagation modeling

A crucial area of study for determining a disease's severity or intensity is plant disease modeling. It's a management system that can predict when plant diseases will occur and how their severity or intensity will evolve. predicting the development of diseases within a given region and time; yet potential yield can be obtained by taking the necessary preventive and control measures beforehand (Slaughter, 2006). By anticipating disease and making management measures cost-effective, they minimize crop and chemical waste at the right moment. To create disease forecasting algorithms, basic plant disease components must be investigated (Baştanlar, and Ozuysal, 2014). To successfully anticipate diseases, there are seven fundamental conditions. The three main disease measuring terminology utilized in disease forecasting research are intensity, incidence, and severity. Field observations, disease data, favorable weather for the transmission of disease, mathematical formulas, and computer utilization are all part of modeling. In mathematical equations displayed as straightforward statements, tables, or graphs, the mathematical and analytical relationship provides information about the interactions between the host, pathogen, and weather variables. Based on relevant factors, such as the inoculum, meteorological variables, and comparative data, disease prediction can be formed. Growers can better understand the impact of epidemic components and subcomponents on yield loss by using computer simulation (Haq *et al.*, 2022).

Limitation of AI and machine learning to detection of fungal pathogen mycotoxin detection and management

Image processing and artificial intelligence approaches provide considerable gains in plant disease diagnosis and classification, but they are not without limits. picture processing algorithms may recognize and isolate affected portions within a picture, but new approaches are needed to handle noise control and unwanted background elements (Mohan *et al.*, 2023).

Real-time machine learning-based disease detection technologies are uncommon in agriculture. Investigating appropriate chemical solutions and their ideal proportions for disease prevention is critical, since inappropriate or inadequate formulations can have a detrimental influence on crop output and nutritional value (Goirand *et al.*, 2021). Farmers frequently require more extensive studies when combining chemicals, resulting in chemical interactions that pose substantial environmental dangers. Leaf pictures may also identify nutritional shortages and water scarcity in plants by carefully observing the leaves. There is an urgent need for modern, hybridized, automated systems capable of meeting these difficulties (Baştanlar, and Ozuysal, 2014).

Technical barriers for AI implementation

Though revolutionary advances in phytopathology are promised by artificial intelligence, careful measures must be taken before incorporating it into agricultural systems. The development of strong, dependable AI solutions for mycotoxin and plant disease situations in the real world is still hindered by technical obstacles. The fact that many machine learning models available today only focus on particular crops, diseases, and controlled environments is one of their main limitations (Goirand *et al.,* 2021). It is common for algorithms that have been trained on small datasets to not generalise to different agricultural environments. The vast differences in soil characteristics, temperatures, pathogen strains, growth stages, and crop cultivars make it difficult to develop AI systems that are flexible enough for in-situ usage (Mohan *et al.,* 2023).

Many concurrent diseases

The notion that each image contains just one disease is not always correct, since many diseases, nutritional deficits, and pests might coexist in the same image. This makes recognizing and monitoring a specific disease more difficult, and the expression of symptoms might differ depending on the geographic area. As a result, it is critical to consider these elements while interpreting photographs.

Ethical concern in AI-driven phytopathology and mycotoxin detection and management

The use of artificial intelligence and machine learning in agriculture also brings up urgent ethical issues that need to be carefully considered, including data privacy, accountability, labour effects, and environmental sustainability (Mohan *et al.,* 2023). Critics warn that AI-enabled crop management systems may strengthen unsustainable industrial farming at the expense of smallholder farmers' food sovereignty, localised expertise, and rural livelihoods. As a result, inclusive stakeholder participation is required for the creation of phytopathology AI systems, focusing on human needs and values (Goirand *et al.,* 2021).

Identification and isolation of disease symptoms

Digital image processing is essential in agricultural research, particularly for recognizing and isolating comparable signs of diverse diseases. It is critical to segment symptoms into diseases that share comparable characteristics in order to improve performance. However, this effort gets difficult when multiple diseases share symptoms and environmental variables. Alternative classification approaches must be investigated in order to detect vegetable diseases with distinct symptoms.

Digital image processing is essential in agricultural research, particularly for recognizing and isolating comparable signs of diverse diseases. It is critical to segment symptoms into diseases that share comparable characteristics in order to improve performance. However, this effort gets difficult when multiple diseases share symptoms and environmental variables. Alternative classification approaches must be investigated in order to detect vegetable diseases with distinct symptoms (Hammond *et al.,* 2016).

Limitations of efficiency of AI in mycotoxins detection

It is evident that machine learning and artificial intelligence has a lot of potential for the agricultural sector, particularly for the detection of mycotoxin and fungal plant diseases. But there are a number of problems that must be identified and resolved before these models can detect function for identification of fungal pathogens and mycotoxins (Sparrow *et al.,* 2021). Images of plant diseases used in agricultural research may contain extraneous noise, colorful backgrounds, and other elements like roots, grass, soil, etc. It is important to identify and exclude these elements. One technique for separating obstructing areas from the captured images is segmentation (Sparrow *et al.,* 2021). In order to efficiently identify infections in the fields, the suggested automatic system has to eliminate unnecessary components from the image and isolate only the necessary segment. This would enable real-time identification of fungal plant pathogens (Inglis *et al.,* 2024).

Environment influence

The majority of the photos in the existing datasets are captured under controlled conditions, usually in lab settings. However, getting an identical image can be difficult because of factors such as light intensity, moisture content, and ambient conditions. Obtaining visual representations of the same leaf specimen from several angles, time periods, and environmental situations is essential to achieve research objectives. The performance of the system is greatly influenced by the tools used for acquiring images (Chhaya *et al.,*2022). The accuracy of forecasts is influenced by a number of parameters, such as the type of sample-taking tool, the amount of moisture, light intensity, and time of day. It is imperative to combine training with the automated disease prediction model's quick deployment to effectively address these problems (Purushotham *et al.,* 2024). Most researchers train their models using a few thousand photos, which emphasises the need for further data for certain diseases. This is addressed by the DL-based data augmentation method, which improves total training images. In this case, the difference between the training data that is used to acquire the model and model data is applied results in a covariate shift. Huge data sets can lead to better model performance, but they can also cause computational complexity (Mohan *et al.,* 2023). Unbalanced patterns in the training dataset present difficulties for automated detection methods. There is a lack of evidence and variation among the classes for a number of fungal diseases. In order to prevent bias, maintaining an impartial and balanced dataset for prediction and precise analysis requires that diseases be represented by disease samples of comparable size, both infected and healthy ones (Inglis *et al.,* 2024)

Conclusion

Artificial intelligence and machine learning have the impact to significantly improve mycotoxin detection and management. Notable accomplishments have been achieved in using AI for automated disease (Goirand *et al.,* 2021). Detection and diagnosis with image recognition techniques such as convolutional neural network analysis. According to studies, artificial intelligence can diagnose diseases in plants with 95-97% accuracy exceeds human visual examination. Artificial intelligence and machine learning can anticipate disease outbreaks using weather, soil, and agricultural factors with 81-95% accuracy, up to 3 weeks in advance. This allows for proactive and tailored pesticide protection tactics, reducing overall consumption. Real-world research shows that AI-enabled solutions may improve mycotoxin management, crop resilience, and provide new insights via complicated agricultural streams of data (Mohan *et al.,* 2023).ML algorithms such as SVM and RF have demonstrated significant success in disease diagnosis and identification, but CNNs have excelled in deep learning. Given the substantial progress made in plant disease prediction using image-based approaches, it is critical to prioritize reliability enhancement, actual time testing, and implementation. Exploring possible chemical and pesticide recommendations for detected illnesses is a promising area of agricultural research.

Artificial intelligence and machine learning are self-improving and generalizable qualities make it ideal for addressing mycotoxin detection, management and plant health concerns related to changing climates, the process of globalization and intensification of agricultural practices.

References

Abd-Elghany, S. M. and Sallam, K. I. (2015). Rapid determination of total aflatoxins and ochratoxins A in meat products by immuno-affinity fluorimetry. *Food chemistry*, 179: 253-256.

Agarwal, A., Sarkar, A. and Dubey, A. K. (2019). Computer vision-based fruit disease detection and classification. In *Smart Innovations in Communication and Computational Sciences: Proceedings of ICSICCS-2018* (pp. 105-115). Springer Singapore.

Alpaydin, E. (2020). *Introduction to machine learning*. MIT press.

Alshannaq, A. and Yu, J. H. (2017). Occurrence, toxicity, and analysis of major mycotoxins in food. *International journal of environmental research and public health*, 14(6): 632.

Babu, M. M., Luscombe, N. M., Aravind, L., Gerstein, M. and Teichmann, S. A. (2004). Structure and evolution of transcriptional regulatory networks. *Current opinion in structural biology*, 14(3): 283-291.

Balakrishna, K. and Rao, M. (2019). Tomato plant leaves disease classification using KNN and PNN. *International Journal of Computer Vision and Image Processing (IJCVIP)*, 9(1): 51-63.

Baştanlar, Y. and Ozuysal, M. (2014). Introduction to machine learning. *miRNomics: MicroRNA biology and computational analysis*, 105-128.

Bernardes, R. C., De Medeiros, A., da Silva, L., Cantoni, L., Martins, G. F., Mastrangelo, T. and Mastrangelo, C. B. (2022). Deep-learning approach for fusarium head blight detection in wheat seeds using low-cost imaging technology. *Agriculture*, 12(11): 1801.

Camardo Leggieri, M., Mazzoni, M. and Battilani, P. (2021). Machine learning for predicting mycotoxin occurrence in maize. *Frontiers in microbiology*, 12: 661132.

Chhaya, R. S., O'Brien, J. and Cummins, E. (2022). Feed to fork risk assessment of mycotoxins under climate change influences-recent developments. *Trends in Food Science & Technology*, 126: 126-141.

Eaton, D. L. and Groopman, J. D. (2013). *The toxicology of aflatoxins: human health, veterinary, and agricultural significance*. Elsevier.

Eskola, M., Kos, G., Elliott, C. T., Hajšlová, J., Mayar, S. and Krska, R. (2020). Worldwide contamination of food-crops with mycotoxins: Validity of the widely cited 'FAO estimate'of 25%. *Critical reviews in food science and nutrition*, 60(16): 2773-2789.

Frisvad, J. C., Hubka, V., Ezekiel, C. N., Hong, S. B., Novßkovß, A., Chen, A. J. and Houbraken, J. (2019). Taxonomy of Aspergillus section Flavi and their production of aflatoxins, ochratoxins and other mycotoxins. *Studies in mycology*, 93(1): 1-63.

Goirand, M., Austin, E. and Clay-Williams, R. (2021). Implementing ethics in healthcare AI-based applications: a scoping review. *Science and engineering ethics*, 27(5): 61.

González-Rodríguez, V. E., Izquierdo-Bueno, I., Cantoral, J. M., Carbú, M. and Garrido, C. (2024). Artificial intelligence: A promising tool for application in phytopathology. *Horticulturae*, 10(3):197.

Hammond, J. L., Formisano, N., Estrela, P., Carrara, S. and Tkac, J. (2016). Electrochemical biosensors and nanobiosensors. *Essays in biochemistry*, 60(1): 69-80.

Haq, I. U., Khan, N. A. and Sarwar, M. K. (2022). Predictive Models for Plant Disease Assessment. *Trends in Plant Disease Assessment*, 225-239.

Inglis, A., Parnell, A. C., Subramani, N. and Doohan, F. M. (2024). Machine Learning Applied to the Detection of Mycotoxin in Food: A Systematic Review. *Toxins*, 16(6): 268.

Inglis, A., Parnell, A. C., Subramani, N., & Doohan, F. M. (2024). Machine Learning Applied to the Detection of Mycotoxin in Food: A Systematic Review. *Toxins*, *16*(6), 268.

Ismaiel, A. A. and Papenbrock, J. (2015). Mycotoxins: producing fungi and mechanisms of phytotoxicity. *Agriculture*, 5(3): 492-537.

Jafar, A., Bibi, N., Naqvi, R. A., Sadeghi-Niaraki, A. and Jeong, D. (2024). Revolutionizing agriculture with artificial intelligence: plant disease detection methods, applications, and their limitations. *Frontiers in Plant Science*, 15: 1356260.

Jafar, A., Bibi, N., Naqvi, R. A., Sadeghi-Niaraki, A. and Jeong, D. (2024). Revolutionizing agriculture with artificial intelligence: plant disease detection methods, applications, and their limitations. *Frontiers in Plant Science*, 15:1356260.

Liakos, K. G., Busato, P., Moshou, D., Pearson, S. and Bochtis, D. (2018). Machine learning in agriculture: A review. *Sensors*, 18(8): 2674.

Liu, J. and Wang, X. (2021). Plant diseases and pests detection based on deep learning: a review. *Plant Methods*, 17: 1-18.

Liu, Y., & Wu, F. (2010). Global burden of aflatoxin-induced hepatocellular carcinoma: a risk assessment. *Environmental health perspectives*, 118(6): 818-824.

Marroquín-Cardona, A. G., Johnson, N. M., Phillips, T. D. and Hayes, A. W. (2014). Mycotoxins in a changing global environment–a review. *Food and Chemical Toxicology*, 69:220-230.

Mavrommatis, A., Giamouri, E., Tavrizelou, S., Zacharioudaki, M., Danezis, G., Simitzis, P. E. and Feggeros, K. (2021). Impact of mycotoxins on animals' oxidative status. *Antioxidants*, 10(2): 214.

Mesterházy, Á., Oláh, J. and Popp, J. (2020). Losses in the grain supply chain: Causes and solutions. *Sustainability*, 12(6): 2342.

Mitra, D. (2021). Emerging plant diseases: research status and challenges. *Emerging trends in plant pathology*, 1-17.

Mohan, S. S., Venkat, R., Rahaman, S., Vinayak, M. and Babu, B. H. (2023). Role of AI in agriculture: applications, limitations and challenges: A review. *Agricultural Reviews*, 44(2): 231-237.

Montanari, C., Bargossi, E., Gardini, A., Lanciotti, R., Magnani, R., Gardini, F. and Tabanelli, G. (2016). Correlation between volatile profiles of Italian fermented sausages and their size and starter culture. *Food Chemistry*, 192: 736-744.

Naik, B. N., Malmathanraj, R. and Palanisamy, P. (2022). Detection and classification of chilli leaf disease using a squeeze-and-excitation-based CNN model. *Ecological Informatics*, 69: 101663.

Ostry, V., Malir, F., Toman, J. and Grosse, Y. (2017). Mycotoxins as human carcinogens—the IARC Monographs classification. *Mycotoxin research*, 33: 65-73.

Patil, V. U., Vanishree, G., Pattanayak, D., Sharma, S., Bhardwaj, V., Singh, B. P. and Chakrabarti, S. K. (2017). Complete mitogenome mapping of potato late blight pathogen, Phytophthora infestans A2 mating type. *Mitochondrial DNA Part B*, 2(1): 90-91.

Purushotham, P., Rakhonde, G., Reddy, N. K., Nayak, A. M., Pravalika, K. M. and Sahana, G. (2024). Harvesting the Future: Navigating the Challenges and Limitations of AI in Agriculture. In *Artificial Intelligence and Smart Agriculture: Technology and Applications* (pp. 547-564). Singapore: Springer Nature Singapore.

Qiu, Z., Verma, J. P., Liu, H., Wang, J., Batista, B. D., Kaur, s. and Singh, B. K. (2022). Response of the plant core microbiome to Fusarium oxysporum infection and identification of the pathobiome. *Environmental Microbiology*, 24(10): 4652-4669.

Radivojevic, T., Costello, Z., Workman, K. and Garcia Martin, H. (2020). A machine learning Automated Recommendation Tool for synthetic biology. Nat. Commun. 11: 4879

Ramesh, K. V., Rakesh, V. and Rao, E. V. S. (2020). Application of big data analytics and artificial intelligence in agronomic research. *Indian Journal of Agronomy*, 65(4):383-395.

Rasel, R. I., Sultana, N. and Meesad, P. (2018). An application of data mining and machine learning for weather forecasting. In *Recent Advances in Information and Communication Technology 2017: Proceedings of the 13th International Conference on Computing and Information Technology (IC2IT)* (pp. 169-178). Springer International Publishing.

Sarker, I. H. (2021). Deep learning: a comprehensive overview on techniques, taxonomy, applications and research directions. *SN computer science*, 2(6): 420.

Shaikh, T. A., Rasool, T. and Lone, F. R. (2022). Towards leveraging the role of machine learning and artificial intelligence in precision agriculture and smart farming. *Computers and Electronics in Agriculture*, 198: 107119.

Shoaib, M., Hussain, T., Shah, B., Ullah, I., Shah, S. M., Ali, F. and Park, S. H. (2022). Deep learning-based segmentation and classification of leaf images for detection of tomato plant disease. *Frontiers in plant science*, 13: 1031748.

Slaughter, G. (2006). Current advances in biosensor design and fabrication. *Encyclopedia of Analytical Chemistry: Applications, Theory and Instrumentation*, 1-25.

Sparrow, R., Howard, M. and Degeling, C. (2021). Managing the risks of artificial intelligence in agriculture. *NJAS: Impact in Agricultural and Life Sciences*, 93(1): 172-196.

Steinkellner, H., Binaglia, M., Dall'Asta, C., Gutleb, A. C., Metzler, M., Oswald, I. P. and Alexander, J. (2019). Combined hazard assessment of mycotoxins and their modified forms applying relative potency factors: Zearalenone and T2/HT2 toxin. *Food and Chemical Toxicology*, 131: 110599.

Streit, E., Schwab, C., Sulyok, M., Naehrer, K., Krska, R. and Schatzmayr, G. (2013). Multi-mycotoxin screening reveals the occurrence of 139 different secondary metabolites in feed and feed ingredients. *Toxins*, 5(3): 504-523.

Tapia, M. S., Alzamora, S. M. and Chirife, J. (2020). Effects of water activity (aw) on microbial stability as a hurdle in food preservation. *Water activity in foods: Fundamentals and applications*, 323-355.

Thapa, S. and Thapa, S. (2019). Scope of value-addition in potato. *International journal of Horticulture, Agriculture and Food science*, 3: 132-465.

The World Health Organization (WHO) (2023), Food Safety. The World Health Organization, https://www.who.int/news-room/fact-sheets/detail/mycotoxins.

Umereweneza, D., Kamizikunze, T. and Muhizi, T. (2018). Assessment of mycotoxins types in some foodstuff consumed in Rwanda. *Food Control*, 85: 432-436.

Vila-Donat, P., Marín, S., Sanchis, V. and Ramos, A. J. (2018). A review of the mycotoxin adsorbing agents, with an emphasis on their multi-binding capacity, for animal feed decontamination. Food and chemical toxicology, 114: 246-259.

Wani, J. A., Sharma, S., Muzamil, M., Ahmed, S., Sharma, S. and Singh, S. (2022). Machine learning and deep learning based computational techniques in automatic agricultural diseases detection: Methodologies, applications, and challenges. *Archives of Computational methods in Engineering*, 29(1): 641-677.

Wu, F. (2015). Global impacts of aflatoxin in maize: trade and human health. *World Mycotoxin Journal*, 8(2): 137-142.

Yeasin, M., Haldar, D., Kumar, S., Paul, R. K. and Ghosh, S. (2022). Machine learning techniques for phenology assessment of sugarcane using conjunctive SAR and optical data. *Remote Sensing*, 14(14): 3249.

Zhang, P., Yang, L. and Li, D. (2020). EfficientNet-B4-Ranger: A novel method for greenhouse cucumber disease recognition under natural complex environment. *Computers and Electronics in Agriculture*, 176: 105652.

Zhang, S., Wu, X., You, Z. and Zhang, L. (2017). Leaf image based cucumber disease recognition using sparse representation classification. *Computers and electronics in agriculture*, 134:135-141.

Chapter 13

Biological Strategies to Manage Fungal Pathogens and Mycotoxins

Akurathi. Ramya[1*], ***Emmadi. Venu***[2], ***Sathiyaseelan. K***[2], ***Bhukya. Srinivas***[3], ***Yeluru. Mohan Babu***[2]

[1]*Department of Plant Pathology, Junagadh Agricultural University, Gujarat.*
[2]*Division of Plant Pathology, Indian Agricultural Research Institute, PUSA, New Delhi.*
[3]*Department of Plant Pathology, Professor Jayashankar Telangana State Agricultural University Hyderabad, Telangana.*
E-mail: ramyae0410@gmail.com

Abstract

The number of virulent phytopathogenic fungus is constantly increasing, resulting in large global economic losses. There are already a number of methods for quickly identifying and managing phytopathogenic fungus. Traditionally, phytopathogenic fungi were controlled with conventional and synthetic fungicides. Nevertheless, in addition to their detrimental effects, there has been a noticeable rise in the development of fungal resistance to these fungicides, which has prompted researchers to investigate a variety of novel mycotoxin and fungus control techniques around the globe. Mycotoxin contamination from mycotoxigenic fungi poses a major concern for food safety and quality is the presence of significant threats. The growing resistance to fungi and difficulties with traditional methods call for the creation of novel biological approaches for quick removal that have effect on quality. In this chapter we discuss the modern biological control strategies which are being applied to manage fungal pathogens for reducing mycotoxin in food include the use of Natural essential oils (NEOs) to control mycotoxins.

Key words: Phytopathogenic fungus, Mycotoxins, Control strategies, Biological control

Introduction

Commercial agriculture heavily relies on chemical fungicides to provide protection of crops from the pathogens of fungi by targeting also destroying their cell and spore structures. However, the widespread and repeated use of these fungicides often leads to their overuse, causing detrimental implications for organisms that are useful, human and animal well-being, along with the ecosystem (Youssef *et al.*, 2019). Additionally, the appearance of fungal strains that are resistant has made managing plant fungal diseases increasingly difficult. To address these issues, it is crucial to develop healthier, non-toxic, and eco-friendly alternatives to chemical fungicides. Infections (Muhammad et al., 2022). Several strategies are being explored, including microbial (Moreno-Gavíra et al., 2021) and botanical fungicides, (Yoon et al., 2013), biological control of phytopathogenic fungus, (Nega *et al.*, 2014) These approaches offer safer and more effective means of managing plant fungal infections while minimizing negative impacts.

Mycotoxins are a major worldwide worry as a result of their detrimental effects on food safety and their widespread presence in food crops. These low molecular weight toxic metabolites are produced by mycotoxigenic fungi, including species such as Aspergillus, Alternaria, Fusarium, and Penicillium. They can contaminate a variety of foods and feeds throughout the entire food chain, from pre-harvest through to postharvest and storage. This contamination poses a significant challenge to ensuring the safety and quality of food products (Emmanuel et al., 2020; Gavahian, Sheu, Magnani, & Mousavi Khaneghah, 2022).

Aflatoxins (AF) and ochratoxins (OT) are among the most toxic mycotoxins. Aflatoxin B1 (AFB1) is particularly notorious for its potent hepatocarcinogenic effects and is frequently present in cereals, grains, animal feed and nuts. When animals that are nursing consume contaminated feed, AFB1 and its derivative AFB2 can be metabolized into hydroxylated forms, AFM1 and AFM2. Ochratoxin A (OTA), which has both hepatotoxic and nephrotoxic properties, is frequently detected in dried fruit, wine, coffee, cereals, and grape juice. (Smith et al., 2016: Bangar et al., 2021). Given the significant health risks posed by these toxins, it is crucial to develop and implement effective strategies and methods to cut down on or get rid of mycotoxin contamination in food products.

Biological Control of Phytopathogenic Fungi

Bio-fumigation (Biological Soil Disinfection)

This control method involves incorporating fresh organic material into the soil and then covering it with plastic tarping (Blok *et al.*, 2000). The organic matter ferments beneath the plastic cover, generating anaerobic conditions. These conditions generate toxic metabolites that effectively inactivate phytopathogenic fungi, thereby controlling their spread.

Plants belonging to the Alliaceae family, which includes onions, produce toxic compounds that have a influence, either intangible or direct, on fungal plant pathogens. These toxic molecules play a crucial role in defending the plants by either directly inhibiting the growth of fungi or by affecting their ability to cause disease.

Degradation of tissues of onions and garlic releases volatile sulfur compounds, including zwiebelanes and thiosulfinates, that are subsequently transformed into disulfides biocidal that target fungal phytopathogens. (Arnault et al., 2004); Blok *et al.* (2000) demonstrated that incorporating recently harvested organic materials, such ryegrass or cabbage, into soil and following up with plastic tarping effectively controls phytopathogenic fungus found in soil like Fusarium oxysporum and *Rhizoctonia solani*. This approach offers a viable alternative to the prohibited disinfection with methyl bromide, which poses risks to respiratory and central nervous systems in humans. Recently, a study conducted in Egypt explored the using Indian mustard (*Brassica juncea*) for biofumigation manage *Rhizoctonia solani* infections in common beans. The research tested various forms of *Brassica juncea*, including both dried and damp vegetation, methanol extracts, and seed powder or meal, demonstrating its potential as a control measure (Abdallah et al., 2020).

The Role of Microbial Antagonists in Disease-Suppressive Soils

Suppressive soils, rich in antagonistic microorganisms, are particularly effective in controlling plant diseases, with fungi often being the dominant agents. For example, to manage due to *Phytophthora palmivora*-caused papaya foot rot, papaya seedlings were initially grown in virgin, limiting soil before being transferred to gaps in the contaminated planting soil (Nega et al., 2014). This method is particularly useful during the early growth stages of the plant when it is most vulnerable. Virgin soil, collected from areas never previously used for papaya cultivation, is typically free from *Phytophthora palmivora*. This strategy proved effective, as 42% of seedlings planted in infected soil without the addition of virgin soil perished within three months, whereas all seedlings planted with virgin soil survived. Similarly, suppressive soils have also been utilized to combat wilt disease caused by the phytopathogenic fungus *Fusarium oxysporu* (Alabouvette et al., 2009). Research into soils that inhibit Fusarium wilting from Chateaurenard has uncovered novel fungal and bacterial taxa that have a critical role in suppressing this disease (Siegel-Hertz et al., 2018).

In certain instances, maintaining single-crop monocultures in soil prone to pathogens can actually help minimize fungus in plants diseases subsequent to prolonged periods a serious outbreak (Kim et al., 2019). This is because, over time, the antagonistic microflora that naturally combats fungal pathogens can become more prevalent in such environments. For example, continuous monocultures of cucumber or wheat have been shown to decrease the incidence of diseases like cucumber damping-off and wheat take-all, both of which are caused by the fungus *Rhizoctonia*. This increase in beneficial microflora helps suppress the fungal pathogens, leading to a reduction in disease levels (Nega *et al.*, 2014).

Another effective method for controlling phytopathogenic fungi involves the use of soil suppressiveness through the cultivation of specific crops as soil amendments

(Inderbitzin *et al.*, 2018, Panth *et al.*, 2020). Cruciferous vegetables, in particular, play a significant role in this approach by enhancing the soil's resident antagonistic microflora, which helps to suppress pathogens. For example, incorporating broccoli into the soil has been shown to provide biocontrol against *Sclerotinia sclerotiorum*, the fungus responsible for lettuce drop. This practice leverages the natural ability of these crops to foster beneficial microorganisms that can protect other plants from disease (Agrios et al., 2005).

Controlling Plant Diseases: Microbial Strategies against Phytopathogenic Fungi

Plant fungal infections are controlled by microbial fungicides can be broadly categorized into two comprehensive types. The initially identified kind involves direct interaction with the intended fungi through diverse processes, including parasitic relationships, the process of of antibiotics that specifically target pathogens, or competition for resources such as food, water, and space. These fungicides effectively combat pathogens by holding the same place in the ecosystem. The second type works obliquely by inducing resistance in plants combating virulent infections. This inducer can be a less pathogenic variation of the same plant pathogen, a different microbial species, or natural products derived from these organisms. Both approaches leverage the power of microorganisms to manage plant fungal diseases in complementary ways.

Microbial antagonists, such as *Trichoderma* species, employ a variety of mechanisms to control plant fungal diseases. These beneficial fungi combat soil-borne phytopathogens through several strategies: they parasitize the pathogens directly, produce enzymes that break down fungal cell walls and antibiotic, and compete for essential nutrients like nitrogen and carbon. Additionally, *Trichoderma* species can stimulate plant growth by producing auxin-like compounds. This multifaceted approach makes them effective agents in biocontrol strategies (Vinale et al., 2008; Degani, et al., 2021).

Certain *Trichoderma species*, such as *Trichoderma harzianum*, are among the greatest commonly used Microbiological antagonists for the biocontrol of fungal infections in plants. They effectively combat pathogens like *Fusarium*, *Rhizoctonia*, and various phytopathogenic fungi that are found in soil (Harman et al., 2006). Another effective biocontrol agent is *Coniothyrium minitans*, targets infections in crops such as brassicas, lettuce, oilseed rape, beans, and carrots affected by *Sclerotinia sclerotiorum*. Additionally, the bacterium *Paenibacillus jamilae* HS-26 demonstrates strong antagonistic properties against numerous soil-borne fungal species by releasing the antifungal metabolites found outside of cells and also synthesizing Enzymes that hydrolyze (Wang et al., 2019). Furthermore, Actino 48 formulations of *Streptomyces cellulosae*, which generates chitinase, have been cited in manage *Sclerotium rolfsii*, a pathogen responsible for soil-borne diseases in peanuts (Abo-Zaid et al., 2021).

Innovative methods are being explored for plant disease management, including the use of honeybees in blueberry cultivation. During their pollination activities, bees can inadvertently carry the phytopathogenic fungus *Monilinia vaccinii-corymbosi*, which causes mummy blueberry disease, from one flower to another (Kim et al., 2019). However, bees also have the potential to act as 'flying doctors' by delivering bacterial fungicides, such as Bacillus subtilis, to the blueberry flowers, helps in control the disease development (Dedej *et al.*, 2004). The endophytic bacterium *Bacillus mojavensis* has been identified as a potent fungicide against several harmful fungal pathogens, consisting of *Fusarium oxysporum*, *Rhizoctonia solani*, and *Sclerotinia sclerotiorum*, further enhancing the potential for biological control in agriculture (Camele *et al.*, 2019).

Botanical-based Fungicides

Plants have developed a variety of antimicrobial agents to defend themselves against fungal infections. Certain organic fungicides, which include substances such as phytoanticipins and phytoalexins, vary widely in structural, weight on molecular basis, and function (Ribera et al., 2012). Among the Secondary metabolism products that plants generate that serve like organic fungicides are phenolics, fatty acids, Edris and Farrag (2003) demonstrated combining peppermint and sweet basil essential oily substances possess antifungal properties, effectively combating certain phytopathogenic fungi. Specifically, the key aroma components of these oils—menthol in peppermint oil and linalool/eugenol in basil oil—contribute to their ability to inhibit the development of fungus such as *Rhizopus stolonifer* and *Sclerotinia sclerotiorum*. Flavonoids, alkaloids, glycosides, terpenoids, and tannins. Each of these compounds plays a unique role in protecting plants from phytopathogenic fungi, reflecting the diverse strategies plants employ to safeguard themselves.

Milsana

The plant-based fungicide Milsana derived from the plant *Reynoutria sachalinensis* and extracted with ethanol, is utilized primarily as a preventative measure against powdery mildew in greenhouse-grown plants (Konstantinidou et al., 2006). It is especially effective when applied to young seedlings in glasshouses, reducing powdery mildew infection by approximately 97%. For optimal results, Milsana should be sprayed to the point of runoff 48 h before planting (Bokshi et al., 2008; Reglinski et al., 2009). Its efficacy stems from both its direct antifungal properties and its ability to induce plant resistance. Milsana functions as an organic inducer of phytoalexins, which are antibacterial substances that plants produce in response to pathogen attacks, thus enhancing the plant's natural immune system and providing a robust defense against the disease (Copping et al., 2004).

Jojoba Oil

Jojoba oil, taken out of the beans of the jojoba plant, has proven to be an effective fungicide for managing powdery mildew on grapevines and decorative landscaping

(Copping et al., 2004; Abdu-Allah et al., 2017). The reliability of it even at high temperatures, makes it a versatile solution for various climatic conditions (Copping et al., 2004; Abdu-Allah et al., 2017). Typically, jojoba oil is applied as a spray at a final concentration of approximately 1%, demonstrating its broad functionality and reliability as a botanical fungicide (Copping et al., 2004).

Essential Oils from Plants

Edris and Farrag (2003) demonstrated peppermint and sweet basil essential oily substances possess antifungal properties, effectively combating certain phytopathogenic fungi. Specifically, the key aroma components of these oils-menthol in peppermint oil and linalool/eugenol in basil oil—contribute to their capacity to impede the development of fungi such as *Rhizopus stolonifer* and *Sclerotinia sclerotiorum*.

Calocedrus macrolepis essential oil, along with its individual elements, has demonstrated notable antimicrobial action against a number of harmful fungi. Specifically, it has been efficient at preventing the development of *Fusarium oxysporum, Fusarium solani, Rhizoctonia solani, and Colletotrichum gloeosporioides* and *Pestalotiopsis funerea* (Chang et al., 2008).

Cinnamaldehyde

Considering that cinnamaldehyde was first isolated through cinnamon essential oil in 1834 (Dumas *et al.,* 1834). It had been produced chemically in order to use in crop defense. This compound essentially combats various fungus that are phytopathogenic, including *Sclerotinia homoeocarpa*, which causes dollar spot in turfgrasses; *Verticillium fungicola*, responsible for dry bubble disease in *Agaricus bisporus*; and *Fusarium moniliforme*, which leads to pitch canker infections. Additionally, cinnamon has been reported to possess antifungal properties against fungi that cause wood decay (Wang et al., 2005).

Cinnamaldehyde has been demonstrated to have significant antifungal properties, particularly against wood-rot fungi (Cheng et al., 2008). Research by Khan *et al.* (2011) highlighted cinnamaldehyde as the most effective antifungal agent among several essential oil components, including citral, eugenol, and geraniol. Cinnamaldehyde exhibited superior inhibitory effects on the growth of *Trichophyton rubrum* and *Aspergillus fumigatus*. Additionally, it adversely impacted on the virulence factors and ultrastructure of fungal hyphae, further underscoring its potent antifungal activity.

Cinnamaldehyde, when composed as a Wettable powder is marketed extensively through the trade names VERTIGO and Proguard. As such, combination is designed to be water-resistant, ensuring that it does not dissolve in water but can still be effectively used as a fungicide. Additionally, cinnamaldehyde in this form is quickly degraded in soil, minimizing environmental impact and posing no significant risk with relation to an unintended recipient creature.

Phenolic Compounds

Investigation has shown that the phenolic compounds produced by medicinal plants have antifungal properties, which are categorized into anthraquinones, flavonoids, coumarins, and simple phenolics (Pizzolitto et al., 2015). Among these, curcuminoids-polyphenolic pigments found in *Curcuma longa* rhizomes-have demonstrated notable anti-fungus activity in various *in vivo* tests (Yoon et al., 2013). Extracted from the rhizomes of *Curcuma longa* using methanol, these curcuminoids include curcumin, bisdemethoxycurcumin, and demethoxycurcumin. Of these, demethoxycurcumin has proven to be the most effective against late blight in tomatoes and blast in rice, followed by curcumin. Additionally, curcuminoids from *Curcuma longa* have shown the ability to inhibit germination of spores and mycelial development in *Colletotrichum coccodes*, the fungus responsible for anthracnose in red pepper.

Alkaloid Compounds

Plants produce alkaloids had been identified as effective fungicides against Plant-pathogenic fungi. For instance, a piperidine alkaloid isolated from *Piper longum* has demonstrated notable The fungicidal action of certain pathogenic fungus (Lee et al., 2001). Furthermore, two other alkaloids extracted from the seeds of *Chimonanthus praecox* (Japanese allspice) exhibit significant fungicide properties against five different Plant-pathogenic fungi (Zhang et al., 2009).

Various natural compounds have demonstrated promising antifungal properties. Allosecurinine, found in the root of *Phyllanthus amarus*, has shown effective fungicidal activity *in vitro*, specifically inhibiting the germination of fungal spores from five different phytopathogenic fungi. Additionally, a dihydropyrrole that *Datura metel* produced has exhibited significant fungicidal *in vitro* activity effects against several *Aspergillus* species, including *Aspergillus fumigatus*, *A. flavus*, and *A. niger*. These findings suggest that these natural compounds could be potential candidates for developing new antifungal treatments.

Impact of Arbuscular Mycorrhizal Fungi on Plant Defense Mechanisms against Fungal Diseases

Symbiotic interactions are formed between plant roots and arbuscular mycorrhizal fungus, benefiting a wide range of horticultural, agricultural, and hardwood crops. These fungi enhance plant resistance to soil-borne pathogens, though their impact on shoot diseases can vary depending on the pathogen's strategies (Pozo et al., 2007). During mycorrhizal formation, plant defense mechanisms are modulated, potentially through interactions between routes for salicylic acid and jasmonate signaling. This modulation helps plants better prepare for and activate their defense responses (Jung et al., 2012). For example, *Glomus mosseae*, an arbuscular mycorrhizal fungus has been shown to stimulate both resistance both locally and systemically against *Phytophthora parasitica* in tomato roots (Pozo et al., 2002). Locally, this fungus stimulates the production of antimicrobial enzymes like beta-1,3-glucanase, chitosanase, and chitinase. Systemically, it enhances root protein extracts' ability to break down the pathogen's cell membrane by lysis, effectively reducing disease symptoms through these combined mechanisms.

Controlling Phytopathogenic Fungi: The Effects of Homeopathy and Herbal Flower Teas

Diseases in living things are treated with homeopathy by believed to arise from suppressive processes that disrupt the vital principle or life force. When these suppressive forces are introduced, they trigger symptoms by getting rid of factors that affect the essential balance (Castro et al., 2013). This imbalance in essential force can lead to ailments in plants, ultimately therefore producing reduced yields or even death.

Chamomile tea has been effectively accustomed to manage damping-off disease in seedlings, which is brought resulting from various fungi that cause disease. The blooms of the chamomile plant are rich in sulfur, a natural fungicide. Elemental sulfur has a long history of use as an organic fungicide, known for its ability to kill fungal spores and prevent fungal diseases (Sokovic et al., 2013).

Managing the Virulence of Phytopathogenic Fungi using Mycoviruses

Mycoviruses, which infect fungi, hold potential for controlling fungal diseases by inducing hypovirulence. Despite this potential, the survival of strains of mycovirus in the field remains uncertain, and no mycovirus has yet been used in practical crop protection. Zhang *et al.* (2020) explored this potential by using a mycovirus to transform the pathogenic fungus *Sclerotinia sclerotiorum* into a beneficial endophyte. Their research demonstrated that the mycovirus reduces the expression of key genes associated with pathogenic factors in *S. sclerotiorum* while an infection is present. This finding suggests that mycoviruses could influence the development of endophytism. The study proposes a novel disease control strategy that involves using mycovirus-infected fungal strains to improve plant well-being and to introduce enter the realm of mycoviruses. Furthermore, strains of mycovirus that were shown to modulate plant transcription of genes related to defense, circadian rhythm, and hormone signaling pathways, which in turn enhances plant development and resistance to illness.

Polyphenols and flavonoids

Phytonutrients, particularly both flavonoids and polyphenols, have gained significant attention inside food chains as a result of their beneficial biological activities, such as the characteristics that are anti-inflammatory, antiviral, antibacterial, and antioxidant (Mehany et al., 2021). These compounds interact with mycotoxins through various mechanisms. Firstly, their capacity for antioxidants and lipophilicity are crucial to their bioactivity. Secondly, they inhibit mycotoxin production by altering the fungal membrane structure. Thirdly, they downregulate genes involved in mycotoxin synthesis. Finally, they inhibit enzymatic activities related to mycotoxin production (Ahmed et al., 2022). Hamad et al. (2021) highlighted the utilization of organic plant extracts together with polyphenols in them as effective strategies for controlling proliferation of fungi and the formation of mycotoxin. For instance, polyphenols such as umbelliferone and quercetin have been shown to down-regulate genes involved in patulin biosynthesis, while chlorogenic and gallic acids have demonstrated inhibitory effects on AFB1 in edible beans (Sanzani et al., 2009; Telles et al., 2017). Additionally, flavanones like naringin, hesperetin glycoside, neohesperidin, and hesperidin, derived from citrus leftovers, achieved up to a 95% reduction in patulin accumulation (Salas et al., 2012). Furthermore, Makhuvele et al. (2020) found that phenols with bioactive properties combined with β-cyclodextrin in a nanosponge technique showed prospective capacity in order to prevent fungal invasion and eliminate mycotoxin. Overall, these studies suggest that polyphenols and dietary fiber interventions hold promises for mitigating the detrimental consequences of fungal mycotoxins in the future (Sharma & Patial, 2021).

Natural essential oils

The beneficial compounds found in natural essential oils (NEOs) are gaining recognition as eco-friendly and effective alternatives for combating fungus development and mycotoxin accumulation in food products. (Chaudhari et al., 2019; Mutlu-Ingok, Devecioglu, Dikmetas, Karbancioglu-Guler, & Capanoglu, 2020). These oils, such as those derived from *Origanum majorana* L., turmeric, garlic, wild oregano, sage, and mint, demonstrate significant antifungal and mycotoxin inhibitory properties, with minimal risk of drug resistance. (Tian et al., 2011). Their mechanisms of action involve several pathways: influencing fungal gene expression, disrupting cell walls and membranes, and directly damaging cellular structures (Cai et al., 2022). For instance, *Origanum majorana* L. essential oil, when encapsulated in nanoemulsion, has shown promise as a novel food preservative by effectively inhibiting fungal growth, AFB1 mycotoxin, and exhibiting antioxidant activity. Capsaicin has been reported to reduce OTA production by *Aspergillus species* by up to 78.1% and 61.5%, respectively, while essential oils from garlic and wild oregano have completely prevented OTA production. Additionally, turmeric oil has shown high antifungal actions against *A. flavus* and aflatoxin pollutants found in corn, and *Mentha spicata* essential oil has been effective in inhibiting toxin production in chickpeas for up to a year. Vital oils sourced from Palmarosa, eucalyptus, lemon, and grapefruit have also been reported to influence zearalenone (ZEN) levels, with the whole essential oil of *Curcuma longa* repressing *Fusarium graminearum* growth and reducing zearalenone levels too little concentrations. These findings highlight the potential of NEOs as effective and sustainable solutions for improving food safety (Naveen Kumar et al., 2016; Perczak et al., 2016).

Conclusion

Recent scientific literature highlights a range of innovative and environmentally friendly strategies for controlling phytopathogenic fungi that offer promising alternatives to traditional chemical fungicides. These biological strategies include biological and microbial control methods. Although some of these technologies are still in development, they show great promise for protecting human and animal health, preserving soil biodiversity, and mitigating the adverse environmental impacts of chemical fungicides. As these methods advance, they could revolutionize agricultural practices by providing safer and more sustainable solutions for managing plant fungal diseases.

References

Abdallah, I., Yehia, R., & Kandil, M. A. H. (2020). Biofumigation potential of Indian mustard (Brassica juncea) to manage Rhizoctonia solani. *Egyptian Journal of Biological Pest Control*, 30(1), 99.

Abdu-Allah, G. A., & Abo-Elyousr, K. A. (2017). Effect of certain plant extracts and fungicides against powdery mildew disease of Grapevines in Upper Egypt. *Archives of Phytopathology and Plant Protection*, 50(19-20), 957-969.

Abo-Zaid, G., Abdelkhalek, A., Matar, S., Darwish, M., & Abdel-Gayed, M. (2021). Application of bio-friendly formulations of chitinase-producing Streptomyces cellulosae Actino 48 for controlling peanut soil-borne diseases caused by Sclerotium rolfsii. *Journal of Fungi*, 7(3), 167.

Agrios, G.N. Plant disease epidemiology (Chapter 8). In Plant Pathololology, 5th ed.; Agrios, G.N., Ed.; Elsevier: Burlington, NJ, USA, 2005; pp. 265–291.

Ahmed, O. S., Tardif, C., Rouger, C., Atanasova, V., Richard-Forget, F., & Waffo-Téguo, P. (2022). Naturally occurring phenolic compounds as promising antimycotoxin agents: Where are we now?. *Comprehensive reviews in food science and food safety*, 21(2), 1161-1197.

Alabouvette, C., Olivain, C., Migheli, Q., & Steinberg, C. (2009). Microbiological control of soil-borne phytopathogenic fungi with special emphasis on wilt-inducing Fusarium oxysporum. *New Phytologist*, 184(3), 529-544.

Arnault, I., Mondy, N., Diwo, S., & Auger, J. (2004). Soil behaviour of sulfur natural fumigants used as methyl bromide substitutes. *International Journal of Environmental Analytical Chemistry*, 84(1-3), 75-82.

Bangar, S. P., Sharma, N., Kumar, M., Ozogul, F., Purewal, S. S., & Trif, M. (2021). Recent developments in applications of lactic acid bacteria against mycotoxin production and fungal contamination. *Food Bioscience*, 44, 101444.

Bokshi, A. I., Jobling, J., & McConchie, R. (2008). A single application of Milsana® followed by Bion® assists in the control of powdery mildew in cucumber and helps overcome yield losses. *The Journal of Horticultural Science and Biotechnology*, 83(6), 701-706.

Cai, J., Yan, R., Shi, J., Chen, J., Long, M., Wu, W., & Kuca, K. (2022). Antifungal and mycotoxin detoxification ability of essential oils: A review. *Phytotherapy Research*, 36(1), 62-72.

Camele, I., Elshafie, H. S., Caputo, L., Sakr, S. H., & De Feo, V. (2019). *Bacillus mojavensis*: Biofilm formation and biochemical investigation of its bioactive metabolites. *Journal of Biological Research-Bollettino della Società Italiana di Biologia Sperimentale, 92(1).*

Castro, D. M. (2013, September). Homeopathy: principles and applications. In *International Conference on Homeopathy in Agriculture* (Vol. 2).

Chang, H. T., Cheng, Y. H., Wu, C. L., Chang, S. T., Chang, T. T., & Su, Y. C. (2008). Antifungal activity of essential oil and its constituents from Calocedrus macrolepis var. formosana Florin leaf against plant pathogenic fungi. *Bioresource technology*, 99(14), 6266-6270.

Chaudhari, A. K., Dwivedy, A. K., Singh, V. K., Das, S., Singh, A., & Dubey, N. K. (2019). Essential oils and their bioactive compounds as green preservatives against fungal and mycotoxin contamination of food commodities with special reference to their nanoencapsulation. *Environmental Science and Pollution Research*, 26, 25414-25431.

Cheng, S. S., Liu, J. Y., Chang, E. H., & Chang, S. T. (2008). Antifungal activity of cinnamaldehyde and eugenol congeners against wood-rot fungi. *Bioresource technology*, 99(11), 5145-5149.

Copping, L. G. (2004). The manual of biocontrol agents. *Hampshire*, 702.

Dedej, S., Delaplane, K. S., & Scherm, H. (2004). Effectiveness of honey bees in delivering the biocontrol agent Bacillus subtilis to blueberry flowers to suppress mummy berry disease. *Biological Control*, 31(3), 422-427.

Degani, O., Rabinovitz, O., Becher, P., Gordani, A., & Chen, A. (2021). Trichoderma longibrachiatum and Trichoderma asperellum confer growth promotion and protection against late wilt disease in the field. *Journal of Fungi*, 7(6), 444.

Dumas, M.M.J.; Peligot, E. Recherches de chimie organique. – Sur l'huile de cannelle, l'acid hippurique et l'acide sebacique. Ann. Chim. Phys., 1834, 57, 305-334.

Edris, A. E., & Farrag, E. S. (2003). Antifungal activity of peppermint and sweet basil essential oils and their major aroma constituents on some plant pathogenic fungi from the vapor phase. *Food/Nahrung*, 47(2), 117-121.

Gavahian, M., Sheu, S. C., Magnani, M., & Mousavi Khaneghah, A. (2022). Emerging technologies for mycotoxins removal from foods: Recent advances, roles in sustainable food consumption, and strategies for industrial applications. *Journal of Food Processing and Preservation*, 46(10), e15922.

Hamad, G. M., Mohdaly, A. A. A., El-Nogoumy, B. A., Ramadan, M. F., Hassan, S. A., & Zeitoun, A. M. (2021). Detoxification of aflatoxin B1 and ochratoxin A using

Salvia farinacea and Azadirachta indica water extract and application in meat products. *Applied Biochemistry and Biotechnology*, 193(10), 3098-3120.

Harman, G. E. (2006). Overview of Mechanisms and Uses of Trichoderma spp. *Phytopathology*, 96(2), 190-194.

Inderbitzin, P., Ward, J., Barbella, A., Solares, N., Izyumin, D., Burman, P., ... & Subbarao, K. V. (2018). Soil microbiomes associated with Verticillium wilt-suppressive broccoli and chitin amendments are enriched with potential biocontrol agents. *Phytopathology*, 108(1), 31-43.

Jung, S. C., Martinez-Medina, A., Lopez-Raez, J. A., & Pozo, M. J. (2012). Mycorrhiza-induced resistance and priming of plant defenses. *Journal of chemical ecology*, 38, 651-664.

Khan, M. S. A., & Ahmad, I. (2011). In vitro antifungal, anti-elastase and anti-keratinase activity of essential oils of Cinnamomum-, Syzygium-and Cymbopogon-species against Aspergillus fumigatus and Trichophyton rubrum. *Phytomedicine*, 19(1), 48-55.

Kim, D. R., Cho, G., Jeon, C. W., Weller, D. M., Thomashow, L. S., Paulitz, T. C., & Kwak, Y. S. (2019). A mutualistic interaction between *Streptomyces* bacteria, strawberry plants and pollinating bees. *Nature Communications*, 10(1), 4802.

Kim, D. R., Jeon, C. W., Shin, J. H., Weller, D. M., Thomashow, L., & Kwak, Y. S. (2019). Function and distribution of a lantipeptide in strawberry Fusarium wilt disease–suppressive soils. *Molecular Plant-Microbe Interactions*, 32(3), 306-312.

Konstantinidou-Doltsinis, S., Markellou, E., Kasselaki, A. M., Fanouraki, M. N., Koumaki, C. M., Schmitt, A., ... & Malathrakis, N. E. (2006). Efficacy of Milsana®, a formulated plant extract from *Reynoutria sachalinensis*, against powdery mildew of tomato (*Leveillula taurica*). *Biocontrol*, 51, 375-392.

Lee, S. E., Park, B. S., Kim, M. K., Choi, W. S., Kim, H. T., Cho, K. Y., ... & Lee, H. S. (2001). Fungicidal activity of pipernonaline, a piperidine alkaloid derived from long pepper, Piper longum L., against phytopathogenic fungi. *Crop Protection*, 20(6), 523-528.

Makhuvele, R., Naidu, K., Gbashi, S., Thipe, V. C., Adebo, O. A., & Njobeh, P. B. (2020). The use of plant extracts and their phytochemicals for control of toxigenic fungi and mycotoxins. *Heliyon*, 6(10).

Mehany, T., Khalifa, I., Barakat, H., Althwab, S. A., Alharbi, Y. M., & El-Sohaimy, S. (2021). Polyphenols as promising biologically active substances for preventing SARS-CoV-2: A review with research evidence and underlying mechanisms. *Food Bioscience*, 40, 100891.

Moreno-Gavíra, A., Diánez, F., Sánchez-Montesinos, B., & Santos, M. (2021). Biocontrol effects of Paecilomyces variotii against fungal plant diseases. *Journal of Fungi*, 7(6), 415.

Muhammad, M., Abdul Wahab, R., Huyop, F. Z., Rusli, M. H., Syed Yaacob, S. N., & Teo, H. L. (2022). An overview of the potential role of microbial metabolites as greener fungicides for future sustainable plant diseases management. *Journal of Crop Protection, 11*(1), 1-27.

Nega, A. (2014). Review on concepts in biological control of plant pathogens. *Journal of Biology, Agriculture and Healthcare*, 4(27), 33-54.

Panth, M., Hassler, S. C., & Baysal-Gurel, F. (2020). Methods for management of soilborne diseases in crop production. *Agriculture*, 10(1), 16.

Pietrzak, K., Twarużek, M., Czyżowska, A., Kosicki, R., & Gutarowska, B. (2015). Influence of silver nanoparticles on metabolism and toxicity of moulds. *Acta Biochimica Polonica,* 62(4), 851-857.

Pizzolitto, R. P., Barberis, C. L., Dambolena, J. S., Herrera, J. M., Zunino, M. P., Magnoli, C. E., ... & Dalcero, A. M. (2015). Inhibitory effect of natural phenolic compounds on Aspergillus parasiticus growth. *Journal of Chemistry*, 2015(1), 547925.

Pozo, M. J., & Azcón-Aguilar, C. (2007). Unraveling mycorrhiza-induced resistance. *Current opinion in plant biology*, 10(4), 393-398.

Pozo, M. J., Cordier, C., Dumas-Gaudot, E., Gianinazzi, S., Barea, J. M., & Azcón-Aguilar, C. (2002). Localized versus systemic effect of arbuscular mycorrhizal fungi on defence responses to Phytophthora infection in tomato plants. *Journal of experimental botany*, 53(368), 525-534.

Reglinski, T., & Walters, D. (2009). Induced resistance for plant disease control. In *Disease control in crops.*

Ribera, A. E., & Zuñiga, G. (2012). Induced plant secondary metabolites for phytopathogenic fungi control: a review. *Journal of soil science and plant nutrition*, 12(4), 893-911.

Salas, M. P., Reynoso, C. M., Céliz, G., Daz, M., & Resnik, S. L. (2012). Efficacy of flavanones obtained from citrus residues to prevent patulin contamination. *Food Research International*, 48(2), 930-934.

Sanzani, S. M., Schena, L., Nigro, F., De Girolamo, A., & Ippolito, A. (2009). Effect of quercetin and umbelliferone on the transcript level of Penicillium expansum genes involved in patulin biosynthesis. *European Journal of Plant Pathology*, 125, 223-233.

Sharma, V., & Patial, V. (2021). Food mycotoxins: Dietary interventions implicated in the prevention of mycotoxicosis. *ACS Food Science & Technology*, 1(10), 1717-1739.

Siegel-Hertz, K., Edel-Hermann, V., Chapelle, E., Terrat, S., Raaijmakers, J. M., & Steinberg, C. (2018). Comparative microbiome analysis of a Fusarium wilt suppressive soil and a Fusarium wilt conducive soil from the Châteaurenard region. *Frontiers in Microbiology*, 9, 568.

Smith, M. C., Madec, S., Coton, E., & Hymery, N. (2016). Natural co-occurrence of mycotoxins in foods and feeds and their in vitro combined toxicological effects. *Toxins*, 8(4), 94.

Soković, M. D., Glamočlija, J. M., & Ćirić, A. D. (2013). Natural products from plants and fungi as fungicides. *Fungicides-showcases of integrated plant disease management from around the world*, 185-232.

Telles, A. C., Kupski, L., & Furlong, E. B. (2017). Phenolic compound in beans as protection against mycotoxins. Food Chemistry, 214, 293-299.

Tian, J., Ban, X., Zeng, H., He, J., Huang, B., & Wang, Y. (2011). Chemical composition and antifungal activity of essential oil from *Cicuta virosa* L. var. *latisecta Celak*. *International Journal of Food Microbiology*, 145(2-3), 464-470.

Vinale, F., Sivasithamparam, K., Ghisalberti, E. L., Marra, R., Woo, S. L., & Lorito, M. (2008). Trichoderma–plant–pathogen interactions. *Soil biology and Biochemistry*, 40(1), 1-10..

Wang, S. Y., Chen, P. F., & Chang, S. T. (2005). Antifungal activities of essential oils and their constituents from indigenous cinnamon (*Cinnamomum osmophloeum*) leaves against wood decay fungi. *Bioresource technology*, 96(7), 813-818.

Wang, X., Li, Q., Sui, J., Zhang, J., Liu, Z., Du, J., ... & Liu, X. (2019). Isolation and Characterization of Antagonistic Bacteria Paenibacillus jamilae HS-26 and Their Effects on Plant Growth. *BioMed Research International*, 2019(1), 3638926.

Yoon, M. Y., Cha, B., & Kim, J. C. (2013). Recent trends in studies on botanical fungicides in agriculture. *The plant pathology journal*, 29(1), 1.

Zhang, J. W., Gao, J. M., Xu, T., Zhang, X. C., Ma, Y. T., Jarussophon, S., & Konishi, Y. (2009). Antifungal activity of alkaloids from the seeds of Chimonanthus praecox. *Chemistry & biodiversity*, 6(6), 838-845.

Zhang, Y., Li, M., Liu, Y., Guan, E., & Bian, K. (2020). Reduction of aflatoxin B1 in corn by water-assisted microwaves treatment and its effects on corn quality. *Toxins*, 12(9), 605.

Chapter 14

Post- Harvest Management of Mycotoxins and Fungal Pathogens

***Priyanka Bhardwaj**[*1], **Nivedita Singh**[1], **Deepika Sharma**[1], **Kumud Jarial**[2] and **RS Jarial**[2]*

[1] *Department of Plant Pathology, College of Horticulture, Nauni, Solan (HP) -173230 India*

[2]*Department of Plant Pathology, College of Horticulture and Forestry, Neri, Hamirpur (HP) -177001 India*

**Corresponding Author Email id: bhardwajpriyanka939@gmail.com*

Abstract

Postharvest diseases reduce the quality as well as quantity of agriculture commodities. In case of fruits and vegetables, due to firm texture, various microorganisms attack the fruit during ripening stages and cause losses. In some cases, as for latent infections, inoculation occurs prior to harvest but the disease does not develop until the postharvest period. The major post-harvest pathogens include *Aspergillus, Alternaria, Fusarium, Penicillium, Colletotrichum, Botrytis, etc.* and most of these pathogens produce secondary metabolites which are toxic to the plant host and aid in spreading the disease. These metabolites are called mycotoxins. Ergotism in the past is the classical example. Important mycotoxins such as aflatoxins, orchratoxins, citrinin, fumonisins, zearalenone and trichothecenes contaminate agricultural commodities. It is practically impossible to completely eliminate or exclude mycotoxin contamination, but integrated management may help to reduce the potential effect. Mycotoxins can be managed by physical (include sorting, processing, storage, radiation), biological (biological agents such as *Trichoderma*), chemical (bases, chitosan, ozone treatment) and enzymatic detoxification (Nano- particles, plant extracts) etc. Physical technique like removal of damaged or incomplete kernels/seeds, drying can be done. Antifungal agents like propionic acid and acetic acid can be used. Mycotoxins are not destroyed during food manufacturing; thus, thermal inactivation is a good choice. Fungal pathogens can be managed to a certain extent by physical treatment of the fruit, certain chemicals, plant extract, essential oils, biological treatment etc. To ensure food and feed safety, it's crucial to take post - harvest steps, such as proper storage, detecting and cleaning up contamination and continuous monitoring during processing and marketing of agricultural products.

Keywords: Post harvest rots, mycotoxins, Fungus, *Fusarium*, *Aspergillus*, management

Introduction

The Greek word "mykes" refers to "fungi" and the Latin word "toxicum" means "poison" (Pittet et al., 1998). The term mycotoxin was introduced in 1962 following an unusual veterinary crisis near London, England, where around 1,00,000 turkeys died. This "turkey X disease" was traced back to a peanut meal contaminated with secondary metabolites from *Aspergillus flavus* called aflatoxins (Bennet and Klich, 2003). Mycotoxins are the toxins produced by fungi. These are secondary metabolites belonging to the phylum Ascomycota and fungi such as *Aspergillus, Fusarium, Penicillium* produce them (Umereweneza et al., 2018). According to Bennett (1987), these metabolites form a toxic and chemically diverse group, categorized together primarily because they can induce disease and death in humans and other vertebrates. It is not surprising that many mycotoxins exhibit overlapping toxicities across invertebrates, plants, and microorganisms. Not all the toxic compounds produced by fungi called mycotoxins. The target and the concentration of the metabolite is important. Mycotoxins are substances produced by fungi that are toxic to vertebrates and other animal groups, even in low concentrations (Bennett, 1987). Mycotoxin contamination poses a worldwide threat to various agriculture commodities including cereals, lentils, fruits, peanuts, almonds, walnuts, pistachios, coffee, cotton seeds and spices etc. (Pleadin et al., 2019). Mycotoxins can contaminate food at different stages throughout the food chain (Bennet and Klich, 2003). The occurrence of mycotoxins can begin as early as the field stage or harvesting, handling, storage and processing (Neme and Mohammad, 2017). Mycotoxins are harmful to animal and human health as they cause cancer, allergies and organ toxicity (Alshannaq and Yu 2017).

While determination of the exact extent is possible, it is suggested that mycotoxin contamination leads to considerable economic losses, along with posing risks to food safety and public health (Oguz, 2017). Fungal pathogens causing postharvest contributes significantly to the postharvest losses of fruits. However, mycotoxins produced compromise the safety and quality of fruits (Godana et al., 2023). However, according to Streit et al. (2013), around 25 per cent of global grain products are contaminated with mycotoxins. It has been reported the loss of roughly 1.3 billion metric tons of food each year due to mycotoxin contamination, representing about one-third of the world's total food production (Mesterhazy et al., 2020). Storage fungi, mainly *Penicillium, Aspergillus* and *Fusarium* causes 25-40 per cent losses of cereal grains (Kumar et al., 2007). Most developing countries are in the world's tropical zones and are subjected to monsoons and high temperature and humidity levels, which contribute to large postharvest crop losses

(Wild et al., 2015). The key postharvest factors that contribute to grain mycotoxin contamination include mechanical injury, insect infestation, timing of harvesting, drying methods, types of storage structures and conditions as well as handling and processing practices (Neme and Mohammad, 2017).

Some notable classes of mycotoxins are aflatoxins, ochratoxin, Alternaria toxins, alternariol, monomethylether, tentoxin, fumonisins, trichothecenes, deoxynivalenol, nivalenol, citrinin, zearalenone, ergot alkaloids etc. Aflatoxins are produced by fungi such as *Aspergillus flavus* and *Aspergillus parasiticus*, typically thriving in hot and humid environments found in tropical and subtropical regions (Frisavad et al., 2019). Ochratoxins are produced by *Aspergillus ochraeus, A. niger, A. alliaceus* and *A. glaucus*. These affect cereals, cheese, grapes, coffee, cocoa and their by-products (Malir et al., 2016). Fumonisins are produced by *Fusarium moniliforme, F. verticilioides* and *F. oxysporum*. They contaminate corn and grapes and cereals (Marin et al., 2013). Trichothecenes are synthesized by *Fusarium* and *Trichothecia* are typically found in cereals, such as oats, corn, barley etc. (Bhat et al., 2010). *Fusarium* species such as *F. moniliforme, F. verticillioides, F. oxysporum* and *F. equiseti* produce zearalenone. These contaminate wheat, rice, corn and oats. *Penicillium* species such as *P. expansum, P. crustosum, P. patulum* and *A. clavatus* produce Patulin and citrinin which contaminates apples and their products (Tapia et al., 2020). *Claviceps* sp. produces ergot alkaloids (Ukwuru et al., 2017). Aflatoxins contaminate one-fourth of the world's food grains, affecting 4.5 billion lives in developing nations (Dhakal et al., 2022).

Mycotoxin contamination in agricultural products results in the rejection of exports, which poses a significant economic concern (Samyal and Sharma, 2023). Some mycotoxins are heat stable during food manufacturing; thus, heat methods may not effectively reduce their levels (Peng et al., 2018). Postharvest mitigation strategies are crucial and cost-effective for controlling mycotoxin contamination. Post-harvest contamination of crops presents a major challenge and is often associated with improper storage conditions, such as insect activity and excessive moisture (Amaike et al., 2011). For example, patulin is absent in undamaged fruit; it is the damage to the fruit's surface that allows *Penicillium* to infect it (Sewram et al., 2000). Aflatoxin contamination of crops is common in the field before harvesting and is frequently associated with drought stress (Diener et al., 1987). In most cases, the key factors in storage are the moisture content of the substrate and the relative humidity of the surrounding environment (Detroy et al., 1971). Ochratoxin is primarily concentrated in the husks of grains, and removing the pericarp, or outer layer, from grains can reduce the OTA concentration by at least 50 per cent (Petzinger et al., 2000). Factors like relative humidity, moisture and temperature that influence Fusarium head blight development also affect deoxynivalenol accumulation (Wegulo et al., 2012). To prevent its production, storing grains at moisture content below 14 per cent and controlling insect activity are key strategies (Awad et al., 2013). Therefore, the key stage for controlling fruit quality is when the fruit enters the processing line. Key interventions include rapid and proper drying, postharvest insect control, adequate transportation and packaging, optimal storage conditions, and the use of natural and chemical agents as well as irradiation. Additionally, grain processing techniques such as sorting, cleaning, milling, fermentation, baking, roasting, flaking, and extrusion cooking have also been reported to reduce mycotoxin levels (Neme and Mohammad, 2017).

Mycotoxin contamination can lead to significant economic losses for farmers, food producers and the entire agricultural sector. Contaminated crops often have to be destroyed or sold at reduced prices, resulting in financial losses. Moreover, costs associated with testing, monitoring, and implementing control measures add to the economic burden. Keeping in view the economic importance of mycotoxins and the losses caused by them, the present chapter focuses on the main strategies for the managing the fungal pathogens and their mycotoxins in agricultural commodities. For mycotoxins to occur in the produce after harvest the fungal growth in the commodity is a pre- requisite.

Management strategies

The steps to avoid the hazardous effects of mycotoxin contamination in the agricultural commodities like food and beverages, animal feed etc., include:

(i) Prevention of contamination in before and after harvesting

(ii) Removal of contaminated food material

(iii) Detoxification of contaminated material

(iv) Timely and effective treatment of the affected individuals

The first step, Prevention starts right at the field level and includes the fungal pathogen from infecting and invading the crop during the growth period and storage and transport. It includes all the disease management practices utilized during the crop cultivation. Crop rotation, soil tillage, use of chemical fungicides, biopesticides, resistant varieties, etc., are some of the majorly used methods in the field. This step is crucial as, a lower fungal count will thereby have less mycotoxin production and in the post-harvest stage. The accumulation of mycotoxins can generally be avoided in the post-harvest stage by providing ideal storage and transport conditions to the agricultural produce. A few of them are discussed in detail under physical methods.

The prevention of fungal pathogens from causing the infection and thereby the disease and mycotoxin production is not always successful for all of the produce. Furthermore, even small quantities of infectious propagules reaching storage stage under optimum growth conditions can lead to accumulation of mycotoxins in the produce. Therefore, active removal or decontamination of the mycotoxin contaminated grains is essential. This step comprises of all the physical, chemical and biological methods to reduce the levels of mycotoxin and the fungal pathogens producing them (Karlovsky et al., 2016).

Physical Methods

Physical methods include sorting, soaking, washing, adsorption, temperature control, modified atmosphere treatment, irradiation, high pressure processing, pulsed electric fields, ozone, pulsed light and cold plasma (Wang et al., 2022a; Pandey et al., 2023). These detoxification methods are economical and feasible, with high detoxification efficiency, do not produce toxic by-products, and have no negative impact on food nutrition and quality. They therefore have great application potential.

1. Temperature and moisture control:

Fungi require a particular range of temperature and moisture to survive and any modifications in the ecosystem can significantly influence its growth. The agricultural commodities in storage also constitute a mini ecosystem with crop and micro-organisms as the biotic factors and moisture, temperature, gases etc. as the abiotic factors. (Liu et al., 2020). Often the moisture and the temperature are maintained to be outside the optimum range of the fungal pathogens to hinder their growth and spread. In industries and domestic storage, the water content and temperature can be easily controlled during food storage. Although both relative humidity (RH) and moisture content (MC) are used to reflect the water content of food, it is better to use equilibrium RH, because the impact of equilibrium RH on spoilage organisms is consistent across different foods, regardless of their composition (Bradford et al., 2018). The food is safe to keep for over a year when the equilibrium RH below 65% at ambient temperature. Similarly, the storage temperature can also be controlled for preventing the growth of fungal pathogens.

Modified Atmosphere

Modified atmosphere (MA) approaches modify the gas composition considering temperature, relative humidity, moisture content and competing microorganisms (Liu et al., 2020). Controlled atmosphere could reduce mycotoxins content and counteract the vitality of deleterious organisms in foodstuff (Pisuttu et al., 2023). The difference in sensitivity towards the immediate atmosphere in fungi and that required for the storage of the agricultural produce is the main factor used in this method. For instance, high CO_2 and low O_2 content inhibit foodborne fungi. This technique has proved to be useful in reducing the levels of *Penicillium roqueforti* and *A. flavus* (Taniwaki et al., 2009). In a study, corn and wheat grains were stored in a highly purified N_2 (98.5%), controlled atmosphere. The growth and sporulation by the four pathogens (*Fusarium graminearum, Fusarium langsethiae, Aspergillus flavus* and *Fusarium verticillioides)* and aflatoxins production by *Aspergillus* were significantly reduced (Lorenzo et al., 2020). Reibeiro et al. (2021) used vacuum and static controlled atmosphere with low oxygen partial pressure (pO_2) and high carbon dioxide (pCO_2) to manage the fungal microorganisms in pecan nuts. At 10 °C *Aspergillus* sp. and *Penicillium* sp. contamination was lowered or absent for after 2 months of storage, and *Cladosporium* sp. up to 4 months of storage. At 20 °C, high pCO_2 storage led to lower contamination than the low pO_2 condition. The nuts stored in vacuum packaging at 20 °C maintained lower contamination levels by *Aspergillus* in all periods for which they were evaluated.

Sorting

Broken and damaged kernels (usually 3-6 % of bulk load) usually contain most of mycotoxin contamination (Johansson et al., 2006). Such grains can be identified and taken out during the initial processing steps sorting and washing (Grenier et al., 2012). Centrifugation force and flotation in air flow were one of the earliest used methods to sort out bulk grains. Optical sorting in 1960s was used direct streams of grains along an array of optical sensors. On detecting a different coloured grain, the detector triggers a magnetic valve, and it is removed by a jet of pressurized air (Fraenkel, 1962). Sorting proves to be useful method for in aflatoxin contaminated grains as it is usually heterogeneous and separating damaged kernels can effectively reduce contamination (Kabak et al., 2006). Whereas in case of mycotoxins that accumulate without producing visible symptoms in the agricultural produce it is rather useless (Afolabi et al., 2007). This may explain why no reduction of aflatoxin content by sorting was found in a recent study (Mutiga et al., 2014).

Washing and floatation

It is used in the case of water-soluble mycotoxins present on surface of grains. Sodium carbonate solutions are often used as a washing solution in DON and ZEN contaminated barley and corns (Karlovsky et al., 2016). Floatation is dependent on the difference between density gradients of the uninfected and mould-damaged kernel. The mould damaged kernels float in the solution and can be removed. This technique is used in removal of aflatoxin infected corns. 87 % aflatoxin reduction was obtained by using a 30 % sucrose solution with the removal of 50 per cent of the material (Huff, 1980). Likewise, the use of saturated sodium chloride solution removed 74 % of the total aflatoxin content in maize (Karlovsky et al., 2016). The grains must be dried properly dried after both of these methods.

Adsorption

Adsorption is a widely utilized technique for detoxification of food and beverages and animal feed, both in the laboratory and in industry. It is based on removing mycotoxins rather than degrading them into smaller substances. It facilitates the irreversible binding of mycotoxins to adsorbents, like clay which prevents the mycotoxins from being adsorbed on from the gastro-intestinal tract. One such material is phyllosilicate clay montmorillonite which binds aflatoxins and fumonisins, and in turn reduce their toxicity in the gastrointestinal tract in animals and in urinary biomarkers in humans (Phillips et al., 2008; Robinson et al., 2012). The use of nanoparticles as adsorbents is a major improvement in this technique (Song and Qin, 2022). For instance, Fe_3O_4 nanoparticles reduce the concentration of AFM1 aflatoxin from 100 to 5 ppt (Javani Jouni et al., 2018).

Thermal Inactivation

Heat treatments of fruits, vegetables and other agricultural products are a part of many processing and value addition methods and additionally it has a deleterious effect on many mycotoxins and the pathogens producing them. Along with the temperature and exposure period the thermal inactivation or degradation of the mycotoxins depend on the moisture content of the food product, toxin concentration, food type matrix, and additives present. Aflatoxin for example can be inactivated at a temp ≥250 °C as they are highly resistant in their pure form (Temba et al., 2016). Further from a study on cotton seed meal it was concluded that the thermal degradation of aflatoxins is dependent on moisture content of the product. The reduction at 100 °C was 85 per cent aflatoxin with 30 per cent MC and only 50 per cent in cotton seed meal with 6.6 per cent (Mann et al., 1967).

Irradiation

Irradiation utilizes ionizing energy/radiation (gamma rays) and non-ionizing energy/radiation (ultraviolet light, sunlight, microwaves) to check the post-harvest fungal pathogens and in turn the mycotoxins produced by them. It is a non-thermal technology which acts by deteriorating the cellular structure and physiological functions which may include disruption of cell membrane integrity, DNA strand breakage, etc. of the microorganisms (Liu et al., 2020). It utilizes energy/radiations that degrade the fungal pathogens by direct or indirect oxidation which facilitates the cleavage of chemical bonds in the molecules giving rise to much smaller, relatively harmless products. Various cereals (e.g., maize, wheat, barley, rice, sorghum) are often treated with γ-radiations to reduce aflatoxin Mohammadi et al., 2022, ZEA (Aziz et al., 2007), Ochratoxin and FB1 (Aziz et al., 2007). Low doses of UV-C radiation as a post-harvest treatment in pear controlled the *Alternaria alternata* pathogen and the mycotoxin produced by the fungus (Sun et al., 2022).

High pressure processing

The three characteristics of HPP treatment are temperature, pressure and exposure time. It is a widely used technique to inhibit growth of fungal pathogens and inactivate mycotoxins by structural modification, reducing their toxicity and conversion to less toxic compounds (Alizadeh et al., 2021). The applied pressure ranges from 400 and 600 MPa for several seconds to minutes at 4-25°C (Palmade et al., 2019). This technique activates or inactivates enzymes at low temperatures, while low molecular weight compounds involved in the flavor and nutritional value remain unaltered (Barba et al., 2012). This technique has often given good results in cereal grains (Stranska et al., 2021).

Pulsed electric fields

Electrical treatments of different electric field strengths (1-40 kV/cm) are applied between two electrodes for short periods of time and the food product is placed between the two electrodes. It is an efficient tool for inactivating fungal microorganisms at low temperatures in foods like vegetables and fruit juices (Pallares et al., 2021). The technique has proved to be effective in reducing the aflatoxin content in foods infected with *Aspergillus* (Bulut et al., 2020; Gavahian et al., 2020). In *Fusarium* infected seeds the when the frequency was increased from 450 to 900 Hz, the number of viable spores gradually decreased (Palicova et al., 2024).

Pulsed light

This technique utilizes high power flashes of ultra-violet, infra-red and visible light in short, high intensity flashes i.e. pulsed light for mycotoxin degradation. This intense light acts by penetrating the microbial cells, and damaging their cell structures, for example their membrane, proteins and nucleic acids, which significantly hinders the microbial growth on the surfaces of foods and packaging materials (Alizadeh et al., 2021; Kim et al., 2021). Unlike other chemicals used to inhibit fungal growth this does not leave any harmful residues and food. Pulsed light has proved to be a potential new technique for reducing mycotoxins in foods and agricultural produce like red pepper (Woldemariam et al., 2021) and cereals (Wang et al., 2022b) against pathogens like *Aspergillus* (Lin et al., 2023), *Fusarium* (Wang et al., 2023).

Cold plasma

Plasma is the fourth state of matter, which contains partially or fully ionized gas. An electric current is applied through neutral gases (like oxygen, nitrogen, or helium) that cause the dissociation of the gaseous molecules producing a highly reactive mixture of electrons, ions, atoms and free radicals. It includes O, O_3, OH, NO, NO_2. These reactive species can degrade mycotoxins and convert them in other less toxic compounds at ambient temperature (Hojnik et al., 2017; Misra et al., 2019). It is a non-thermal treatment (temperature at or below 0 °C) and does not involve chemicals, so there is no heat damage or contamination of food with chemicals. Cold plasma treatment has been widely used to kill many microorganisms, and for mycotoxin detoxification (Chen et al., 2022; Kis et al., 2020; Wang et al., 2022).

Chemical Methods

1. Control by chemical antifungal agents

This type of control is generally effective to inhibit growth of fungal pathogens which will lead to lesser mycotoxin production. The chemicals generally used include sodium hydroxide and potassium hydroxide in the aflatoxin degradation oil, although these chemicals can cause secondary contamination and have harmful effects on the nutritional value of the products (Ji & Xie, 2020).

2. Ozonation

Ozonation causes the progressive degradation of cell components, the primary target being the cell surface. The ozone mainly causes leakage of cell contents and lysis because of its action on the unsaturated lipids in the cell membrane along with oxidation of various internal proteins and rapid cell death (Karaca and Velioglu, 2007; Silva et al., 2010). The method involves either dipping the materials in ozonised water or pumping the ozone gas in dry or moistened form.

This technique has been effective in both food and feed material but mostly reserved to feed because of the presence of degradation products in the food (Temba et al., 2016). It is considered a green technology because it does not leave any residue as the ozone disassociates into O_2 and related radicals (EPA, 1999). In many cases this technique has proved effective for detoxification of mycotoxins for including fumonisins, aflatoxins (AFB1 & AFG1), (Zorlugenc et al., 2008; Proctor et al., 2016). Furthermore, in Kiwi TiO2 photocatalytic ozonation has proved to control fungal growth in post-harvest storage (Hur et al., 2004). It is considered a green technology because it does not leave any residue as the ozone disassociates into O_2 and related radicals.

3. Ammoniation

Ammoniation involves treating material with either gaseous ammonia or ammonium hydroxide. The efficacy of mycotoxin detoxification increases with the quantity of ammonia used the reaction time, and the temperature and pressure levels. It has been shown to degrade ochratoxin, aflatoxin and citrinin in food and animal feeds.

Biological Methods

Biological control using microbial antagonists, either alone or as part of an integrated strategy, has emerged as a promising approach to reducing pesticide use and controlling aflatoxins and other mycotoxins in both pre- and post-harvested crops (Dimitrios et al., 2012). Using biofungicides involves applying various microorganisms, microbial antagonists, or competitors to suppress harmful fungi Bacteria like *Bacillus* and *Pseudomonas*, as well as fungi from the genus *Trichoderma*, are among the most effective biocontrol agents. They combat a wide range of plant pathogens in an environmentally friendly way (Bhattacharjee & Dey, 2014).

Biological control with microbial antagonists, either on its own or as part of a comprehensive control strategy, has emerged as a promising method for managing mycotoxins in crops. Detoxifying agricultural products contaminated with mycotoxins is challenging due to safety concerns, potential loss of nutritional value, limited effectiveness and high costs (Kohl et al., 2011). Use of bacteria, fungi and yeasts have potential feasible substitute to reduce the use of chemical compounds (Medeiros et al., 2012).

Mitigation Strategies

1. **Pre-Harvest:** Employing crop management practices such as rotation, proper irrigation and pest control to reduce fungal infection.
2. **Post-Harvest**: Implement effective drying and storage techniques to minimize fungal growth and mycotoxin production.
3. **Processing**: Use methods such as cleaning, sorting and thermal processing to reduce mycotoxin levels in food products.

4. **Regulation and Monitoring**: Adhere to regulations regarding maximum allowable mycotoxin levels and conduct regular monitoring and testing of food and feed.

Examples

The market was initially graced with the introduction of first commercial product for farmers, such as BINAP T in France in 1976 (Ricard and Ricard, 1997). Twenty BCAs were tested against ten pathogenic fungi. Some of the cell-free supernatants (CFS) tested showed in vitro antifungal activity versus pathogenic fungi. The highest fungicidal activity was observed in the fermented CFS of *Paenibacillus chibensis* CECT 375, *Bacillus amyloliquefaciens* CECT 493, and *Pantoea agglomerans* CECT 850, which showed a minimum inhibitory concentration (MIC) and minimum fungicidal concentration (MFC) values of 125 and 250 g/L, respectively (Illeuca et al., 2021)

New biocontrol agents need to be safe in addition to being effective against the intended plant pathogen. Aspire (*Candida oleophila* strain 182; Bio-Save 10 and 11 (*Pseudomonas syringae* strains ESC10 and ESC 11; AF36 (atoxigenic strain of *A. flavus* and Afla-Guard (atoxigenic strain of *A. flavus* NRRL 21882 are just a few of the commercial products for controlling mycotoxigenic fungi that are available on the market in various countries throughout the world. (Nunes, 2002; Kohl et al., 2011).

Concept of natural buffering

Risks posed by the fungi and their mycotoxins requires a coordinated approach involving agricultural practices, food safety regulations and ongoing research to develop effective prevention and control measures. The identification of the role of antagonistic microorganisms responsible for natural buffering, such as saprophytes inhibiting *Fusarium roseum 'Culmorum'* from colonizing grain stumps, served as one of the foundations for research on biological management (Cook, 2008).

The natural buffering of cropping systems is facilitated by the naturally occurring interference between plant pathogens and beneficial microorganisms, which prevents or limits the development of disease. This is particularly evident in suppressive soils where the pathogen population has developed an antagonistic potential (Weller et al., 2002).

In recent times, diseases affecting seedlings, roots, leaves and fruits have been effectively targeted by antagonists (Paulitz and Bélanger, 2001; Sharma et al., 2009). Commercial businesses in the US advertised 26 microbial products with the proven purpose of controlling biological diseases in 2005 (Fravel, 2005). According to Ehlers (2009), 14 microbiological products for the control of diseases were registered in Europe in 2008 and included in Annex 1 of Directive 91/414 EEC.

Particularly in the last ten years, the biocontrol industry that deals with microorganisms has grown into a professional sector. Manufacturers are now part of the International Biocontrol Manufacturers Association (IBMA), based in

Basele, Switzerland and there are international business meetings like the Annual Biocontrol Industry Meeting (ABIM), held in Lucerne, Switzerland (International Biocontrol Manufacturers' Association, 2010) (Köhl et al., 2011)

Government laws pertaining to usage and registration are implemented in a manner identical to that of artificial chemical plant protection goods. Genetic stability of the antagonist, market size for the biocontrol product and the possibilities of patent protection for the application is a significant step (Whitesides et al., 1994). Isolation of a microorganism showing antagonistic property to the commercial marketing of an economically viable biological control product is of great importance (Blum, 2007).

Biodetoxification

Biodetoxification is an emerging strategy for reducing mycotoxins using non-pathogenic microbes or their enzymes through catabolic processes. These microbes not only reduce or neutralize toxins to less harmful compounds but are also generally considered safe because they produce beneficial end products via biodegradation or bio-adsorption. Probiotics combat toxigenic fungi by competing for space and nutrients essential for growth and metabolism, forming biofilms and triggering defensive responses, including the release of free oxygen radicals (Afshar, 2020).

Bacteria

Flavobacterium aurantiacum, which can detoxify aflatoxins; Phenylobacterium immobile, known for degrading ochratoxin A; and *Gliocladium roseum*, which detoxifies zearalenone through ring opening followed by decarboxylation (Schatzmayr, 2006). Eubacterium BBSH 797 was the first isolated bacterial strain capable of biotransforming the epoxide group in trichothecenes and due to epoxidase enzymatically reduced deoxynivalenol (DON) to the non-toxic metabolite de-epoxy-deoxynivalenol (DOM-1).

Probiotics

Probiotics can eliminate these contaminants through biodegradation or bioadsorption. Biodegradation is a more permanent process that can alter the toxin's structure, potentially producing harmful metabolites (such as aflatoxicol from aflatoxin B1), which may negatively affect the host. In contrast, bioadsorption involves the rapid binding of toxins, which can be easily released depending on the bacterial affinity for the toxin (Solis-Cruz et al., 2019). Probiotic yeast can quickly grow on inexpensive substrates in fermenters, making them suitable for large-scale production. The use of yeasts is safe for humans, animals, host plants and the environment and it is unlikely that the target organisms will develop resistance (Diaz et al., 2020). *Saccharomyces cerevisiae* is regarded as a probiotic yeast that can significantly decompose deoxynivalenol (DON) and reduce the release of lactate dehydrogenase (LDH) (Liu et al., 2019). The ability of *Rhodotorula mucilaginosa*

JM19 to degrade more than 50 per cent of patulin suggests that this yeast could be valuable for improving the safety and quality of foods and raw materials (Li et al., 2019). The primary mechanisms by which bioprotective yeasts exert their effects include competition for nutrients and space, synthesis and secretion of antibacterial compounds, mycoparasitism and the secretion of lytic enzymes. Additionally, these yeasts can form biofilms, engage in quorum sensing, induce systemic resistance in the host fruit and produce reactive oxygen species, all of which contribute to their bioprotective activity. Utilizing antagonistic yeasts as a biological approach helps control fungal decay and sustainably adsorb and/or degrade mycotoxins (Oztekin et al., 2023). Yeasts show significant potential as natural alternatives to chemical additives in various food applications. Yeasts are appealing candidates for biological control because of their broad-spectrum antimicrobial activity, genetic stability, low nutritional requirements, and ability to thrive in low-temperature environments. These characteristics make them suitable for use in various settings, especially in food preservation and safety, where such attributes are crucial for effectively combating spoilage organisms and pathogens. Their ability to inhibit pathogens, degrade toxins, and enhance food preservation through natural mechanisms makes them valuable for improving food safety and extending shelf life without relying on synthetic chemicals (Yan et al., 2024).

Fungi

Fungi, such as *Rhizopus*, *Trichoderma*, *Clonostachys*and *Penicillium* species, also show potential for use in mycotoxin biocontrol strategies (Alberts et al., 2017). *Trichoderma* strains, in particular, have shown significant antibiosis and parasitism capabilities, which makes them suitable as mycoparasites against toxigenic *Fusarium* isolates. They prevent *Fusarium* growth by coiling around its hyphae and penetrating them (Błaszczyk et al., 2017).

Nano particles (NPs)

NPs have emerged as promising agents for mycotoxin detoxification. For example, magnetic carbon nanocomposites have been effective in reducing the toxicity of AFB1, while chitosan-coated Fe3O4 nanoparticles have been utilized for the detoxification of patulin (PAT). Silver nanoparticles have demonstrated inhibitory effects against the growth of *Fusarium* species and have also proven effective in reducing mycotoxin levels (Luo et al., 2018). According to Kumar et al. (2024), natural biopolymers and bioactive compounds can be utilized to create nanoformulations that extend the shelf life of citrus fruits by reducing fungal growth, offering a potential alternative to traditional fungicides and improving post-harvest shelf-life of citrus fruits.

Enzymatic detoxification

It is an effective method for mycotoxin degradation which combines chemical and biological methods with high performance and specification. For instance, β-1,3-gulcanase is an enzyme secreted by *Aspergillus* which has the properties to degrade a mycotoxin from produce particularly fumonosins (Lyagin and Eferemenko, 2019).

Plant extracts

Plant metabolites contain a diverse array of chemicals with different properties and roles making them versatile enough to be used in mycotoxin management. The use of botanicals or plant extracts is economically and environmentally feasible in managing the mycotoxins in the food or agricultural products. Some of these metabolites are involved in plant defense against disease causing pathogens. They are synthesized and stored by the plants in non-toxic form (Chaves Lobon et al., 2019). A number of studies have been put forward in this field. Capsaicin inhibits Ochratoxin (*Aspergillus* and *A. carbonarius*) production in grapes (Kollia et al., 2019). Various aflatoxins are also degraded by plant extracts like turmeric, eugenol, citrol, etc. (Jahanshiri et al., 2015; Hu et al., 2017; Moon et al., 2017).

A major disadvantage in using plant extract in post-harvest includes their highly volatile nature, vulnerability to oxidation reactions. Their organoleptic features are also a drawback as they put off-flavour to the food products. These drawbacks can be overcome by the use of plant extracts in form of sprays (Gao et al., 2015), films (Giteru et al., 2015) or edible coatings (Peretto et al., 2014).

Biological control and the good agricultural practices

Adhering to excellent agricultural practices in the field and storage is the first and most sensible step in managing phytopathogen contamination. Certain of these methods depend on the precise timing of harvest and the kind of crop material to plant (Wilson et al., 2006; Blandino et al., 2009). For example; maize grains, the selection of hybrids, plant density, time of planting and insect control may have an influence on toxin contamination (Magan and Aldred, 2007).

Factors affecting fungi and mycotoxin occurrence in the food chain (Medeiros, 2012)

Biological factors: a compatible toxic fungus and a sensitive crop.

Environmental factors: moisture, temperature, damage from insects and birds, and mechanical injuries.

Harvesting: crop maturity, moisture content, temperature, and contamination identification.

Storage: monitoring of humidity, temperature, and contaminants.

Distribution: identifying and rerouting contaminants

Conclusion

Postharvest diseases reduce the quality as well as quantity of agriculture commodities. The major post-harvest pathogens include *Aspergillus, Alternaria, Fusarium, Penicillium, Colletotrichum, Botrytis, etc.* and most of these pathogens produce secondary metabolites which are toxic to the plant host and aid in spreading the disease. Mycotoxin contamination in food and feed may lead to several health issues. Mycotoxins can contaminate food at various stages of the food chain. Their occurrence can start as early as the field stage or during harvesting, handling, storage, and processing. Effectively managing mycotoxins in agriculture is crucial for protecting human and animal health, preserving economic interests, ensuring food safety and security, adhering to regulatory standards, and supporting sustainable agricultural practices. There are many pre and post-harvest approaches which can help to minimise the contamination.

Important measures to avoid the harmful effects of mycotoxin contamination in agricultural commodities include preventing contamination before and after harvesting, removing contaminated food, detoxifying affected materials and promptly and effectively treating those impacted. Maintaining good storage conditions, using natural and chemical agents, and applying irradiation are also reported to prevent mycotoxins after harvest. Postharvest methods to reduce mycotoxins involve rapid and proper drying, appropriate transportation and packaging, sorting, cleaning, drying, smoking, postharvest insect control and the use of botanical or synthetic pesticides as storage protectants. Various physical, chemical and biological methods are employed to reduce the contamination. The lack of pathogenicity towards host plants and non-toxic metabolic profile further enhances its safety and compatibility with other treatments. Thus, integrated disease management can be adopted to reduce the mycotoxin contamination and fungal pathogens in order to have food security and health safety.

References

Afolabi, C.G., Ojiambo, P.S., Ekpo, E.J.A., Menkir, M., & Bandyopadhyay, R. (2007). Evaluation of maize inbred lines for resistance to Fusarium ear rot and fumonisin accumulation in grain in tropical Africa. *Plant Disease, 91*, 279-286.

Afshar P., Shokrzadeh M., Raeisi S.N., Ghorbani-HasanSaraei A. & Nasiraii L.R. (2020). Aflatoxins biodetoxification strategies based on probiotic bacteria. *Toxicon. 178*:50-58.

Alberts J.F., Lilly M., Rheeder J.P., Burger H.-M., Shephard G.S. & Gelderblom W.C.A. (2017). Technological and community-based methods to reduce mycotoxin exposure. *Food Control, 73*, 101-109.

Alizadeh, A. M., Hashempour-Baltork, F., Khaneghah, A. M., & Hosseini, H. (2021). New perspective approaches in controlling fungi and mycotoxins in food using emerging and green technologies. *Current opinion in food science, 39*, 7-15.

Alshannaq, A. & Yu, J. H. (2017). Occurrence, toxicity, and analysis of major mycotoxins in food. *International journal of environmental research and public health. 14*(6), 632.

Amaike, S., Keller, N. P. (2011). Aspergillus flavus. Annual review of phytopathology. *49*, 107-133. doi: 10.1146/annurev-phyto-072910-095221.

Awad W., Ghareeb K., Bhm, J. & Zentek, J. (2013). The Toxicological Impacts of the Fusarium Mycotoxin, Deoxynivalenol, in Poultry Flocks with Special Reference to Immunotoxicity. *Toxins, 5*, 912-925.

Aziz, N. H., Ferial, M., Shahin, A. A., & Roushy, S. M. (2007). Control of Fusarium moulds and fumonisin B1 in seeds by gamma-irradiation. *Food Control, 18*(11), 1337-1342.

Barba, F. J., Esteve, M. J., & Frígola, A. (2012). High pressure treatment effect on physicochemical and nutritional properties of fluid foods during storage: A review. *Comprehensive Reviews in Food Science and Food Safety, 11*(3), 307-322.

Ben Amara, A., Mehrez, A., Ragoubi, C., Romero-González, R., Garrido Frenich, A., Landoulsi, A., & Maatouk, I. (2022). Fungal mycotoxins reduction by gamma irradiation in naturally contaminated sorghum. *Journal of Food Processing and Preservation, 46*(3), 16345.

Bennett, J. W. (1987). Mycotoxins, mycotoxicoses, mycotoxicology and mycopathology. *Mycopathlogia, 100*, 3-5.

Bennett, J. W. & Klich, M. (2003). Mycotoxins. *Clinical microbiology reviews, 16*, 497-516.

Bhat, R., Rai, R.V. & Karim, A. A. (2010). Mycotoxins in food and feed: present status and future concerns. *Comprehensive review in food science and food Safety, 9*(1), 57-81.

Bhattacharjee R. & Dey U. (2014). An overview of fungal and bacterial biopesticides to control plant pathogens/diseases. *African journal of microbiology research, 8*, 1749-1762.

Blandino, M., Reyneri, A., Vanara, F., Tamietti, G. & Pietri, A., 2009. Influence of agricultural practices on *Fusarium* infection, fumonisin and deoxynivalenol contamination of maize kernels. *World Mycotoxin Journal, 2*(4), pp.409-418.

Blaszczyk, L., Basinska-Barczak A., Cwiek-Kupczynska H., Gromadzka K., Popiel D. & Stępien L. (2017). Suppressive Effect of *Trichoderma* spp. on Toxigenic Fusarium Species. *Polish journal of microbiology, 66*, 85-100. doi: 10.5604/17331331.1234996.

Blum, B. J. (2007). Concepts and Strategies for a successful product development: the industry's development concept. In *Cost Action 850–Conference Schloss Salzau, Germany*.

Bradford, K.J., Dahal, P., Asbrouck, J., Kunusoth, K., Bello, P., Thompson, J. & Wu, F. (2018). The dry chain: Reducing postharvest losses and improving food safety in humid climates, *Trends in Food Science & Technology*, 71, 84-93.

Bulut, N., Atmaca, B., Akdemir Evrendilek, G., & Uzuner, S. (2020). Potential of pulsed electric field to control Aspergillus parasiticus, aflatoxin and mutagenicity levels: Sesame seed quality. *Journal of Food Safety, 40*(6), 12855.

Chaves Lobón, N., Ferrer de la Cruz, I., & Alías Gallego, J. C. (2019). Autotoxicity of diterpenes present in leaves of Cistus ladanifer L. *Plants, 8*(2), 27.

Chen, X., Qiu, Y., Zhang, J., Guo, Y., Ding, Y., & Lyu, F. (2022). Degradation efficiency and products of deoxynivalenol treated by cold plasma and its application in wheat. *Food Control*, 136, 108874.

Cook, R. J. (2007). Tell me again what it is that you do. *Annu. Rev. Phytopathol., 45*(1), 1-23.

DBM, S. (1962). Effect of ozone on survival and permeability of Escherichia coli. *J of Bacteriology*, 85, 567-576.

Dimitrios, T., Dimakopoulou, M., Antoniou, P. & Tjamos, E. (2012). Biological control strategies of mycotoxigenic fungi and associated mycotoxins in Mediterranean basin crops. *Phytopathologica Mediterranean*. 51, 158-174.

Detroy, R. W., Lillehoj, E. B. & Ciegler, A. (1971). Aflatoxin and related compounds. vol. VI: fungal toxins. In: A. Ciegler, S. Kadis, & S. J. Ajl (ed.). Microbial toxins, Academic Press, New York, 3-178.

Dhakal, A., Hashmi, M.F. & Sbar, E. (2022). Aflatoxin Toxicity. In: StatPearls [Internet]. Treasure Island (FL): StatPearls Publishing. PMID: 32491713.

Diaz M.A., Pereyra M.M., Picón-Montenegro E., Meinhardt F. & Dib J.F. (2020). Killer Yeasts for the Biological Control of Postharvest Fungal Crop Diseases. *Microorganisms, 8*, 1680.

Diener, U. L., Cole, R. J. & Sanders, T. H. (1987). Epidemiology of aflatoxin formation by Aspergillus flavus. *Annual review of phytopathology, 25*, 249-270.

Ehlers, R. U. (2009). REBECA–EU-policy support action to review regulation of biological control agents. *recent developments in management of plant diseases*, 147-161.

Eisa, N. A., Ali, F. M., El-Habbaa, G. M., Abdel-Reheem, S. K., & Abou-El-Ela, M. F. (2003). Pulsed electric field technology for checking aflatoxin production in cultures and corn grains.

Fraenkel H (1962) Colour sorting machine. US Patent No. 3,066,797

Fravel, D. R. (2005). Commercialization and implementation of biocontrol. *Annual review of phytopathology, 43*(1), 337-359.

Freitas-Silva, O., & Venâncio, A. (2010). Ozone applications to prevent and degrade mycotoxins: a review. *Drug metabolism reviews, 42*(4), 612-620.

Frisvad, J. C., Hubka, V., Ezekiel, C. N., Hong, S. B., Novbkovb, A. A., Chen, J. & Houbraken, J. (2019). Taxonomy of Aspergillus section Flavi and their production of aflatoxins, ochratoxins, and other mycotoxins. *Studies in mycology, 93*(1), 1-63.

Gao, M., Feng, L., & Jiang, T. (2014). Browning inhibition and quality preservation of button mushroom (Agaricus bisporus) by essential oils fumigation treatment. *Food chemistry, 149,* 107-113.

Gavahian, M., Pallares, N., Al Khawli, F., Ferrer, E., & Barba, F. J. (2020). Recent advances in the application of innovative food processing technologies for mycotoxins and pesticide reduction in foods. Trends in Food Science & Technology, *106,* 209-218. https://doi.org/10.1016/j.tifs.2020.09.018

Giteru, S. G., Coorey, R., Bertolatti, D., Watkin, E., Johnson, S., & Fang, Z. (2015). Physicochemical and antimicrobial properties of citral and quercetin incorporated kafirin-based bioactive films. *Food chemistry, 168,* 341-347.

Godana, EA., Yang, Q., Zhang, X., Zhao, L.,Wang, K., Dhanasekaran, S., Mehari, G & Zhang, H. (2023). Biotechnological and Biocontrol Approaches for Mitigating Postharvest Diseases Caused by Fungal Pathogens and Their Mycotoxins in Fruits: A Review. *Journal of Agricultural and Food Chemistry 71* (46), 17584-17596.

Grenier, B., Bracarense, A. P. F., Schwartz, H. E., Trumel, C., Cossalter, A. M., Schatzmayr, G. & Oswald, I. P. (2012). The low intestinal and hepatic toxicity of hydrolyzed fumonisin B1 correlates with its inability to alter the metabolism of sphingolipids. *Biochemical pharmacology, 83*(10), 1465-1473.

Hojnik, N., Cvelbar, U., Tavcar-Kalcher, G., Walsh, J. L., & Krizaj, I. (2017). Mycotoxin decontamination of food: Cold atmospheric pressure plasma versus "classic" decontamination. *Toxins, 9*(5), 151.

Hu, Y., Zhang, J., Kong, W., Zhao, G., & Yang, M. (2017). Mechanisms of antifungal and anti-aflatoxigenic properties of essential oil derived from turmeric (Curcuma longa L.) on Aspergillus flavus. *Food chemistry, 220,* 1-8.

Huff, W. E. (1980). A physical method for the segregation of aflatoxin-contaminated corn.

Hur, J. S., Oh, S. O., Lim, K. M., Jung, J. S., Kim, J. W., & Koh, Y. J. (2005). Novel effects of TiO2 photocatalytic ozonation on control of postharvest fungal spoilage of kiwifruit. *Postharvest biology and technology, 35*(1), 109-113.

Illueca, F., Vila-Donat, P., Calpe, J., Luz, C., Meca, G., & Quiles, J. M. (2021). Antifungal activity of biocontrol agents in vitro and potential application to reduce mycotoxins (aflatoxin B1 and ochratoxin A). *Toxins, 13*(11), 752.

Jahanshiri, Z., Shams-Ghahfarokhi, M., Allameh, A., & Razzaghi-Abyaneh, M. (2015). Inhibitory effect of eugenol on aflatoxin B 1 production in Aspergillus parasiticus by downregulating the expression of major genes in the toxin biosynthetic pathway. *World journal of microbiology and biotechnology, 31,* 1071-1078.

Javani Jouni, F., Zafari, J., Abdolmaleki, P., Vazini, H., Ghandi, L., & Satari, M. (2018). Aflatoxin M 1 detoxification from infected milk using Fe3O4 nanoparticles attached to specific aptamer. *Journal of nanostructure in chemistry, 8,* 13-22.

Ji, J., & Xie, W. (2020). Detoxification of Aflatoxin B1 by magnetic graphene composite adsorbents from contaminated oils. *Journal of hazardous materials, 381*, 120915.

Johansson, A. S., Whitaker, T. B., Hagler Jr, W. M., Bowman, D. T., Slate, A. B., & Payne, G. (2006). Predicting aflatoxin and fumonisin in shelled corn lots using poor-quality grade components. *Journal of AOAC International, 89*(2), 433-440.

Jouany J. (2007). Methods for preventing, decontaminating and minimizing the toxicity of mycotoxins in feeds. *Animal feed science technology, 137*, 342-362.

Kabak, B., Dobson, A. D., & Var, I. I. L. (2006). Strategies to prevent mycotoxin contamination of food and animal feed: a review. *Critical reviews in food science and nutrition, 46*(8), 593-619.

Karaca, H., & Velioglu, Y. S. (2007). Ozone applications in fruit and vegetable processing. *Food Reviews International, 23*(1), 91-106.

Karlovsky, P., Suman, M., Berthiller, F., De Meester, J., Eisenbrand, G., Perrin, I., & Dussort, P. (2016). Impact of food processing and detoxification treatments on mycotoxin contamination. *Mycotoxin research, 32*, 179-205.

Kim, J. H., Lee, C. J., Min, K. D., Hwang, B. U., Kang, D. G., Choi, D. H., & Jung, S. B. (2021). Intense pulsed light surface treatment for improving adhesive bonding of aluminum and carbon fiber reinforced plastic (CFRP). *Composite Structures, 258*, 113364.

Kiš, M., Milošević, S., Vulić, A., Herceg, Z., Vukušić, T., & Pleadin, J. (2020). Efficacy of low pressure DBD plasma in the reduction of T-2 and HT-2 toxin in oat flour. *Food chemistry, 316*, 126372.

Kohl, J., Postma, J., Nicot, P., Ruocco, M., & Blum, B. (2011). Stepwise screening of microorganisms for commercial use in biological control of plant-pathogenic fungi and bacteria. *Biological control, 57*(1), 1-12.

Kollia, E., Proestos, C., Zoumpoulakis, P. & Markaki, P. (2019). Capsaicin, an inhibitor of Ochratoxin A production by *Aspergillus* section Nigri strains in grapes (*Vitis vinifera* L.). *Food additives and contaminants*, Part. A, 36, 1-13.

Kumar, R., Mishra, A. K., Dubey, N.K. & Tripathi, Y.B. (2007). Evaluation of Chenopodium ambrosioides oil as a potential source of antifungal, antiaflatoxigenic and antioxidant activity. *International journal of food microbiology*. 115, 159-164.

Kumar, N., Upadhyay, A., Shukla, S. *et al.* (2024). Next generation edible nanoformulations for improving post-harvest shelf-life of citrus fruits. *Food Measure* 18, 1825-1856.

Kumera, N. & Mohammed A. (2017). Mycotoxin occurrence in grains and the role of postharvest management as a mitigation strategies. A review. *Food Control, 78*, 412-425.

Li X., Tang H., Yang C., Meng X. & Liu B. (2019). Detoxification of mycotoxin patulin by the yeast *Rhodotorula mucilaginosa*. *Food Control, 96*, 47-52.

Lin, S., Jiang, H., Fu, Q., Huang, S., Tang, L., Li, A., & Liu, M. (2023). Effects of Pulsed Light on Mycelium Growth and Conidiation in Aspergillus oryzae. *Fermentation, 9*(7), 674.

Liu Y., Chang J., Wang P., Yin Q., Huang W., Liu C., Bai X., Zhu Q., Gao T. & Zhou P. (2019). Effects of *Saccharomyces cerevisiae* on alleviating cytotoxicity of porcine jejunal epithelia cells induced by deoxynivalenol. *AMB Express. 9*, 137.

Liu, Y., Galani Yamdeu, J. H., Gong, Y. Y., & Orfila, C. (2020). A review of postharvest approaches to reduce fungal and mycotoxin contamination of foods. *Comprehensive Reviews in Food Science and Food Safety, 19*(4), 1521-1560.

Lorenzo, M., Sabrina, S., Gianpaola, P., Antonio, M., Miriam, H., & Giovanni, V. (2020). N 2 controlled atmosphere reduces postharvest mycotoxins risk and pests attack on cereal grains. *Phytoparasitica, 48*, 555-565.

Luo Y., Liu X. & Li J. (2018). Updating techniques on controlling mycotoxins-A review. *Food Control,* 89, 123-132.

Lyagin, I., Efremenko, E. (2019). Enzymes for detoxification of various mycotoxins: Origins and mechanisms of catalytic action. Molecules, 24: 2362.

Magan, N., & Aldred, D. (2007). Post-harvest control strategies: minimizing mycotoxins in the food chain. *International journal of food microbiology, 119*(1-2), 131-139.

Malir, F., Ostry, V., Pfohl-Leszkowicz, A., Malir, J. & Toman, J. (2016). Ochratoxin A: 50 years of research. *Toxins, 8*(7), 191.

Mann, G. E., Codifer, L. P., & Dollear, F. G. (1967). Effect of heat on aflatoxins in oilseed meals. *Journal of agricultural and food chemistry, 15*(6), 1090-1092.

Marin, S., Ramos, A. J., Cano-Sancho, G. & Sanchis V. (2013). Mycotoxins: occurrence, toxicology, and exposure assessment. *Food and chemical toxicology, 60*, 218-237.

Medeiros, F. H. V. D., Martins, S. J., Zucchi, T. D., Melo, I. S. D., Batista, L. R., & Machado, J. D. C. (2012). Biological control of mycotoxin-producing molds. *Ciência e Agrotecnologia, 36*, 483-497.

Mesterhazy, A., Olah J & Popp, J. (2020). Losses in the grain supply chain: causes and solutions. *Sustainability, 12*(6), 2342

Misra, N. N., Yadav, B., Roopesh, M. S., & Jo, C. (2019). Cold plasma for effective fungal and mycotoxin control in foods: mechanisms, inactivation effects, and applications. *Comprehensive reviews in food science and food safety, 18*(1), 106-120.

Mohammadi, S. B., Rezaie, M. R., & Zareie, N. (2022). Formulation of Aflatoxins B1 & B2 reduction in corn by low level gamma irradiation. *Toxicon, 218*, 83-87.

Moon, Y. S., Lee, H. S., & Lee, S. E. (2018). Inhibitory effects of three monoterpenes from ginger essential oil on growth and aflatoxin production of Aspergillus flavus and their gene regulation in aflatoxin biosynthesis. *Applied Biological Chemistry, 61*, 243-250.

Mutiga, S. K., Were, V., Hoffmann, V., Harvey, J. W., Milgroom, M. G., & Nelson, R. J. (2014). Extent and drivers of mycotoxin contamination: Inferences from a survey of Kenyan maize mills. *Phytopathology, 104*(11), 1221-1231.

Neme, K., & Mohammed, A. (2017). Mycotoxin occurrence in grains and the role of postharvest management as a mitigation strategies. A review. *Food Control, 78*, 412-425.

Nunes, C., Usall, J., Teixido, N., Abadias, M., & Vinas, I. (2002). Improved control of postharvest decay of pears by the combination of Candida sake (CPA-1) and ammonium molybdate. *Phytopathology, 92*(3), 281-287.

Oguz, H. (2017). Mycotoxins and their importance. *Turkish clinics journal of veterinary scinces- pharmacology and toxicology, 3*, 113-119.

Oztekin, S., Dikmetas, DN., Devecioglu, D., Acar, EG., & Karbancioglu-Guler, F. (2023). Recent Insights into the Use of Antagonistic Yeasts for Sustainable Biomanagement of Postharvest Pathogenic and Mycotoxigenic Fungi in Fruits with Their Prevention Strategies against Mycotoxins. *Journal of Agricultural and Food Chemistry, 71* (26), 9923-9950.

Palicova, J., Chrpova, J., Tobolkova, A., Ovesna, J., & Stranska, M. (2024). Effect of pulsed electric field on viability of *Fusarium* micromycetes. *Cereal Research Communications*, 1-7.

Pallares, N., Barba, F. J., Berrada, H., Tolosa, J., & Ferrer, E. (2020). Pulsed electric fields (PEF) to mitigate emerging mycotoxins in juices and smoothies. *Applied Sciences, 10*(19), 6989.

Pandey, A. K., Samota, M. K., Kumar, A., Silva, A. S., & Dubey, N. K. (2023). Fungal mycotoxins in food commodities: present status and future concerns. *Frontiers in Sustainable Food Systems, 7*, 1162595.

Paulitz, T. C., & Belanger, R. R. (2001). Biological control in greenhouse systems. *Annual review of phytopathology, 39*(1), 103-133.

Peng W. X., Marchal, J. L. M. & Van der Poel, A. F. B. (2018). Strategies to prevent and reduce mycotoxins for compound feed manufacturing. *Animal feed science and technology, 237*, 129-153.

Peretto, G., Du, W. X., Avena-Bustillos, R. J., Sarreal, S. B. L., Hua, S. S. T., Sambo, P., & McHugh, T. H. (2014). Increasing strawberry shelf-life with carvacrol and methyl cinnamate antimicrobial vapors released from edible films. *Postharvest Biology and Technology, 89*, 11-18.

Petzinger, E., Ziegler, K. (2000). Ochratoxin A from a toxicological perspective. *Journal of veterinary pharmacology and therapeutics, 23*, 91-98.

Phillips, T. D., Afriyie-Gyawu, E., Williams, J., Huebner, H., Ankrah, N. A., Ofori-Adjei, D., & Wang, J. S. (2008). Reducing human exposure to aflatoxin through the use of clay: A review. *Food Additives & Contaminants: Part A*, 25(2), 134–145. https://doi.org/10.1080/02652030701567467

Picart-Palmade, L., Cunault, C., Chevalier-Lucia, D., Belleville, M. P., & Marchesseau, S. (2019). Potentialities and limits of some non-thermal technologies to improve sustainability of food processing. *Frontiers in nutrition, 5*, 130.

Pisuttu, C., Risoli, S., Moncini, L., Nali, C., Pellegrini, E. & Sarrocco, S. (2023). Sustainable Strategies to Counteract Mycotoxins Contamination and Cowpea Weevil in Chickpea Seeds during Post-Harvest. *Toxins, 15*, 61.

Pittet, A.C. (1998). Natural occurrence of mycotoxins in foods and feeds: An update review. *Revue de medicine veterinaire. 6*:479-492.

Pleadin, J., Frece, J. & Markov K. Mycotoxins in food and feed. (2019). *Advances in food and nutrition research, 89*, 297-345.

Proctor, A. D., Ahmedna*, M., Kumar, J. V., & Goktepe, I. (2004). Degradation of aflatoxins in peanut kernels/flour by gaseous ozonation and mild heat treatment. *Food additives and contaminants, 21*(8), 786-793.

Ribeiro, S. R., Garcia, M. V., Copetti, M. V., Brackmann, A., Both, V., & Wagner, R. (2021). Effect of controlled atmosphere, vacuum packaging and different temperatures on the growth of spoilage fungi in shelled pecan nuts during storage. *Food Control, 128*, 108173.

Ricard, J. L., & Ricard, T. J. (1997). The ethics of biofungicides–A case study: *Trichoderma harzianum* ATCC 20476 on Elsanta strawberries against Botrytis cinerea (gray mold). *Agriculture and Human Values, 14*(3), 251-258.

Robinson, A., Johnson, N. M., Strey, A., Taylor, J. F., Marroquin-Cardona, A., Mitchell, N. J., & Phillips, T. D. (2012). Calcium montmorillonite clay reduces urinary biomarkers of fumonisin B1 exposure in rats and humans. *Food Additives & Contaminants: Part A, 29*(5), 809-818.

Samyal, S. & Sharma, A. (2023). Mycotoxins: Structure, Biosynthesis, Health Effects, and Their Biological Detoxification. In: Singh, I., Rajpal, V.R., Navi, S.S. (eds). Fungal Resources for Sustainable Economy. Springer, Singapore. https://doi.org/10.1007/978-981-19-9103-5_18

Schatzmayr G., Zehner F., Täubel M., Schatzmayr D., Klimitsch A., Loibner A.P. & Binder E.M. (2006). Microbiologicals for deactivating mycotoxins. *Molecular Nutrition & Food Research, 50*, 543-551.

Sewram, V., Nair, J., Nieuwoudt, T., Leggott, N. & Shephard, G. (2000). Determination of patulin in apple juice by high-performance liquid chromatography-atmospheric pressure chemical ionization mass spectrometry. *Journal of chromatography, 897*, 365-74.

Sharma, R. R., Singh, D., & Singh, R. (2009). Biological control of postharvest diseases of fruits and vegetables by microbial antagonists: A review. *Biological control, 50*(3), 205-221.

Solis-Cruz B., Hernandez-Patlan D., Hargis B. & Tellez G. (2019). Control of Aflatoxicosis in Poultry Using Probiotics and Polymers. In: Njobeh P.B., Stepman F., editors. *Mycotoxins-Impact and Management Strategies.* IntechOpen; London, UK.

Song, C., & Qin, J. (2022). High-performance fabricated nano-adsorbents as emerging approach for removal of mycotoxins: a review. *International Journal of Food Science & Technology, 57*(9), 5781-5789.

Spadaro, D., & Gullino, M. L. (2004). State of the art and future prospects of the biological control of postharvest fruit diseases. *International journal of food microbiology, 91*(2), 185-194.

Stranska, M., Prusova, N., Behner, A., Dzuman, Z., Lazarek, M., Tobolkova, A., ... & Hajslova, J. (2023). Influence of pulsed electric field treatment on the fate of *Fusarium* and *Alternaria* mycotoxins present in malting barley. *Food Control, 145*, 109440.

Streit, E., Schwab, C., Sulyok, M., Naehrer, K., Krska, R. & Schatzmayr, G. (2013) Multi-mycotoxin screening reveals the occurrence of 139 different secondary metabolites in feed and feed ingredients. *Toxins, 5*(3), 504-523.

Sun, T., Ouyang, H., Sun, P., Zhang, W., Wang, Y., Cheng, S., & Chen, G. (2022). Postharvest UV-C irradiation inhibits blackhead disease by inducing disease resistance and reducing mycotoxin production in 'Korla'fragrant pear (Pyrus sinkiangensis). *International Journal of Food Microbiology, 362*, 109485.

Taniwaki, M. H., Hocking, A. D., Pitt, J. I., & Fleet, G. H. (2009). Growth and mycotoxin production by food spoilage fungi under high carbon dioxide and low oxygen atmospheres. *International journal of food microbiology, 132*(2-3), 100-108.

Tapia, M. S., Alzamora, S. M & Chirife, J. (2020). Effects of water activity (aw) on microbial stability as a hurdle in food preservation. *Agricultural and Food Sciences, Environmental Science.* pp. 323-355

Temba, B. A., Sultanbawa, Y., Kriticos, D. J., Fox, G. P., Harvey, J. J., & Fletcher, M. T. (2016). Tools for controlling a major global food and feed safety risk: Non-biological post-harvest procedures to decontaminate mycotoxins in food and feeds.

Tsitsigiannis, D. I., Dimakopoulou, M., Antoniou, P. P., & Tjamos, E. C. (2012). Biological control strategies of mycotoxigenic fungi and associated mycotoxins in Mediterranean basin crops. *Phytopathologia Mediterranea*, 158-174.

Ukwuru, M. U., Ohaegbu C. G. & Muritala, A. (2017). An overview of mycotoxin contamination of foods and feeds. *Journal of biochemical & microbial toxicology*, 1, 101.

Umereweneza, D., Kamizikunze, T. & Muhizi T. (2018). Assessment of mycotoxins types in some foodstuff consumed in Rwanda. *Food Control, 85,* 432-436.

Wang, X., Han, Y., Niu, H., Zhang, L., Xiang, Q., & Zong, W. (2022a). Alternaria mycotoxin degradation and quality evaluation of jujube juice by cold plasma treatment. *Food Control, 137,* 108926.

Wang, Y., Li, B., Shang, H., Ma, R., Zhu, Y., Yang, X. & Jiao, Z. (2022b). Effective inhibition of fungal growth, deoxynivalenol biosynthesis and pathogenicity in cereal pathogen Fusarium spp. by cold atmospheric plasma. *Chemical Engineering Journal,* 437, 135307.

Wegulo, S. N. (2012). Factors Influencing Deoxynivalenol Accumulation in Small Grain Cereals. *Toxins, 4,* 1157-1180. doi: 10.3390/toxins4111157.

Weller, D. M., Raaijmakers, J. M., Gardener, B. B. M., & Thomashow, L. S. (2002). Microbial populations responsible for specific soil suppressiveness to plant pathogens. *Annual review of phytopathology, 40*(1), 309-348.

Whitesides, S. K., Daoust, R. A., & Gouger, R. J. (1994). Commercialization of biological control agents: an industry perspective. *Biological Control of Postharvest Diseases of Fruits and Vegetables: Theory and Practice, CL Wilson and ME Wisniewski, Eds. CRC Press, Boca Raton FL.*

Wilson, J. P., Jurjevic, Z., Hanna, W. W., Wilson, D. M., Potter, T. L., & Coy, A. E. (2006). Host-specific variation in infection by toxigenic fungi and contamination by mycotoxins in pearl millet and corn. *Mycopathologia, 161,* 101-107.

Woldemariam, H. W., Harmeling, H., Emire, S. A., Teshome, P. G., Toepfl, S., & Aganovic, K. (2022). Pulsed light treatment reduces microorganisms and mycotoxins naturally present in red pepper (*Capsicum annuum* L.) powder. *Journal of Food Process Engineering, 45*(2), e13948.

Yan He, Pascal Degraeve & Nadia Oulahal. (2024). Bioprotective yeasts: Potential to limit postharvest spoilage and to extend shelf life or improve microbial safety of processed foods, *Heliyon, 10*(3), e24929.

Zorlugenç, B., Zorlugenç, F. K., Öztekin, S., & Evliya, I. B. (2008). The influence of gaseous ozone and ozonated water on microbial flora and degradation of aflatoxin B1 in dried figs. *Food and chemical toxicology, 46*(12), 3593-3597.

Chapter 15

Agronanotechnology: Sustainable Innovations for Managing Fungal Diseases and Mycotoxin Contamination in Plants

Emmadi. Venu[1], *Akurathi. Ramya[2*], Sathiyaseelan. K[1], Bhukya. Srinivas[3], Yeluru Mohan Babu[1]*

[1]Division of Plant Pathology, Indian Agricultural Research Institute, Pusa, New Delhi.

[2]Department of Plant Pathology, Junagadh Agricultural University, Junagadh, Gujarat.

[3]Department of Plant Pathology, Professor Jayashankar Telangana State Agricultural University Hyderabad, Telangana.

E-mail: ramyae0410@gmail.com

Introduction

Nanotechnology, which operates on materials ranging from 0.1 to 100 nanometers in size, holds significant promise for various applications in everyday life. It leverages the unique properties of particles and micron-sized granules, and specific atoms to influence three systems: chemical, biological, and physical. In particular, study on nanotechnology explores how altering the size of current antibacterial substances can enhance or modify their effects. By controlling particle size, researchers are able to rediscover and optimize the biological impacts of these agents. This approach has led to the development of both inorganic and organic antimicrobial nanoparticles designed to effectively combat bacterial, fungal, and viral pathogens (Patel et al., 2014, Sakr et al., 2018; Mohamed et al., 2021).

Nanoparticles synthesized through the green synthesis approach, utilizing primary and secondary metabolites from plant extracts, are recognized for their stability, cost-effectiveness, and environmental friendliness (Shabaaz Begum et al., 2020). These plant-derived metabolites not only support plant development, inhibit fungus-related pathogens, and mitigate crop diseases, but also facilitate the eco-friendly production of nanoparticles. They achieve this by acting as electron shuttles and aiding in the decrease and anchoring of elements (Mittal et al., 2013; Banerjee et al., 2014). This method of nanoparticle synthesis exemplifies green nanotechnology applications in managing fungal pathogens, with various nanoparticles specifically targeting different fungal threats.

3.1. Zinc Oxide Nanoparticles (ZnO NPs)

Senthilkumar and Sivakumar (2014) demonstrated the synthesis of zinc oxide nanoparticles (ZnO NPs) using aqueous leaf extracts of green tea (*Camellia sinensis*) and confirmed their effectiveness in combating fungi such as the three types of *Aspergillus: fumigatus, flavus, and niger*. Similarly, Lakshmeesha *et al.* (2019) reported the biofabrication of ZnO NPs using the extract from the buds of *Syzygium aromaticum* flowers, showing that these nanoparticles were capable of controlling *Fusarium graminearum* by inhibiting the development of mycelium and mycotoxin output. Furthermore, another study highlighted that ZnO NPs synthesized from *Eucalyptus globules* exhibited fungicidal properties against fungus that cause disease affecting apple trees, including *Alternaria mali, Diplodia seriata*, and *Botryosphaeria dothidea* (Ahmad et al., 2020). These findings suggest that ZnO NPs, produced through various plant extracts, offer significant potential for managing fungi disorders and protecting fruit trees.

3.2. Gold Nanoparticles (Au NPs)

The green antibacterial synthesis gold nanoparticles (Au NPs) has been explored using a variety of plant extracts, including new blooms or foliage from *Magnolia kobus* and *Diospyros kaki* (Song et al., 2009), as well as *Azadirachta indica* (Thirumurugan et al., 2010), *M(Liny et al., 2012), entha piperita*, (Mubarak et al., 2011), alfalfa (Montes et al., 2011), *Helianthus annuus* (sunflower) *Moringa oleifera* (Anand et al., 2015), and *Artemisia dracunculus* (Wacławek et al., 2018). In addition to plant-based methods, traditional chemical reducing agents such as sodium borohydride and sodium citrate are commonly employed in the synthesis of Au NPs (Islam et al., 2019).

Gold nanoparticles (Au NPs) have demonstrated promising antifungal properties, offering potential applications in agriculture for managing phytopathogenic fungi. For instance, Au NPs synthesized using an Abelmoschus esculentus seed aqueous extract exhibited strong effects of fungicides against *Aspergillus niger, Puccinia graminis var. tritici, and Aspergillus flavus* (Jayaseelan et al., 2013). Similarly, Au NPs produced from the extract of *Agaricus bisporus* (a mushroom that can be eaten) and $HAuCl_4 \cdot 3H_2O$ solution showed enhanced antimycotic activity against *Aspergillus terreus* (Eskandari-Nojedehi et al., 2018). *in contrast to Aspergillus flavus*. Furthermore, Au NPs generated by exposing Water-soluble gold ions added to Salix alba leaf extract showed effective antimicrobial characteristics towards *Aspergillus niger* and *Alternaria solani* but were less effective against *Aspergillus flavus*. It is also noted that these nanoparticles exhibited relative instability at elevated temperatures (Islam et al., 2019).

3.3. Copper Nanoparticles (Cu NPs)

Green synthesis methods for producing copper nanoparticles have been explored using various plant extracts, including Magnolia leaf extract (Team et al., 2010), Euphorbia nivulia stem latex (Valodkar et al., 2011), Papaya leaf extract from Carica and leaf extract from Aloe Vera (Gunalan et al., 2012). Shende et al. (2015) further advanced this field by demonstrating the effective green synthesis of copper *Citrus medica* as a source of nanoparticles. Their research validated that these nanoparticles exhibit significant inhibitory effects against several phytopathogenic fungi, specifically *Fusarium graminearum, Fusarium oxysporum, and Fusarium culmorum* respectively. Given their established in vitro antifungal potential, copper nanoparticles could be effectively utilized in managing fungal infections of plants (Kanhed et al., 2014, Banik et al., 2017).

4. Deactivating and Evacuating Fungal Cells Using Ghost Techniques

The sponge-like protocol was developed to create microbial ghost cells using inexpensive an as well the secure chemical-based mixtures, including NaOH, SDS, $NaHCO_3$, and H_2O_2 (Amara et al., 2013). This method generates ghost cells from numerous microbes by inducing the formation of evacuation pores in the microbial cell wall (El-Baky et al., 2021; Amara et al., 2014). What sets this protocol apart is its use of both the minimum MIC and MGC, or minimal growth and inhibition concentrations, of these chemicals. MIC is used to kill the cells under minimal conditions, while MGC allows the cells to survive but still impacts their cell walls. The protocol's unique approach involves optimizing the experimental design through either the complete or simplified Plackett-Burman experimental scheme to effectively prepare microbial ghosts.

5. Alternative Emerging Approaches

5.1. Suppressing Plant Fungal Diseases with Ultraviolet Light

Recent advancements in UV-protective substances, such as sheets and nets made of polyethylene, have been evolved to manage transmitters of insects that affect enclosed environment vegetation crops (Diaz et al., 2007). These resources effectively filter UV radiation (280–400 nm), disrupting insect vision and thereby reducing their impact on plants. Research by Roberts and Paul (2006) highlighted that illumination not just influences plant development and biochemical procedures, but it also has a significant impact on enhancing plant resistance to pests. Additionally, the application of UV light has been explored as a method to combat fungal infections of plants, for example powdery mildew on strawberries and apples (Onofre et al., 2019). Studies have shown that divulging foliage of plants to a UV dose of approximately 30 mJ/cm^2 can significantly reduce infection rates without adversely affecting plant health. This technique involves using low doses of UV light to kill fungi, with plants being able to endure much higher UV levels. Consistent use of this method it is better over time for effective disease control.

Enhancing Plant Resistance to Fungal Diseases

Success in plant breeding for disease resistance can be exemplified by the development of barley varieties with long-lasting and all-encompassing management of *Blumeria graminis f. sp. hordei*, the pathogen responsible for powdery mildew. This resistance is largely due to the recessive alleles of the Mildew locus O (Mlo) gene (Acevedo-Garcia et al., 2017). Additionally, plant resistance strategies have been effectively applied to mitigate aflatoxin accumulation in maize caused by Aspergillus flavus. For instance, six maize germplasm lines bred in Africa were evaluated in the US and demonstrated a significant reduction in aflatoxin levels (Brown et al., 2016).

Breeding for disease resistance in ornamental plants has been increasingly emphasized, though not all ornamental species naturally possess such resistance. Consequently, disease control often relies on selecting varieties that are inherently resistant to diseases (Jesenicnik et al., 2019). Understanding the interactions between host plants and pathogens is essential for creating tactics that work to enhance plant-growth resistance (Hammond-Kosack et al., 2003). Traditional breeding methods involve selecting disease-resistant traits through non-transgenic approaches, such as using DNA-based markers and conducting multiple breeding cycles to integrate resistance with desirable ornamental features. In contrast, genetic engineering methods that introduce either broad-spectrum or particular disease resistance through carefully regulated transgenes. Despite these advances, breeding ornamentals with resistance is comparatively underdeveloped, as it typically only becomes a focus in the later stages of breeding (Michelmore et al., 2003). The Rarely does ornamentals have genetic mapping of disease resistance due to their complex genomes. (Arens et al., 2012). However, transferring genes related to pathogenesis, such as those encoding chitinase, glucanase, osmotin, and defensin, has shown promise in achieving tolerance to fungus-related diseases. (Parmar et al., 2017). Additionally, methods for modifying the genome are being employed to build resistance to illness in ornamental plants Mushtaq et al., 2019; Mekapogu et al., 2021).

RNA interference (RNAi) strategies, such as host-induced gene silencing (HIGS) (Nowara et al., 2010), and spray-induced gene silencing (SIGS) (Koch et al., 2016) offer promising alternatives to traditional fungicides for protecting plants from fungal pathogens. These strategies involve applying double-stranded RNA (dsRNA) molecules either by spraying them onto plant foliage or by generating them through genetically engineered plants. Overall potency of RNAi-based crop protection depends heavily on the design of the dsRNA. It is essential to select the right gene and the specific region of the gene to target for optimal results. Enhancing dsRNA uptake can be achieved through various methods, including the use of carriers and formulations that protect the dsRNA from decline caused by environmental elements as well as aid in its movement within the plant. Additionally, it is important to ensure that the target fungal pathogen possesses functional RNAi machinery to achieve successful control (Secic et al., 2021).

6. Factors Affecting the Production and Contamination of Mycotoxin in Food

Mycotoxins, which are toxic compounds produced can infect a variety of substrates via certain fungus agricultural and culinary goods, with cereals, cereal-based products, and animal-derived foods being the most common sources of human exposure (Persi, Pleadin, Kovacevic, Scortichini, & Milone, 2014). It is possible for infection to happen at any stage in the lead-up to the harvest, and while the presence of fungi is a risk factor, it does not always lead to mycotoxin contamination, as the conditions required for mycotoxin production differ from those needed for fungus development (Kochiieru et al., 2020). To address this, the Food Safety Management System (FSMS) has been proposed as an effective strategy for mitigating mycotoxin contamination. The FSMS encompasses a comprehensive approach to food hygiene and safety, including good hygiene practices (GHP), good agricultural practices (GAP), good storage practices (GSP), good manufacturing practices (GMP), and hazard analysis and critical control points (HACCP). By implementing these procedures and management policies, the FSMS aims to control the food production process and environment, thereby ensuring the safety of food products for consumption (Nada, Nikola, Bozidar, Ilija, & Andreja, 2022).

Mycotoxin synthesis and contamination in food are significantly influenced by local conditions such as substrate kind, fungus species, as well as nutrition availability. Mold can grow on a wide variety of food items because the essential nutrients for mycotoxigenic fungi, primarily nitrogen and carbon, are commonly found in carbohydrate-rich foods. While tryptophan can inhibit fungal growth, additional nutrients such as tyrosine, triglycerides, fats, and sugar can enhance mycotoxin production (Gacem & Ould El Hadj-Khelil, 2016). Various factors, including substrate composition, temperature, pH, and osmotic pressure, play crucial roles in proliferation of fungi and the formation of mycotoxin (Kokkonen, Jestoi, & Rizzo, 2005). For instance, increased sugar content in the growth medium has been linked to higher levels of aflatoxin B1 generated by *Aspergillus flavus* (Liu et al., 2016; Uppala, Bowen, & Woods, 2013). Additionally, the specific strain of fungi can affect toxin production, as different fungal species have varying levels of toxicity and growth conditions (Greeff-Laubscher, Beukes, Marais, & Jacobs, 2020). For example, *Aspergillus favus* and *Aspergillus carbonarius* produce aflatoxin B1 and ochratoxin, respectively, under varying temperature conditions, ranging from 15–44°C and 8–40°C (Mannaa & Kim, 2017).

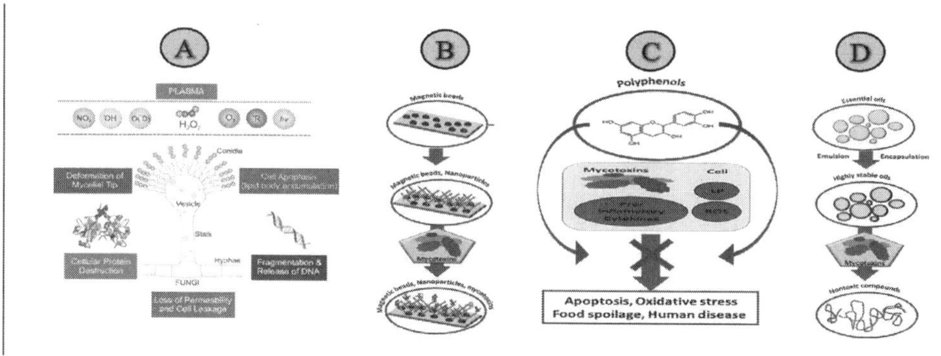

Fig.1. New approaches for managing infestation of food with mycotoxin and its mechanisms include: (A) cold plasma (Adapted from Wu *et al.*, 2021), (B) polyphenols, (C) natural essential oils, and (D) magnetic and nanoparticles.

Current Innovative Techniques for managing contamination with mycotoxin in Food

Recent research into reducing the presence of mycotoxin in food has led to the exploration of several innovative strategies that effectively address toxicity and disease while minimizing toxic residuals and preserving food quality. Among these approaches are magnetic materials, polyphenols and flavonoids, cold atmospheric plasma (CAP), nanoparticles, and natural essential oils (NEOs). (Fig.1). These methods offer various mechanisms for mycotoxin control, each with its unique benefits and limitations. foods (Ahmed et al., 2022; Cai et al., 2022; Chtioui, Balmas, Delogu, Migheli, & Oufensou, 2022; Diao et al., 2022; Gavahian, Pallares, Al Khawli, Ferrer, & Barba, 2020; Guimar˜ aes & Venancio,ˆ 2022; Nasrollahzadeh, Mokhtari, Khomeiri, & Saris, 2022; Sen, Onal-Ulusoy, & Mutlu, 2019; Jafarzadeh, Hadidi, Forough, Nafchi, & Mousavi Khaneghah, 2022; Wang et al., 2022).

7.1. Cold plasma

Plasma, often referred to as the fourth condition of substance, is an entirely or partly ionized gas that can be produced by passing a neutral gas through an electric current, causing the separation of its constituent parts (Kis et al., 2020; Wielogorska et al., 2019). when the breakdown voltage is exceeded by the applied voltage, plasma is created, producing reactive oxygen and nitrogen species (RONS). Notably, cold atmospheric plasma (CAP) is able to produce at pressure in the atmosphere and the surrounding temperature while using a minimal amount of energy (Hojnik et al., 2021; Wielogorska et al., 2019). Water can be treated with CAP either directly or indirectly, buffer solutions, and acids, resulting in broad-spectrum disinfectants (Ali, Sun, Cheng, & Johnson Esua, 2022; Esua, Cheng, & Sun, 2021). Its antimicrobial properties and capability to degrade complex biochemical molecules stem from a unique combination of reactive species, such as O_2–, NO_2–, NO_3–, H_2O_2, OH, ONOO–, and O_3 along with electric fields and ultraviolet photons fields. These elements work together while maintaining near-ambient temperatures, thus characterizing CAP as non-thermal (Esua, Sun, Cheng, Wang, & Chen, 2022; Hojnik et al., 2021; Sen et al., 2019; Wielogorska et al., 2019).

Non-thermal Cold Atmospheric Plasma (CAP) is recognized as an efficient and cost-effective method to remove harmful germs from food and agricultural goods. This innovative strategy not only ensures rapid decontamination but also maintains the quality of the treated items, as noted by Esua, Cheng, and Sun (2020).

The effectiveness of mycotoxin degradation using plasma technology is highly reliant upon the chemical composition of the toxin and the specific circumstances of processing employed. For instance, the rate at which deoxynivalenol (DON) breaks down in wheat is affected by elements like moisture content, matrix properties, processing duration, gas type, and plasma voltage. Studies show that applying 50 kV of atmospheric air plasma for 8 minutes can achieve a 25.82% reduction in DON levels. Additionally, increasing the moisture content from 8% to 20% improves the degradation efficiency from 9.59% to 36.10%, while only a slight increase in wheat whiteness is noted, indicating minimal impact on wheat quality (Chen et al., 2022).

Similarly, the breakdown of HT-2 toxin, the primary metabolite of T-2 toxin, in oat flour is affected by type, humidity, and duration of exposure of gas used in cold plasma treatment. Using nitrogen (N2) as the generating gas, the toxins are effectively oxidized by cold plasma-derived oxygen atoms, leading to 43.25% decreases in the levels of T-2 toxin and 38.54% for HT-2 toxin (Kiš et al., 2020). An further study demonstrates that boosting the power that is provided from 10 to 30 kV results in reductions of 1.3–56.5% for alternariol (AOH) and 8.1–62.8% for alternariol monomethyl ether (AME) in jujube juice. This treatment also significantly affects the juice's quality, resulting in a decrease in Brix, along with improvements in antioxidant activity, physical stability, total acids, and sugar reduction (Wang et al., 2022).

Non-thermal plasma (CAP) inactivates fungi through a mechanism primarily involving oxidative damage. This process starts with the induction of oxidation, which compromises the integrity of the cell walls by causing harm to the cell membranes. As a result, there is an increase in membrane permeability. This damage leads to DNA fragmentation and leakage, the destruction among the proteins found in cells, and ultimately, the death of cells. Additionally, the mycelial spores undergo distortion, further contributing to the overall inactivation of the fungus.

7.4. Magnetic compounds and nanoparticles

Recent research has highlighted the effectiveness of eco-friendly, cost-efficient methods for controlling mycotoxins using magnetic substances and small nano-sized particles. For instance, $Fe3O_4$ coating of chitosan on magnetic particles have been demonstrated to effectively adsorb patulin from fruit juice. Additionally, the reaction between retinoic acid and nanocellulose has been shown to adsorb aflatoxin B1 (AFB1) from various food products free of any detectable toxicity, based on the pH and concentration levels (Jebali, Yasini Ardakani, Sedighi, & Hekmatimoghaddam, 2015; Luo, Zhou, & Yue, 2017). Various nanoparticles that are magnetized, such as Surface-active maghemite, surface-clay, and surface-gel, and materials in the form of nanoparticles, such as selenium (SEN), copper, silver, zinc oxide (ZON), and have proven functional in mycotoxin removal and binding in agricultural feed and food products (Abd-Elsalam, Hashim, Alghuthaymi, & Said-Galiev, 2017; Magro et al., 2016). Specifically, SEN derived from *Trichoderma harzianum* JF309 reduced fumonisin B1 (FB1) and deoxynivalenol (DON) levels by 63% and 76%, respectively (Hu et al., 2019). SLN inhibited the expansion of ochratoxin and aflatoxin-producing fungi, thus preventing ochratoxin A (OTA) and aflatoxin accumulation in corn-based media. Meanwhile, ZON effectively suppressed the expansion of *Penicillium*, *Aspergillus*, and *Fusarium* species and reduced the food model systems' generation of FB1, OTA, and AFB1 (Gomez, Tarazona, Mateo, Jiménez, & Mateo, 2019; Hassan, Howayda, & Mahmoud, 2013).

Made from maize byproducts, magnetic carbon nanocomposites, have demonstrated significant promise for detoxifying aflatoxin B1 (AFB1) in poultry. These composites achieved up to 90% adsorption at pH 7 within 180 minutes of utilization. Additionally a chitosan-glutaraldehyde compound that is cross-linked has been identified as a powerful substance for adsorption for multiple toxins, showing superior adsorption performance for various mycotoxins compared to other components examined (Zahoor & Ali Khan, 2016; Zhao et al., 2015).

Nanoparticles contribute to reducing mycotoxin risk through various molecular mechanisms, including inducing oxidative stress, triggering inflammation, and interacting with nucleic acids to destroy microorganisms in both vegetation and animals (Horky et al., 2018). These relationships often result in apoptosis, Reactive oxygen species (ROS) production suppression of lipid peroxidation and mitochondrial activities, and the autophagous mode. Specifically, the ZON toxin generates ROS that disrupt the fungal hyphae's bilayer of lipid. This damage leads to the breakdown of the fungal cells, which is visible as strange protrusions and deformations at the surface-level of the hyphae when treated containing 12 mmol/L ZON, as noticed using electron microscopy by scanning (Pietrzak et al., 2015).

8. Recent Innovations and Strategies

Pulsed electric fields (PEF) have been increasingly utilized to inactivate fungi and their mycotoxins in food products, particularly targeting aflatoxins B1 and G1 made by Aspergillus (Gavahian et al., 2020). Research indicates that PEF is a highly effective and cost-efficient method for detoxifying various mycotoxins, including aflatoxins, fumonisins, zearalenone, ochratoxin A (OTA), and trichothecenes. However, specific assessments for different foods are recommended before widespread industry integration (Nunes et al., 2021). Additionally, water-assisted microwave treatment (WMT) has been explored for reducing aflatoxin B1 in corn. This method not only effectively eliminates AFB1 but also minimizes heat damage to the corn kernels, offering a novel approach to aflatoxin reduction (Zhang et al.,

2020). High-pressure treatment (HP) has also proven successful in significantly reducing juice samples with mycotoxins (Gavahian et al., 2020). Furthermore, technology for active food packing has been identified like a successful means to manage fungus and detoxify mycotoxins across various foods such as cereal grains, dairy products, fruits, nuts, and bread goods (Jafarzadeh et al., 2022). The use of biopolymers, or their combinations such as polysac economical and ecologically benign charides, lipids, and proteins, or their blends, offers a economical and ecologically benign alternative due to the fact that they degrade in the environment.

Recent research has explored the use of Sodium and calcium bentonite (Ca- and Na-bentonite), as innovative methods for removing ochratoxin A (OTA) from certain types of cheese, including Karish and Roomy cheeses, which exhibited OTA concentrations of 3.399 and 4.138 μg/kg, respectively (Hamad et al., 2022). The study found that Ca-bentonite demonstrated superior adsorption efficiency compared to Na-bentonite, especially at concentrations of 60 and 100 mg/mL and across various pH levels in phosphate-buffered saline (PBS). In vivo tests with rats showed that the enzymatic activities in samples treated with both OTA and bentonite were comparable to those in the control group, suggesting effective detoxification. Additionally, scanning electron microscopy (SEM) and Fourier transform infrared spectroscopy (FTIR) verified the exceptional binding capabilities of bentonites for OTA. Notably, feta cheese supplemented with Ca-bentonite exhibited enhanced sensory attributes, marking this approach as a promising and inventive adsorbent of food grade method for the elimination of OTA in food products.

9. Future Perspectives:

Farmers and agriculturists should focus on adopting methods that are safe, non-toxic, and environmentally sustainable for managing fungal infections in plants. These green approaches are crucial for safeguarding human and animal health as well as preserving soil biodiversity. Implementing these methods in real-world conditions is essential for effective plant disease management. Additionally, there is a need for ongoing research to drive further innovations in sustainable fungal disease control.

Conclusion

Mycotoxin contamination in produce from nutrition, agriculture and feed for livestock is a significant worldwide issue due to its toxicity and the risks it poses to the health of people and animals, in addition to the economic losses associated with it. Despite the use of necessary initiatives like GAP, or Good Agricultural Practices, Good Manufacturing Practices (GMP), Good Storage Practices (GSP), Good Hygiene Practices (GHP), and Hazard Analysis and Critical Control Points (HACCP) at various stages of pre-, post-, and processing; contamination with mycotoxin remains a persistent challenge (Nada et al., 2022). Quick and easy identification is crucial for eliminating these toxins and ensuring food safety. However, traditional pharmaceutical, both genetic and physiological detoxification approaches face limitations due to increasing consumer awareness, regulatory concerns, the creation of toxins that cause cancer, and resistance by new mycotoxin strains. As a result, there is a growing focus on developing creative tactics regarding the swift management, mitigation, and eradication of mycotoxins that minimize impacts on food quality, including its morphological, physicochemical, textural, and structural properties, as well as on the environment.

Recent reviews have highlighted the productivity of creative tactics for managing development of fungi that is toxic and mycotoxin contamination within foodstuff. Techniques such as Cold Atmospheric Plasma (CAP) technology, polyphenol and flavonoid inhibitors, magnetic materials, nanoparticles, and novel emerging options (NEOs) have demonstrated significant potential in reducing and detoxifying mycotoxins. The success of these methods is influenced by both inherent elements such as contamination and kind of substrate level, humidity percentage, and analytical composition, as well as external aspects including decontamination methods and processing conditions. CAP technology, in particular, offers an economically viable and secure substitute for traditional thermal methods, aligning well with sustainable development goals. However, its efficiency is closely linked to the reactive oxygen and nitrogen species (RONS) produced, which depend on the type of gas used. A notable concern is that CAP can introduce nitrites into foods, which may degrade over time into N-nitroso compounds with potential health risks. This issue necessitates careful regulation according to US FDA guidelines for edible products. Additionally, longer exposure times may be required for complete decontamination, and exploring indirect application methods might offer viable alternatives.

NEOs offer several advantages, including adopting practices that minimize environmental impact while optimizing resource use, but they face limitations such as reduced efficacy, high sensitivity, susceptibility to oxidation, and not soluble in nature (Loi, Paciolla, Logrieco, & Mule, 2020). Similarly, while magnetic substances and small particles show significant potential in the sectors of nutrition, crop cultivation, and animal production industries due to their ability to adsorb mycotoxins, their applications are still in the early stages. The mechanisms through which these inhibit fungal growth materials are not fully elucidated, and there persists limited knowledge about their possible unintended consequences, particularly concerning electromagnetic compounds and nanoscale fragments. This lack of understanding could restrict their entire range of possibilities and implementations. Furthermore, the strategies for removing toxins, their ability to preserve dietary aspects and sensory attributes, and assessment of harmful effects from remaining compounds require detailed analysis to prevent food from being contaminated again.

Despite the significant progress made, current innovative strategies for mycotoxin decontamination have not yet achieved 100% efficiency and are not universally effective across all matrices. However, as environmental protection awareness grows, these strategies offer valuable understanding to fast-track the development of emerging eco-friendly technologies in this field. The multidimensional approach to decontamination, combined with technological advancements, suggests that integrating various approaches combining traditional and modern techniques could enhance detoxification effectiveness and flexibility to various culinary matrices. Nonetheless, most research investigations remain at the research setting, highlighting the requirement for collaborative efforts across various fields that includes economic studies and engineering practices to scale up and commercialize these technologies in a cost-effective and sustainable manner.

References

Abd El-Baky, N., Sharaf, M. M., Amer, E., Kholef, H. R., Hussain, M. Z., Ahmed, R., ... & Amara, A. A. (2018). The Minimum Inhibition and Growth Concentrations for Controlling Fungal Infections as well as Ghost Cells Preparation: *Aspergillus flavus* as a Model. *Biomedical Journal*, 1, 5.

Abd-Elsalam, K. A., Hashim, A. F., Alghuthaymi, M. A., & Said-Galiev, E. (2017). Nanobiotechnological strategies for toxigenic fungi and mycotoxin control. In *Food preservation* (pp. 337-364). Academic Press.

Acevedo-Garcia, J., Gruner, K., Reinstädler, A., Kemen, A., Kemen, E., Cao, L., ... & Panstruga, R. (2017). The powdery mildew-resistant *Arabidopsis* mlo2 mlo6 mlo12 triple mutant displays altered infection phenotypes with diverse types of phytopathogens. *Scientific reports*, 7(1), 9319.

Ahmad, H., Venugopal, K., Rajagopal, K., De Britto, S., Nandini, B., Pushpalatha, H. G., ... & Jogaiah, S. (2020). Green synthesis and characterization of zinc oxide nanoparticles using Eucalyptus globules and their fungicidal ability against pathogenic fungi of apple orchards. *Biomolecules*, 10(3), 425.

Ali, M., Sun, D. W., Cheng, J. H., & Esua, O. J. (2022). Effects of combined treatment of plasma activated liquid and ultrasound for degradation of chlorothalonil fungicide residues in tomato. *Food chemistry*, 371, 131162.

Amro, A. A., Salem-Bekhit, M. M., & Alanazi, F. K. (2014). Plackett–Burman randomization method for Bacterial Ghosts preparation form E. coli JM109. *Saudi Pharmaceutical Journal*, 22(3), 273-279.

Anand, K., Gengan, R. M., Phulukdaree, A., & Chuturgoon, A. (2015). Agroforestry waste Moringa oleifera petals mediated green synthesis of gold nanoparticles and their anti-cancer and catalytic activity. *Journal of Industrial and Engineering Chemistry*, 21, 1105-1111

Atiq, M., Naeem, I., Sahi, S. T., Rajput, N. A., Haider, E., Usman, M., ... & Qayyum, A. (2020). Nanoparticles: A safe way towards fungal diseases. *Archives of Phytopathology and Plant Protection*, 53(17-18), 781-792.

Banerjee, P., Satapathy, M., Mukhopadhyay, A., & Das, P. (2014). Leaf extract mediated green synthesis of silver nanoparticles from widely available Indian plants: synthesis, characterization, antimicrobial property and toxicity analysis. *Bioresources and Bioprocessing*, 1, 1-10.

Banik, S., & Pérez-de-Luque, A. (2017). In vitro effects of copper nanoparticles on plant pathogens, beneficial microbes and crop plants. *Spanish Journal of Agricultural Research*, 15(2), e1005-e1005.

Bart, H., Evelyne, D., & Els, D. (2020). Carry-over of some Fusarium mycotoxins in tissues and eggs of chickens fed experimentally mycotoxin-contaminated diets. *Food and Chemical Toxicology*, 145, 111715.

Begum, J. S., Manjunath, K., Pratibha, S., Dhananjaya, N., Sahu, P., & Kashaw, S. (2020). Bioreduction synthesis of zinc oxide nanoparticles using *Delonix regia* leaf extract (Gul Mohar) and its agromedicinal applications. Journal of Science: Advanced Materials and Devices, 5(4), 468-475.

Brown, R. L., Williams, W. P., Windham, G. L., Menkir, A., & Chen, Z. Y. (2016). Evaluation of African-bred maize germplasm lines for resistance to aflatoxin accumulation. *Agronomy*, 6(2), 24.

Chen, X., Qiu, Y., Zhang, J., Guo, Y., Ding, Y., & Lyu, F. (2022). Degradation efficiency and products of deoxynivalenol treated by cold plasma and its application in wheat. *Food Control*, 136, 108874.

Cho, J. Y., Choi, G. J., Lee, S. W., Lim, H. K., Jang, K. S., Lim, C. H., ... & Kim, J. C. (2006). In vivo antifungal activity against various plant pathogenic fungi of curcuminoids isolated from the rhizomes of *Curcuma longa*. *The Plant Pathology Journal*, 22(1), 94-96.

Chtioui, W., Balmas, V., Delogu, G., Migheli, Q., & Oufensou, S. (2022). Bioprospecting phenols as inhibitors of trichothecene-producing Fusarium: Sustainable approaches to the management of wheat pathogens. *Toxins*, 14(2), 72.\

Diao, E., Ma, K., Qian, S., Zhang, H., Xie, P., Mao, R., & Song, H. (2022). Removal of patulin by thiol-compounds: A review. *Toxicon*, 205, 31-37.

Díaz Desani, B. M., & Fereres, A. (2007). Ultraviolet-blocking materials as a physical barrier to control insect pests and plant pathogens in protected crops.

Eskandari-Nojedehi, M., Jafarizadeh-Malmiri, H., & Rahbar-Shahrouzi, J. (2018). Hydrothermal green synthesis of gold nanoparticles using mushroom (*Agaricus bisporus*) extract: physico-chemical characteristics and antifungal activity studies. *Green Processing and Synthesis*, 7(1), 38-47.

Esua, O. J., Cheng, J. H., & Sun, D. W. (2020). Antimicrobial activities of plasma-functionalized liquids against foodborne pathogens on grass carp (*Ctenopharyngodon Idella*). *Applied Microbiology and Biotechnology*, 104, 9581-9594.

Esua, O. J., Cheng, J. H., & Sun, D. W. (2021). Functionalization of water as a nonthermal approach for ensuring safety and quality of meat and seafood products. *Critical reviews in food science and nutrition*, 61(3), 431-449.

Esua, O. J., Sun, D. W., Cheng, J. H., Wang, H., & Chen, C. (2022). Hybridising plasma functionalized water and ultrasound pretreatment for enzymatic protein hydrolysis of Larimichthys polyactis: Parametric screening and optimization. *Food Chemistry*, 385, 132677.

Gacem, M. A., & El Hadj-Khelil, A. O. (2016). Toxicology, biosynthesis, bio-control of aflatoxin and new methods of detection. *Asian Pacific Journal of Tropical Biomedicine*, 6(9), 808-814.

Gavahian, M., Pallares, N., Al Khawli, F., Ferrer, E., & Barba, F. J. (2020). Recent advances in the application of innovative food processing technologies for mycotoxins and pesticide reduction in foods. *Trends in Food Science & Technology*, 106, 209-218.

Gavahian, M., Pallares, N., Al Khawli, F., Ferrer, E., & Barba, F. J. (2020). Recent advances in the application of innovative food processing technologies for mycotoxins and pesticide reduction in foods. *Trends in Food Science & Technology*, 106, 209-218.

Gómez, J. V., Tarazona, A., Mateo, F., Jiménez, M., & Mateo, E. M. (2019). Potential impact of engineered silver nanoparticles in the control of aflatoxins, ochratoxin A and the main aflatoxigenic and ochratoxigenic species affecting foods. *Food Control*, 101, 58-68.

Greeff-Laubscher, M. R., Beukes, I., Marais, G. J., & Jacobs, K. (2020). Mycotoxin production by three different toxigenic fungi genera on formulated abalone feed and the effect of an aquatic environment on fumonisins. *Mycology*, 11(2), 105-117.

Guimarães, A., & Venâncio, A. (2022). The potential of fatty acids and their derivatives as antifungal agents: *A review. Toxins*, 14(3), 188.

Gunalan, S., Sivaraj, R., & Venckatesh, R. (2012). Aloe barbadensis Miller mediated green synthesis of mono-disperse copper oxide nanoparticles: optical properties. Spectrochimica Acta Part A: *Molecular and Biomolecular Spectroscopy*, 97, 1140-1144.

Hamad, G. M., El-Makarem, H. A., Elaziz, A. A., Amer, A. A., El-Nogoumy, B. A., & Abou-Alella, S. A. (2022). Adsorption efficiency of sodium & calcium bentonite for ochratoxin A in some Egyptian cheeses: An innovative fortification model, in vitro and in vivo experiments. *World Mycotoxin Journal*, 15(3), 285-300.

Hammond-Kosack, K. E., & Parker, J. E. (2003). Deciphering plant–pathogen communication: fresh perspectives for molecular resistance breeding. *Current opinion in biotechnology*, 14(2), 177-193.

Hassan, A. A., Howayda, M. E., & Mahmoud, H. H. (2013). Effect of zinc oxide nanoparticles on the growth of mycotoxigenic mould. *Studies in Chemical Process Technology* (SCPT) Volume, 1(4).

Hojnik, N., Modic, M., Walsh, J. L., Žigon, D., Javornik, U., Plavec, J., ... & Cvelbar, U. (2021). Unravelling the pathways of air plasma induced aflatoxin B1 degradation and detoxification. *Journal of Hazardous Materials*, 403, 123593.

Hu, D., Yu, S., Yu, D., Liu, N., Tang, Y., Fan, Y., ... & Wu, A. (2019). Biogenic Trichoderma harzianum-derived selenium nanoparticles with control functionalities originating from diverse recognition metabolites against phytopathogens and mycotoxins. *Food Control*, 106, 106748.

Islam, N. U., Jalil, K., Shahid, M., Rauf, A., Muhammad, N., Khan, A., ... & Khan, M. A. (2019). Green synthesis and biological activities of gold nanoparticles functionalized with Salix alba. *Arabian Journal of Chemistry*, 12(8), 2914-2925.

Islam, N. U., Jalil, K., Shahid, M., Rauf, A., Muhammad, N., Khan, A., ... & Khan, M. A. (2019). Green synthesis and biological activities of gold nanoparticles functionalized with Salix alba. *Arabian Journal of Chemistry*, 12(8), 2914-2925.

Jafarzadeh, S., Hadidi, M., Forough, M., Nafchi, A. M., & Mousavi Khaneghah, A. (2023). The control of fungi and mycotoxins by food active packaging: A review. Critical Reviews in Food Science and Nutrition, 63(23), 6393-6411.

Jayaseelan, C., Ramkumar, R., Rahuman, A. A., & Perumal, P. (2013). Green synthesis of gold nanoparticles using seed aqueous extract of Abelmoschus esculentus and its antifungal activity. *Industrial Crops and Products*, 45, 423-429.

Jebali, A., Yasini Ardakani, S. A., Sedighi, N., & Hekmatimoghaddam, S. (2015). Nanocellulose conjugated with retinoic acid: its capability to adsorb aflatoxin B1. *Cellulose*, 22, 363-372.

Jeseničnik, T., Štajner, N., Radišek, S., & Jakše, J. (2019). RNA interference core components identified and characterised in *Verticillium nonalfalfae*, a vascular wilt pathogenic plant fungi of hops. *Scientific reports*, 9(1), 8651.

Kanhed, P., Birla, S., Gaikwad, S., Gade, A., Seabra, A. B., Rubilar, O., ... & Rai, M. (2014). In vitro antifungal efficacy of copper nanoparticles against selected crop pathogenic fungi. *Materials Letters*, 115, 13-17.

Kiš, M., Milošević, S., Vulić, A., Herceg, Z., Vukušić, T., & Pleadin, J. (2020). Efficacy of low pressure DBD plasma in the reduction of T-2 and HT-2 toxin in oat flour. *Food chemistry*, 316, 126372.

Koch, A., Biedenkopf, D., Furch, A., Weber, L., Rossbach, O., Abdellatef, E., ... & Kogel, K. H. (2016). An RNAi-based control of *Fusarium graminearum* infections through spraying of long dsRNAs involves a plant passage and is controlled by the fungal silencing machinery. *PLoS pathogens*, 12(10), e1005901.

Kochiieru, Y., Mankevičienė, A., Cesevičienė, J., Semaškienė, R., Dabkevičius, Z., & Janavičienė, S. (2020). The influence of harvesting time and meteorological conditions on the occurrence of *Fusarium* species and mycotoxin contamination of spring cereals. *Journal of the Science of Food and Agriculture*, 100(7), 2999-3006.

Kumar, K. N., Venkataramana, M., Allen, J. A., Chandranayaka, S., Murali, H. S., & Batra, H. V. (2016). Role of Curcuma longa L. essential oil in controlling the growth and zearalenone production of *Fusarium graminearum*. *LWT-Food Science and Technology*, 69, 522-528.

Lakshmeesha, T. R., Kalagatur, N. K., Mudili, V., Mohan, C. D., Rangappa, S., Prasad, B. D., ... & Niranjana, S. R. (2019). Biofabrication of zinc oxide nanoparticles with *Syzygium aromaticum* flower buds extract and finding its novel application in controlling the growth and mycotoxins of *Fusarium graminearum. Frontiers in microbiology*, 10, 1244.

Liny, P., Divya, T. K., Barasa Malakar, B. M., Nagaraj, B., Krishnamurthy, N. B., & Dinesh, R. (2012). Preparation of gold nanoparticles from *Helianthus annuus* (sun flower) flowers and evaluation of their antimicrobial activities.

Lisboa, S. P., Cupertino, M. D. C., Arruda, V. M., & Casali, V. W. D. (2005). Nova visão dos organismos vivos e o equilíbrio pela homeopatia: de acordo com "O Organon" de Samuel Hahnemann.

Liu, J., Sun, L., Zhang, N., Zhang, J., Guo, J., Li, C., ... & Qi, D. (2016). Effects of nutrients in substrates of different grains on aflatoxin B1 production by *Aspergillus flavus.* BioMed Research International, 2016(1), 7232858..

Luo, Y., Zhou, Z., & Yue, T. (2017). Synthesis and characterization of nontoxic chitosan coated Fe3O4 particles for patulin adsorption in a juice-pH simulation aqueous. *Food Chemistry*, 221, 317–323.

Mannaa, M., & Kim, K. D. (2017). Influence of temperature and water activity on deleterious fungi and mycotoxin production during grain storage. *Mycobiology*, 45(4), 240-254.

Mittal, A. K., Chisti, Y., & Banerjee, U. C. (2013). Synthesis of metallic nanoparticles using plant extracts. *Biotechnology advances*, 31(2), 346-356.

Mohamed, A. A., Elshafie, H. S., Sadeek, S. A., & Camele, I. (2021). Biochemical characterization, phytotoxic effect and antimicrobial activity against some phytopathogens of new Gemifloxacin schiff base metal complexes. Chemistry & Biodiversity, 18(9), e2100365.

Montes, M. O., Mayoral, A., Deepak, F. L., Parsons, J. G., Jose-Yacamán, M., Peralta-Videa, J. R., & Gardea-Torresdey, J. L. (2011). Anisotropic gold nanoparticles and gold plates biosynthesis using alfalfa extracts. *Journal of Nanoparticle Research,* 13, 3113-3121.

Mushtaq, M., Sakina, A., Wani, S. H., Shikari, A. B., Tripathi, P., Zaid, A., ... & Salgotra, R. K. (2019). Harnessing genome editing techniques to engineer disease resistance in plants. *Frontiers in plant science*, 10, 550.

Mutlu-Ingok, A., Devecioglu, D., Dikmetas, D. N., Karbancioglu-Guler, F., & Capanoglu, E. (2020). Antibacterial, antifungal, antimycotoxigenic, and antioxidant activities of essential oils: *An updated review. Molecules*, 25(20), 4711.

Nada, S., Nikola, T., Bozidar, U., Ilija, D., & Andreja, R. (2022). Prevention and practical strategies to control mycotoxins in the wheat and maize chain. *Food Control*, 136, 108855.

Nowara, D., Gay, A., Lacomme, C., Shaw, J., Ridout, C., Douchkov, D., ... & Schweizer, P. (2010). HIGS: host-induced gene silencing in the obligate biotrophic fungal pathogen Blumeria graminis. *The Plant Cell*, 22(9), 3130-3141.

Nunes, V. M., Moosavi, M., Khaneghah, A. M., & Oliveira, C. A. (2021). Innovative modifications in food processing to reduce the levels of mycotoxins. *Current Opinion in Food Science*, 38, 155-161.

Onofre, R. B., Gadoury, D. M., Stensvand, A., Bierman, A., Rea, M., & Peres, N. A. (2021). Use of ultraviolet light to suppress powdery mildew in strawberry fruit production fields. *Plant Disease*, 105(9), 2402-2409.

Patel, N., Desai, P., Patel, N., Jha, A., & Gautam, H. K. (2014). Agronanotechnology for plant fungal disease management: a review. Int. J. Curr. Microbiol. App. Sci, 3(10), 71-84.

Perczak, A., Juś, K., Marchwińska, K., Gwiazdowska, D., Waśkiewicz, A., & Goliński, P. (2016). Degradation of zearalenone by essential oils under in vitro conditions. *Frontiers in microbiology*, 7, 1224.

Roberts, M. R., & Paul, N. D. (2006). Seduced by the dark side: integrating molecular and ecological perspectives on the influence of light on plant defence against pests and pathogens. *New Phytologist*, 170(4), 677-699.

Sakr, S. H., Elshafie, H. S., Camele, I., & Sadeek, S. A. (2018). Synthesis, spectroscopic, and biological studies of mixed ligand complexes of gemifloxacin and glycine with Zn (II), Sn (II), and Ce (III). Molecules, 23(5), 1182

Šečić, E., & Kogel, K. H. (2021). Requirements for fungal uptake of dsRNA and gene silencing in RNAi-based crop protection strategies. *Current Opinion in Biotechnology*, 70, 136-142.

Sen, Y., Onal-Ulusoy, B., & Mutlu, M. (2019). Detoxification of hazelnuts by different cold plasmas and gamma irradiation treatments. *Innovative Food Science & Emerging Technologies*, 54, 252-259.

Senthilkumar, S. R., & Sivakumar, T. (2014). Green tea (*Camellia sinensis*) mediated synthesis of zinc oxide (ZnO) nanoparticles and studies on their antimicrobial activities. *Int. j. pharm. pharm. sci*, 6(6), 461-465.

Shende, S., Ingle, A. P., Gade, A., & Rai, M. (2015). Green synthesis of copper nanoparticles by Citrus medica Linn.(Idilimbu) juice and its antimicrobial activity. *World Journal of Microbiology and Biotechnology*, 31, 865-873.

Song, J. Y., Jang, H. K., & Kim, B. S. (2009). Biological synthesis of gold nanoparticles using *Magnolia kobus* and *Diopyros kaki* leaf extracts. *Process Biochemistry*, 44(10), 1133-1138.

Team, N. S. T. I., Fellows, N. S. T. I., Network, K., & Briefs, T. (2010). Nanotechnology 2010: Advanced materials, CNTs, particles, films and composites. *Tech Connect. Briefs*, 1, 234-237.

Thirumurugan, A., Ramachandran, S., Tomy, N. A., Jiflin, G. J., & Rajagomathi, G. (2012). Biological synthesis of gold nanoparticles by Bacillus subtilis and evaluation of increased antimicrobial activity against clinical isolates. *Korean Journal of Chemical Engineering*, 29, 1761-1765.

Uppala, S. S., Bowen, K. L., & Woods, F. M. (2013). Pre-harvest aflatoxin contamination and soluble sugars of peanut. *Peanut Science*, 40(1), 40-51.

Valodkar, M., Jadeja, R. N., Thounaojam, M. C., Devkar, R. V., & Thakore, S. (2011). Biocompatible synthesis of peptide capped copper nanoparticles and their biological effect on tumor cells. *Materials Chemistry and Physics*, 128(1-2), 83-89.

Wacławek, S., Gončuková, Z., Adach, K., Fijałkowski, M., & Černík, M. (2018). Green synthesis of gold nanoparticles using Artemisia dracunculus extract: control of the shape and size by varying synthesis conditions. *Environmental Science and Pollution Research*, 25, 24210-24219.

Wang, X., Han, Y., Niu, H., Zhang, L., Xiang, Q., & Zong, W. (2022). Alternaria mycotoxin degradation and quality evaluation of jujube juice by cold plasma treatment. *Food Control*, 137, 108926.

Wielogorska, E., Ahmed, Y., Meneely, J., Graham, W. G., Elliott, C. T., & Gilmore, B. F. (2019). A holistic study to understand the detoxification of mycotoxins in maize and impact on its molecular integrity using cold atmospheric plasma treatment. *Food chemistry*, 301, 125281.

WJ, B. (2000). Control of soil-borne plant pathogens by incorporating fresh organic amendments followed by tarping. *Phytopathology*, 90, 253-259.

Wu, Y., Cheng, J. H., & Sun, D. W. (2021). Blocking and degradation of aflatoxins by cold plasma treatments: Applications and mechanisms. Trends in Food Science & Technology, 109, 647–661.

Zahoor, M., & Ali Khan, F. (2016). Aflatoxin B1 detoxification by magnetic carbon nanostructures prepared from maize straw. *Desalination and water treatment*, 57(25), 11893-11903.

Zhang, H., Xie, J., Fu, Y., Cheng, J., Qu, Z., Zhao, Z., ... & Jiang, D. (2020). A 2-kb mycovirus converts a pathogenic fungus into a beneficial endophyte for Brassica protection and yield enhancement. *Molecular plant*, 13(10), 1420-1433.

Chapter 16

Fungal Pathogen and Mycotoxin Control in Organic Agriculture

Dipayan Hazra, Waghamare Minal Bhujangrao, Sudipta Mahato

Ph.D. Research Scholar, Uttar Banga Krishi Vishwavidyalaya, Pundibari, Cooch Behar, West Bengal -736165, India

[1]Corresponding author email - dipayanhazra78@gmail.com*

Abstract

In recent days to regulate the excessive use of chemicals in field new legislative measures are taken. It is an incentive for developing various biocontrol strategies to prevent the pathogenic fungi and overall organic agriculture. Mycotoxigenic fungal diseases have received particular attention because they can directly harm crop productivity and have the potential to produce mycotoxins, which can be harmful to both human and animal health. Mycotoxins are toxic, non-volatile, relatively low- molecular- weight, fungal secondary metabolic products. Mycotoxins are a much- diversified group of toxic compounds produced by five kinds of spore-forming fungi. When crops are contaminated by phytopathogenic species like *Fusarium*, *Aspergillus*, *Alternaria*, and *Penicillium*, the stored crops or the end products (bread, beer, etc.) typically include mycotoxins. Food security is regularly risked by mycotoxins appearing in food. These fungal species and their mycotoxins contaminate harvested seeds causing losses of agricultural commodities in many zones of the world. Such contaminants are fearsome, since they affect the seeds before harvest time, stored condition, leading to alteration of the germination quality of these seeds. It makes reasonable to create disease-control plans with an ecological foundation in an organic system. Antifungal materials are often used during growing in the field or storage to minimise damage and inhibit fungal growth. However, these antifungal antibiotics are equally dangerous to people and animals. For this reason, we need more eco-friendly and healthier bioweapons. Organic agriculture is farming strategy that supports ecosystem, soil health and humans, it is social requirement for sustainable agriculture.

Keywords- Biocontrol, Mycotoxin, Mycotoxigenic fungi, Organic agriculture and Management strategy

Introduction

Mycotoxin is a wide term for the hazardous substance produced by molds. Mycotoxins are the toxic chemical substances found in food and feed, posing the biggest risk to human and health. Every year, a significant number of crops are lost before and after harvest as a result of pathogen infestation. It involves a broad range of diseases, including prokaryotic bacteria, eukaryotic fungi, viruses, nematodes, and viroids. These Plant diseases attack with great persistence, causing both direct and indirect losses around 40 billion dollars globally. It is very much critical to control plant pathogens and the diseases they cause in order to increase productivity for food as well as other plant-derived materials like fibre, wood, oils, medicines, etc., and to meet the world's rapidly expanding food demand. Many of these disease-causing fungi are phytopathogenic and/or mycotoxigenic. Their presence causes numerous soil-borne diseases in plants, which in turn cause a variety of food and feed-borne mycotoxicosis in humans and animals when infected items are consumed. As a result, these filamentous fungi harm not only the health of humans and animals but also the global economy. Cereals, grains, fruits, and vegetables are the primary targets of moulds. Mycotoxin contamination is a serious issue for food security. More than 400 metabolites are classified as mycotoxins, and more than 300 species of fungi with the capacity to produce mycotoxins have been found (Gareis, 1999). Major toxins linked to the contamination include Aflatoxins, Zearalenone, Fumonisins, Ochratoxin A, deoxynivalenol (DON). These are the most significant mycotoxins, mostly generated by *Aspergillus*, *Fusarium*, and *Penicillium* species (Pitt, 2000). Since organic farming prohibits the use of synthetic fungicides, it is common to assume that fungal populations on crops are higher. As a result, there has been much speculation regarding the frequency of organic food and feed being contaminated by mycotoxin (Tinker, 2001). Others contended that eliminating fungicides might increase the diversity of microbial populations and, as a result, restrict the proliferation of certain producers of mycotoxin. Higher accumulation of mycotoxins in grains grown through conventional, organic, and "low input" farming have been the subject of contentious reports (Marx *et al.*, 1995; Berleth *et al.*, 1998; Birzele *et al.*, 2002; Meier *et al.*, 1999; Stahle *et al.*, 1998; Doll *et al.*, 2002). As a result, there has been much speculation regarding the probablity of mycotoxin contamination of organic food and feed (Tinker, 2001; Trewavas, 2001, 2004). It is appropriate to develop disease-control strategies that have an ecological basis in organic farming. In organic farming we utilise the benefits of a variety of naturally occurring plant protection resources such as micro- and macroflora and fauna found in soils, protective agents made from plant extracts, physical techniques like weeding and mulching, cultivar selection, etc.

Mycotoxin

The word "mycotoxin" refers to a large category of dangerous compounds that moulds create. The most dangerous chemicals present in food and feed are called mycotoxins, and they represent the biggest danger to both human and animal health. Mycotoxin is driven from the term "myco" meaning fungus and toxin meaning poison. Mycotoxins causes illness which is known as "Mycotoxicosis". Mycotoxins are the secondary metabolites that fungi make throughout the process of digestion. These biochemicals are produced by fungi for a variety of causes, many of which are yet unclear. As fungi approach hibernation and their growth rates decrease, mycotoxin production typically rises. Mycotoxin production seems to be a defensive response in such cases. Mycotoxins may have the effect of counteracting the elements that slow down the development rate of fungi. As an alternative, fungus may create mycotoxins to shield bacteria and other surviving fungal species from dormant moulds and fungal spores. It's possible that mycotoxins shield mould from unfavourable environmental factors (such as excessive cold or dryness) or from the absence of certain vital nutrients in the substrate that the mould is growing on. Research has identified a number of commercially significant mycotoxins, including aflatoxin (AF), ochratoxins (OTA), fumonisins, trichothecenes (TH), patulin (PAT), citrinin (CIT), and zearalenone (ZEA). Mycotoxins affect crop growers and consumers equally in terms of profitability and health. The most susceptible crops to mycotoxins are maize, millet, wheat, sorghum, soybean, peanut, and their derivatives, as well as byproducts made from affected primary ingredients. Mycotoxins can be found in food waste, mouldy bread, cottonseed, spices, and other foods, as well as in cereals, legume seeds, oilseeds, and industrial byproducts fed to animals (Coppock *et al.*, 2018). Five types of mycotoxins or categories of mycotoxins prevalently present in food. These types are Deoxynivalenol/Nivalenol, Zearalenone, Ochratoxin, Fumonisins, and Aflatoxins (Tola and Kebede, 2016).

Table 1: Mycotoxin and their effects

Mycotoxin	*Pathogen producing toxin*	*Major Food*	*Toxic effects and diseases*
Alfatoxin	*Aspergillus flavus* *A. Parasiticus* *A. Nomius* *Penicillium*	Cereals, feeds, oilseeds and pulp, coconut	Carcinogenicity, hepatotoxicity, teratogenicity, suppression of immune systems, alteration of DNA structure, hepatitis, hemorrhage, renal lesions.
Fumonisin	*Fusarium verticillioides* *F. culmorum*	Cereals, corn	Human encephalomalacia, pulmonary edema, carcinogenicity, neurotoxicity, liver damage, heart failure and esophageal cancer.

Mycotoxin	*Pathogen producing toxin*	*Major Food*	*Toxic effects and diseases*
Ochratoxin (OTA)	*Aspergillus penicillium* *A. ochraceus* *Penicillium nordicum* *P. verrucosum*	Cereals, herbs, oilseeds, figs, beef jerky, fruits, and wine	Kidney and liver damage, loss of appetite, nausea and vomiting, immune system suppression, carcinogenic.
Patulin	*Penicillium carneum* *P. clavigerum* *P. griseofulvum*	Silage, wheat, feeds, apples, grapes, peaches, pears, apricots, olives, cereals	Brain haemorrhage, neural disorders, skin lesions, skin cancer, lung, mutagenicity, antibacterial impact.
Trichothecenes (T2, DON, DAS, HT2)	*Fusarium oxysporum* *Fusarium sp.* *Cephalosporium sp* *Trichoderma sp*	Cereals, feeds, silage, legumes, fruits, and vegetables	Immune suppression, cytotoxicity, skin necrosis, hemorrhage, anemia, granulocytopenia, oral epithelial lesions, GIS lesions, hematopoietic, alimentary toxic aleukia (ATA), hypotension, coagulopathy.
Zearalenone	*Fusarium sp* *F. graminearum* *F. culmorum*	Cereals, feeds, silage, legumes, fruits, and vegetables	Carcinogenic, hormonal imbalance estrogenic effect, reproductive problems, teratogenic.

(Source: Oguz et al., 2017)

Fungal pathogen producing mycotoxins:

"Mycotoxins produced by fungi play a major role in the biochemical warfare that unfolds among competing species in virtually every environment on Earth"

Organisms and tissues, they target and how, fungi differ widely from one another. Mycotoxins are the byproduct of fungal growth and interaction between fungi and host. The synthesis and potential of mycotoxin varies greatly throughout fungal groups. Mycotoxins and moulds do not specifically target higher species like humans; however, mycotoxins create hindrance to animals from consuming contaminated food. However, both farm animals and humans are occasionally the targets of the biochemical conflict between different species of fungi and bacteria battling for the nutrients contained in decomposing and dead organic materials. Different fungi, depending on the substrate the mould is growing on, create varying kinds and concentrations of mycotoxins. All that mould is actually masses of fungus that are growing. Mycotoxins are often produced and spread via the spores that moulds produce. Mycotoxin producing fungi broadly divided into two groups –

(a) Field fungi: They invade plant products before harvest

(b) Storage fungi: They invade crop products after harvest

There are three types of toxicogenic field fungi found-

(a) Plant pathogens: *Fusarium graminearum* producing deoxynivalenol, nivalenol. Toxin production may be a crucial part in pathogenicity.

(b) Fungi growing on senescent or stressed plants: *Fusarium moniliforme* producing fumonisin, *Aspergillus flavus* producing aflatoxin

(c) Fungi that initially colonize the plant before harvest and predispose mycotoxin contamination after harvest: *Penicillium verrucosum* producing ochratoxin, *A. flavus* producing aflatoxin (Tola and Kebede 2016).

Table 2: Mycotoxin produced by different genera of fungal pathogen

Aspergillus	*Penicillium*	*Fusarium*	*Alternaria*	*Claviceps*
Alfatoxin B1	Patulin	Deoxynivalenol (DON)	Tenuazonic acid	Ergot alkaloids
Alfatoxin G1	Ochratoxin A (OTA)	Nivalenol	Alternariol	
Alfatoxin M1	Citrinin	Diacetoxyscripenol	Alternariol methyl ester	
Ochratoxin A (OTA)	Penitrem A (PA)	T-2		
Sterigmatocystein	Cyclopiazonic acid	Fumonisins		
Cyclopiazonic acid		Moniliformin		
		Zearalenone		

(Source: Bennett et al., 2003; Steyn, 1995)

Different Major Mycotoxins:

1. Alfatoxin:

Since its discovery in 1960, Aflatoxin—the most hazardous mycotoxin to humans—has gained international attention. The word "aflatoxin" was obtained from "a" from the genus *Aspergillus*, the "fla" from the species flavus, and "toxin" meaning toxic. It is a potent toxin that damages DNA and can cause cancer and liver damage. Different strains of aflatoxin can be produced by *Aspergillus flavus* and *Aspergillus parasiticus*, two different forms of mould. It has been possible to isolate and identify many aflatoxins. 16 chemicals were reported by Cole and Cox in the aflatoxin grouping, but the only aflatoxins in foods and feeds sold in commerce that are regularly examined are B1, B2, G1, G2, and MI. Of the aflatoxins that have been examined to date, alfatoxin B1 is the most frequently occurring strain and the most carcinogenic (Steyn, 1995). Many experts believe it to be the most potent carcinogen yet discovered by humans (Bennett *et al.,* 2003). A wide range of agricultural commodities, including maize, rice, wheat, spices, and nuts, have been reported to be contaminated by aflatoxins (Nonaka *et al.*, 2009).

Aspergillus flavus is a common fungus found in soil, and it frequently infects nuts and mouldy grains. Toxins known to be produced by *Aspergillus flavus* comprise around half of all strains. Moist conditions and high temperatures promote the formation of aflatoxin. Different fungus species produce at least 13 different forms of aflatoxin.

2. *Fusarium* mycotoxins:

There are hundreds of species of fungi in the genus *Fusarium*, which are extensively dispersed in both temperate and tropical climates in plant materials and soil from where isolation is generally done. These species are significant cereal pathogens that cause illnesses including Fusarium head blight (FHB), which is one of the most significant fungal diseases that affects wheat, maize, barley including other several cereal crops, as well as seedling blight and Fusarium root rot (Nicholson *et al.,* 2004). Thankfully, only a few numbers of *Fusarium* species are known to produce more types varieties of mycotoxins which pose risk regarding the safety of food (Moss, 2002), however, there have been reports of over 100 *Fusarium* toxins.

(a) **Trichothecene:** The two most important mycotoxin types among the four fundamental trichothecene groups (A, B, C, and D) are types A and B. T-2 toxin, which can harm bone marrow and weaken the immune system, is a component of type A toxins. Approximately 14 times more poisonous than deoxynivalenol (DON), formerly known as vomitoxin, the most common type B mycotoxin derived from *Fusarium*. DON, a protein synthesis inhibitor of low molecular weight with haemolytic and cell membrane activity, is present in complex animal feed derived from wheat, barley, maize, and safflower in addition to these crops themselves. Some *Fusarium* species seem to be distinguished by the synthesis of the two kinds toxins. *Fusarium graminearum* and *F. culmorum* are the main hosts of type B trichothecenes, whereas *Fusarium sporotrichoides* and probably *F. poae* are the dominant hosts of type A trichothecenes (D'Mello, 2003).

(b) Zearalenone: One of the secondary metabolites of Fusarium graminearum species is zearalenone. Zearalenone, in contrast to other mycotoxins, has an acute LD-50 of more than 10,000 ppm in rats, meaning that it is almost non-toxic to mammals after acute intake (i.e., 1 percent of the diet). It has been demonstrated that zearalenone at as little as 0.0001 ppm can cause a measurable, hormone-related uterogenic response in female pigs (Bennett *et al.,* 2003). Stated differently, exposure to zearalenone has been demonstrated to affect reproductive processes at a dose 100 million times lower than the lethal level in a rat LD-50 investigation.

(c) **Fuminosins and moniliformin:** Fumonisin B1 (FB1), one of the most prevalent mycotoxin in the world's food supply chain is produced by *Fusarium moniliforme syn. verticillioides*, which also causes the disease known as "ear rot" in maize (Steyn, 1995). Particularly in the liver, FB1 can prevent the production of lipids. It is only recently that fumonisins have been identified and studied. Numerous structurally related fumonisins (FBs) have been linked to a variety of issues in cattle fed FB1-contaminated feed, as well as to human cancer (e.g., FB1 with oesophageal cancer) (D'Mello, 2003; Benbrook, 2005).

3. Ochratoxin:

Mycotoxin known as ochratoxin A (OTA) can be produced by specific *Penicillium* and *Aspergillus* species. *Penicillium nordicum*, which is primarily obtained from meat and cheese, and *Penicillium verrucosum*, which is discovered to impure grain, are the two principal *Penicillium species* that are currently recognised to produce OTA (Larsen *et al.,* 2001; Seifert and Levesque, 2004). *Aspergillus niger* (not all strains), which is believed to be in charge of wine's OTA production and *Aspergillus ochraceus*, which primarily produces OTA in coffee and cocoa (Frank, 1999; Frisvad and Thrane, 2000) are among the *Aspergillus species* that are also an OTA producer (Krogh, 1978; Frisvad and Thrane, 2000; Seifert and Levesque, 2004).

Ochratoxin is mostly present in grains, beans, nuts, and dried fruits; it is typically linked to the storage of these foods. Ochratoxin can also be found in some wines, but barley is especially susceptible to it. One of the main dietary sources of ochratoxin is pork, and animal feed is how pigs acquire the disease (Bennett *et al.,* 2003; Fink-Gremmels, 1999). The primary fungus species that produce ochratoxin is Aspergillus, which can harm kidneys, induce cancer, and weaken the immune system. The gastrointestinal tract can easily absorb it (Fung *et al.,* 2004). Because of this, it is one of the mycotoxins that are most frequently discovered in human and cow's breast milk (Marquardt *et al.,* 1992). Ochratoxin enters human breast milk through a variety of human foods and beverages and finds its way into cow's milk through the consumption of grain-based animal feeds and concentrates (Miraglia *et al.,* 1995; Turconi *et al.,* 2004).

4. Ergot Alkaloids:

Claviceps purpurea is the source of ergot alkaloids, and it has polluted rye flour for millennia and continues to be an issue in many global regions. Since the fifth century AD, the effects of ergot alkaloids (EAs), a form of mycotoxin have been recognised. The oldest documented mycotoxicosis is ergotism, which is a disease that can infect both humans and animals when they consume food or feed contaminated with EAs (Agriopoulou, Stamatelopoulou and Varzakas, 2020) The first recorded case of ergotism occurred in France in 944–945 AD, killing about 10,000 people (Frebortova, *et al.*, 2013). In eastern France, Germany, and Russia, ergotism was widespread between the ninth and the fourteenth centuries (Janik *et al.,* 2020). The word "ergot" comes from the old French word "argot," which translates to "cock's spur" (Krska and Crews, 2008).

Claviceps purpurea, C. fusiformis, C. africana, Epichlo spp., and *Neotyphodium spp.* are a few EA producers (Holderied, Rychlik and Elsinghorst, 2019). Moreover, *Aspergillus, Penicillium*, and *Blansia* all produce EAs (Shahid *et al.*, 2020). The above-mentioned species have the potential to seriously damage and reduce cereal yields. Based on their structural characteristics, EAs are divided into three classes: peptides (ergopeptines), lysergic acid amides, and clavines (Shahid *et al.*, 2020). An alkaloid produced from tryptophan called EAs was the subject of an analytical categorisation study conducted by Arroyo-Manzanares *et al.*, The most well-known produced extravertine amino acids (EAs) from the *Claviceps* species are: "the ergometrine (Em), ergotamine (Et), ergocristine (Ecr), ergokryptine (Ekr), ergosine (Es), and ergocornine (Eco), and their inine-epimeric forms ergotaminine (Etn), egometrinine (Emn), egocristinine (Ecrn), ergokryptinine (Ekrn), ergocroninine (Econ), and ergosinine (Esn)" (Lee and Ryu, 2017).

Even when ergot bodies were removed from grains by manual cleaning, EAs were still found in grains (Shi *et al*; 2019). Sclerotria can contaminate cereal-based food and feed products with EAs since they are collected alongside cereals or grass (Storm *et al.,* 2008). Furthermore, sclerotia are broken and transported to the final product, flour, during cereal milling, rather than being easily isolated from healthy grains (Storm *et al.,* 2008). At least 400 Poaceae species are parasitised by *C. purpurea*, the most damaged crops are weedy grasses and cereals such rye, wheat, barley, millet, and oats. Compared to other cereals like maize, rye is more prone to contracting *Claviceps* infection (Debegnach *et al.,* 2019).

Management of mycotoxin in Organic Agriculture:

None of the environmental elements that are so important in determining the hazards of mycotoxin are radically changed by either organic or conventional farming practices. Alternative farming practices, such as organic farming and the use of genetically modified seeds, can occasionally change the ecology of farming systems to the point where changes occur in the composition and abundance of bacteria and fungi competing in particular ecological niches. These changes may then have an effect on the quantity and presence of mycotoxins in food at harvest.

By encouraging diversity in the fungus and bacteria that colonise plant tissues and live in the soil, organic farming systems lower the prevalence of major fungal infections and, consequently, the dangers of mycotoxin exposure. Given the variety of these species in an ecosystem, there is still a greater chance that there will be intense competition between species, despite varying climatic and meteorological factors. Adaptability within populations of bacteria and fungi reduce the likelihood that fungi that produce mycotoxin may take over and can accumulate to points where mould and mycotoxin causing disease could become dangerous. Numerous initiatives have been made to lower the level of mycotoxin contamination in agricultural products through preventive measures. These can be divided into three main categories:

1. Application of Biocontrol Agent
2. Pre-harvest Agronomic Strategies
3. Post-harvest Strategies

1. Application of Biocontrol Agent

In plant pathology, the term "biocontrol" refers to the use of microorganisms to inhibit the growth and/or stop the colonisation of dangerous plant pathogens (Pal and Mc Spadden, 2006). In this chapter, BCAs are defined as advantageous microorganisms with the capacity to oppose plant diseases and safeguard the plant (DM, 1998; ZK and RS, 2003).

Antibiosis, competition, mycoparasitism, and stimulation or improvement of plant defence are the primary mechanisms of action of BCAs (Pal and Mc Spadden, 2006). In order to combat the pathogen, BCAs typically use multiple modes of action; hence, the existence of one predominant mode of action does not rule out the others. There are several mechanisms found in different biocontrol agents –

Antibiosis: The process of producing secondary metabolites that can inhibit the growth, lessen the virulence, or eradicate pathogenic fungus is known as antibiosis. Examples of these compounds include lytic enzymes (El-Tarabily and Sivasithamparam, 2006), antibiotics (Wang *et al.,* 2007; El-Tarabily and Sivasithamparam, 2006), and various proteins (Daguerre *et al.,* 2014).

Competition: When two or more fungi fight for the same vital nutrients, competition takes place necessary for their development and growth (Benitez *et al.,* 2014; Alabouvette *et al.,* 2009). Exclusionary competition is a different kind through filling the same role (Ehrlich, 2014; Abbas *et al.,* 2006).

Mycoparasitism or hyperparasitism: Direct parasitic attack of one fungus by another is known as mycoparasitism or hyperparasitism which one that ultimately results in the host pathogen's mortality (Kim and Vujanovic, 2016; Howell, 2013).

Colonization of the pant: The plant's colonisation by advantageous microorganisms may result in localised or systemic defence mechanisms, strengthening resistance to plant diseases (Shoresh, Harman and Mastouri, 2010; Lugtenberg, Caradus and Jihnson, 2016).

Biocontrol and mycotoxin

Trichothecene

Trichothecene is the dangerous toxin produced by *Fusarium sp* having four basic groups (A, B, C, D). *Trichoderma gamsii* 6085 was chosen as a possible antagonist against *F. graminearum* and *F. culmorum*. The strain showed the ability to suppress both pathogens' production of DON by up to 92% (Matarese et al.,

2012). In order to assess the effectiveness of various BCAs against ear blight and the occurrence of related DON, a two-season field experiment was carried out on winter wheat. The two strains of *F. equiseti* that performed the best reduced the amount of mycotoxins that *F. graminearum* and *F. culmorum* produced by 70 and 94%, respectively. Nonetheless, it was shown that the cereals treated with *F. equiseti* had low levels of NIV (Dawson *et al.,* 2004). *Piriformospora indica* has recently demonstrated its promising capacity to improve the total grain weight of *F. graminearum* inoculated samples by 54% and lower the severity of the disease induced by *F. graminearum* and mycotoxin DON contamination in wheat by 70–80% (Rabiey and Ullah, 2016). In vitro, novel bacterial endophytes such as *Citrobacter* and *Paenibacillus polymyxa* were able to detoxify DON; however, the effectiveness of some of these isolated strains in field or greenhouse conditions has not yet been described (Mousa *et al.,* 2015).

In addition to fungal and yeast BCAs, bacteria have also been employed to manage DON in maize (Mousa *et al.,* 2015) and wheat (Hu *et al.,* 2014; Palazzini *et al.,* 2007) that is generated by *F. graminearum*. When *B. subtilis* RC 218 and *Bacillus subtilis* RC 263 were applied at anthesis for two seasons, a complete reduction in DON content was achieved (Palazzini *et al., 2016)*. This was consistent with earlier findings under greenhouse conditions by the same authors (Palazzini *et al.,* 2007), though the disease incidence did not consistently decrease. In contrast, Khan and Doohan conducted a small-scale field experiment in wheat and barley to evaluate three strains of *Pseudomonas spp.*—two strains of fluorescens and one strain of *frederiksbergensis*—against *F. culmorum* and DON production. The findings demonstrated that DON was decreased in wheat (12%) and Barley (21%).

Zearalenone

Zearalenone (ZEN) is a significant mycotoxin found in numerous cereals, although its management has received less attention. It is also produced by *Fusarium sp.* It has recently been shown that ZEN can be detoxified by *Trichoderma* isolates by being reduced and sulfated (Tian *et al.,* 2018). This was consistent with earlier findings by Gromadzka *et al.* who examined several isolates of *Clonostachys* and two isolates of *Trichoderma* against two isolates each of *F. graminearum* and *F. culmorum* in vitro. The performance of these isolates under greenhouse or field tests was not validated, despite the great rate of ZEN decrease (above 96%) (Gromadzka *et al.,* 2009).

By using lactonohydrolase to catalyse an enzymatic alkaline hydrolysis in vitro, *C. rosea* transforms ZEN into less harmful molecules (Kosawang et al., 2014;). This has been demonstrated by cloning the relevant gene, zhd 101, and expressing it in *Escherichia coli* and *Schizosaccharomyces pombe*, but not in *Saccharomyces cerevisiae*, which showed only little detoxification activity against ZEN (Takahashi-Ando *et al., 2004)*. BCA resistance to the mycotoxin itself is a crucial component of this strategy, which entails a direct interaction between BCAs and the pathogen toxin, in order to guarantee the efficacy and durability. Furthermore, it has been demonstrated that C. rosea can withstand exposure to ZEN because of its abundance in ATP-binding cassette transporters (Karlsson *et al.,* 2015).

Fumonisin

The concentration of FB1 remained unaffected by a combination of *B. subtilis* and *P. solanacearum* (Cavaglieri, Orlando and Etcheverry, 2005). In field testing on maize, fumonisin B1 content may be effectively decreased by treating seeds with *B. amyloliquefaciens* Ba-S13 (Pereira, Nesci and Etcheverry, 2007). That was verified in a two-year field research using the same bacteria, *B. amyloliquefaciens,* following the application of two distinct treatments: pre-sowing seed inoculation and flowering maize ear inoculation (Pereira *et al.*, 2010). During a three-year investigation, *P. fluorescens* isolated from the maize rhizosphere by Nayaka et al. shown a significant reduction in FB1 content and the disease incidence following challenge with *F. verticillioides* (Nayaka *et al.,* 2009). The incidence of fumonisins was reduced by 88% by seed treatment followed by spray treatment with a pure culture of *P. fluorescens* (Nayaka *et al.,* 2009). One practical way to stop the fungi from spreading vertically is to utilise the endophytic bacterium *B. subtilis* to regulate the synthesis of FB1. This was proposed by Bacon et al. 50% less FB1 was found in greenhouse settings (Bacon *et al.*, 2001).

T. viride and *F. verticillioides* co-inoculating maize kernels resulted in a 72–85% drop in FB1, depending on when the inoculation was done (Yates *et al.,* 1999). In order to stop the buildup of toxins during storage, the fungus was also suggested as a postharvest agent (Yates *et al.,* 1999; Bacon *et al.*, 2001). It has been demonstrated that while *C. rosea* does not degrade fumonisins produced by *F. verticillioides,* it can impede their synthesis (Chatterjee *et al.,* 2016). Constant 60–70% reduction in FB1 while applying a 50:50 mixture of the pathogen and *C. rosea* 016 at different stages of maize cob ripening, depending on the temperature. These studies were carried out because, in the right environmental circumstances, *F. verticillioides* may target maize throughout its ripening stage (Samsudin *et al.,* 2017). Similar outcomes on milled maize agar at the same concentration (50:50/ pathogen: C. rosea 016) have also been reported in the past (Samsudin *et al.,* 2017). It is possible to draw the conclusion that controlling FB1 in vitro and during field experiments was more successful when bacterial BCAs relied on antibiosis.

Alfatoxin

Generally speaking, non-toxic *A. flavus* strains have been used to reduce AFs in a variety of crops (Abbas et al., 2006; Cotty, 1989; Atehnkeng, 2008). A few of these strains—AF36, for instance—are offered for sale on the market (Cotty and Bhatnagar, 1994; Abbas *et al.,* 2011). The exact mechanism of action for the reduction of aflatoxin (AFs) by non-toxigenic *A. flavus* BCAs is still unknown, but two theories have been put forth:

(i) reduction due to competitive exclusion on the toxic wild *A. flavus* population and

(ii) inhibition of biosynthetic pathways involved in aflatoxin production (Hruska *et al.,* 2014).

Using *A. flavus* strain AF36 as a BCA, Doster *et al.,* were able to reduce AF levels by 20–45% in pistachio orchards over the course of four seasons (2008–2011) (Doster, Cotty and Michailides, 2014). Additional studies on ground nuts have been conducted in the field [58–60] and in vitro [61, 66]. After treating groundnut seeds and rhizospheric and geocarpospheric soil with two strains isolated from India, Mallikarjunaiah *et al.,* (2017) measured total AFs in the soil. It was possible to significantly lower the mycotoxin concentration below the highest amounts that are allowed for ground nuts (Hulikunte *et al.,* 2017).

As possible substitute BCAs to control *Aspergillus spp*., researchers have also examined various *Trichoderma* species, including *T. viride, T. harzianum*, and *T. asperellum* (Calistru, Mclean and Berjak, 1997), bacteria (Palumbo, Okeffe and Abbas, 2007); yeast (Niknejad *et al.,* 2012); and algae (Gong *et al.,* 2015). However, not all of these studies have examined mycotoxins. Aflatoxin synthesis in maize and peanuts held at varying water activity was 100% inhibited by *Shewanella algae* strain YM8 when two volatile chemicals, dimethyl trisulfide and 2,4-bis (1,1 dimethylethyl)-phenol, were produced (Gong *et al.,* 2015). It has also been previously shown that *B. subtilis* RCB 90 fully inhibits AFB1 in vitro (Nesci, Bluma and Etcheverry, 2005). Aflatoxin generation was also significantly reduced (by 90–99%) by the yeast *Candida parapsilosis* IP1698 at various pH and temperature ranges (Niknejad *et al.,* 2012). This was consistent with the reduced percentage of *Bacillus spp*. P1 and P11 that was previously obtained (Veras *et al.*, 2016). Aiyaz et al. investigated in the field, four BCAs and all the formulations, by maize seeds treatment application, had a significant reduction in AFs level.

2. Pre-harvest Agronomic Strategies:

Several agronomic practices promote some beneficial microbes in crop rhizosphere. Some are there to create distance between host, not suitable environment for pathogen, exclude pathogens. In the following segment of the chapter such practices described in detail.

Crop Rotation

Crop rotation excludes over-seasoning pathogens from the crop and soil ecology. Because crop wastes include a lot of nutritional residues, they make a great home for fungus. In the absence of crop rotation, fungus can spread to the following crop seeded in the field by living on the leftovers of prior crops. Crop rotation may be able to reduce *Fusarium spp*. growth and, consequently, the mycotoxin levels in grains (Mielniczuk and Skwaryło-Bednarz, 2020; Janssen *et al.,* 2019). *Fusarium spp.* develop and spread more readily when cereals are grown on the same piece of land through several years, especially after wheat and maize (*McMullen et al.*, 2012; Hofgaard *et al.,* 2016; Weber and Kita, 2010). Fields where wheat was grown following maize had greater amount of DON in the grain (Schaafsma *et al.,* 2001), whereas ZEA was found in 45% of the samples (Qiu *et al.,* 2016). Forecrops that serve as barriers against *Fusarium spp.* include legumes and root crops (Fernandez and Conner, 2011).

Tillage

Tillage is the process of cultivating soil by turning over the topsoil up to 30 cm, or shallowing it up to 20 cm. This also has an impact on the decrease of mycotoxins in the grain. This also has an impact on cereals' ability to reduce mycotoxins (Bernhoft, 2012). One of the key strategies for reducing Fusarium Head Blight. But as low tillage becomes more common in conventional farming, DON levels and Fusarium disease situations rise (Krebs *et al.,* 2000).

Choice of crop species

The sensitivity of different grain species to fungi that produce mycotoxin and/ or levels of mycotoxin contamination varies significantly. For instance, rye often contains more ergot contamination than wheat, while wheat is typically infected with DON more often (Tanaka *et al.,* 1988; Lepschy *et al.,* 1989; Doll *et al.,* 2002). The Ochratoxin contamination susceptibility of cereal species varies as well.

Cultivar Choice

In general, mycotoxin contamination is not so much in crops that are resistant to fungal infection during the growth season. There aren't any highly resistant cultivars on the market right now against *Fusarium*, and the best way to characterise the cultivars that are available in Europe that have some resistance or tolerance is as "less susceptible." Doll *et al.*, (2002) have reported distinct variances in cultivar susceptibilities to *Fusarium* infection and DON concentration. These authors conducted a comparison between twelve wheat cultivars grown in winter season) or 120 accessions in all, with a minimum of five samples per cultivar that were traditionally cultivated in Germany during the high-rainfall 1998 growing season on various locations.

Regarding resistance to *Penicillium verrucosum* infection and OTA development, there is least of knowledge on cultivar differences (Axberg *et al.,* 1997). The latter used three wheat and six barley cultivars that had been intentionally contaminated in a lab setting. Following 23 weeks of incubation at roughly 19% moisture, the barley and wheat cultivars had OTA g−1 contents ranging from 6 to 350 and 25 to 890 ng, respectively. In the same field, two of the four spring spelt cultivars had 18 and 92 ng OTA g−1, accordingly, according to a different study by Elmholt and Rasmussen (2005), whereas the other two had less than 0.5 ng OTA g−1. Variations in OTA were most likely the result of varying cultivar sensitivity because they did not correlate with harvest moisture content or *Penicillium verrucosum* levels. That said, more research should be done on this.

Variations in ergot susceptibility have been reported for cultivars of barley (Cunfer *et al.,* 1975; Pageau *et al.,* 1994), durum wheat (Platford and Bernier, 1976), and spring wheat (Platford and Bernier, 1976; Bretag and Merriman, 1981). It was proposed that choosing plants with larger anthers and copious amounts of pollen might enhance resistance to *C. purpurea*. It has been demonstrated that hybrid rye cultivars are more susceptible than non-hybrid cultivars, however non-hybrid cultivars are still more widely utilised in organic farming.

Seed and Sowing:

Using high-quality seed is crucial for preventing the growth of pathogenic fungus. Desirable seed material is healthy, undamaged seed that has the right amount of moisture and vitality (Edwards, 2009). The date of sowing has a big impact on crop output. Winter cereals are less sensitive to *Fusarium* infection since the risk of infection is higher during the flowering season (Champeli, 2004; Jouany, 2007).

In a temperate area, an early sowing date for maize prevent fungal infection (Magan and Aldred, 2007). According to (Magan and Aldred, 2007), throughout the blossoming to the maize maturity stage, there was a high mycotoxin contamination of the grains while there was also a high precipitation and low temperature. The buildup of mycotoxins or fumonisins and maize infection are most noticeable during dry spells (Ferrigo, Raiola and Causin, 2016) although they are easily preventable by using an irrigation system. This minimises the stress that the drought is putting the plant under, which in turn lowers the rate of *F. verticillioides* infection and mycotoxin contamination. Irrigation for tiny cereal can increase the risk of Fusarium Head Blight (Torelli, 2012; Gautam and Dill-Macky, 2012).

Seed Grading

Grading of seeds is thought to be a crucial technique in enhancing the quality of contaminated seed lots since Fusarium head blight (FHB) frequently causes a greater proportion of tiny or dried grains. However, seed grading is insufficient and imprecise to eliminate all contaminated grains and/or DON. According to other writers, there isn't a direct correlation between DON levels and grain size (Lepschy, 1992). Perkowski (1998), however, found that over 75% of the total DON was found in the percentage of smallest barley kernels. However, compared to smaller grains, Chelkowsky and Perkowski (1992) found that grains larger than 2.5 mm had increased DON concentrations. Compared to grains that are usually affected by *Fusarium* (1–600 mg/kg), shrivelled grains had less DON (< 5 mg/kg) according to Sinha and Savard (1997). Additionally, it was demonstrated that even in ears without FHB symptoms, normal-sized grains could be infected (Rintelen, 1995). Therefore, visual grading for FHB symptoms may not be very accurate in predicting the production of FHB and DON (Herrmann *et al.*, 1998; Aufhammer *et al.*, 1999; Lemmens *et al.*, 2004).

Seed Treatment:

The goal of seed treatments is to decrease or eradicate inoculum and/or disease development caused by seed-borne fungi, without compromising the seed's ability to germinate or emerge. One of the early seed treatments applied in agriculture was thermal treatment. They could be based on hot water treatments (52 °C, 10 min), warm water treatments (45 °C, 2 h), or hot air/steam treatments. Warm or hot water treatments have been demonstrated to successfully prevent or limit the development of *Fusarium* disease (Jahn, 2002), frequently demonstrating an efficacy comparable to that of chemical seed dressing (Winter *et al.*, 1997; Forsberg *et al.*, 2005). However, agricultural practice has not yet used hot air or hot steam method of management (Jahn, 2002).

In order to directly improve seed quality and eliminate or minimise inoculum for seedborne diseases, Lindner *et al.*, (1996) and Schauder (2003) devised and tested a seed dressing approach based on "electrons." In field testing, "Electron" seed dressing efficiently eliminated common bunt spores (*Tilletia caries*) and decreased bunt infections in comparison to untreated seeds.

Use of 'organic', 'home-grown' and/or 'farm-saved' seed

When comparing conventionally produced seeds to those grown on organically certified farms, a number of studies conducted in the early 1990s revealed decreased *Fusarium* infection (Piorr, 1992; Dornbusch *et al.*, 1993). This was ascribed to several methods employed in organic systems, including (i) not using maize as a pre-crop for cereals in the rotation, (ii) fertilisation regimes that result in less availability of plant-available nitrogen and (iii) soil management techniques that lead to increased biological activity and the prevention of diseases (Knudsen *et al.*, 1995, 1999). The fungicide treatments used in conventional seed production were also thought to be relatively ineffective in this regard. However, the issues found in traditional seed production systems might be resolved by the effective application of well-planned fungicide treatments (Stahle *et al.*, 1998).

Elmholt (2003) notes that the low presence of *P. verrucosum* in the majority of soils suggests that seed contamination could be a significant cause of infection. Few research, meanwhile, have looked into the significance of seed inoculum on *P. verrucosum* preharvest contamination. When compared to home-grown or farm-saved seed, Elmholt (2003) provided samples of commercially produced seed with significantly lower contamination levels. It should be highlighted, nonetheless, that a large number of the homegrown samples were from farms with inadequate drying facilities, where *P. verrucosum* infestation affected up to 100% of the kernels. Future research should ascertain whether the possibility of *P. verrucosum* infection of "home-grown" seed may be eliminated by quickly and effectively drying the seed.

Fertilization

Remarkably, applying mineral fertilisers in the field may increase the rate of *Fusarium spp.* infection (Gajecki *et al.*, 2010). Namely, the increased incidence of *Fusarium* fungal infections in grains is caused by the elevated nitrogen content in the soil. DON levels are not as impacted by fertiliser type, despite the fact that urea, ammonium nitrate, or calcium nitrate can change the pace at which grain becomes affected with *Fusarium spp.* (Yi *et al.,* 2001). In contrast to wheat treated with 120 kg N ha 1, winter wheat fertilised with a greater nitrogen dose of 200 kg N ha 1 showed distinct mycotoxins, according to a study. It was also established that there is a statistically significant correlation between the quantity of nitrogen fertiliser applied and the wheat cultivar and the mycotoxin concentration (Podolska *et al.,* 2010).

Breeding and Selection:

Agronomic and environmental production circumstances, as well as diverse genetic pools from breeding programs across national borders, supply varying genetic material (Edwards, 2004). Genetic modification has produced cultivars thus far that are either partially or completely resistant to *Fusarium spp.* This has been shown to be the most effective strategy for suppressing *Fusarium* infections (Tekauz *et al.,* 2008; Lanubile *et al.,* 2017; Buerstmayr *et al.,* 2002). Cereals have developed five different types of defence mechanisms against *Fusarium spp.*: type I resistance is to infection, and type II resistance is to the pathogen's spread within the head (Schroeder and Christensen, 2017; Steiner, Buerstmayr and Michel, 2017). Type III refers to the so-called capacity to break down or resist DON. Type IV refers to grain's resistance to mycotoxin buildup and degradation by converting them into non-toxic derivatives or by preventing the biosynthesis of toxic metabolites (Boutigny, Richard-Forget and Barreau, 2008; Kluger, 2015). Type IV denotes the presence of DON and other related secondary metabolites as well as the plants' resistance to infection (Mesterhazy, 1995). Understanding the fundamental molecular relationships between the host-pathogen and plant defence systems is crucial for successful breeding and transgenesis (Perincherry, Lalak-Kanczugowska and Stepien, 2019). Barley becomes less prone to FHB infection when the antifungal gene HvNEP-1 is overexpressed in the endosperm. Lower mycotoxin levels in the grain result from this (Bekalu *et al.,* 2020). Targeted gene silencing is a crucial method for controlling *Fusarium spp.* in cereal crops. While host-induced gene silencing (HIGS) is a transgenic technique used to silence fungal genes on plants during attempted infection with successful reduction of infection, RNA interference (RNAi) is a natural mechanism that controls gene expression (Machado *et al.,* 2018). This technique depends on the host plant's capacity to manufacture mobile, short interfering RNA molecules that are complementary to the targeted fungal genes and function as effectors and regulators of the plant's defence against infections. These molecules are created from lengthy double-stranded RNA. These chemicals are to be transmitted from the plant to fungi in order to achieve and induce gene silencing (Majumdar *et al.,* 2017). By selecting a resistant cultivar and reducing mycotoxin deposition and biosynthesis, mycotoxin levels in grains can be decreased. Mycotoxin synthesis can be regulated by phenolic substances, peptides or carotenoids, and pro-oxidative chemicals like hydrogen peroxide (Boutigny, Richard-Forget and Barreau, 2008).

3. Post-harvest Strategies

Microbial agents

This firstly refers to hostile microorganisms as a natural alternative to chemicals. Microorganisms and cereal plants have been shown to interact, and these interactions have been identified as potentially advantageous since they help strengthen plants' defence mechanisms (Kuzniar, Wlodarczyk and Wolinska, 2019). Nonetheless, fungi possess the capacity to produce several secondary metabolites, or antibiotics, which function as antifungal, antibacterial, and insecticidal agents, thereby hindering the development and spread of other microbes (Dutta *et al.,* 2014). Applying *Trichoderma harzianum* T22 to maize seeds has the potential to inhibit *F. verticillioides* growth and the associated build-up of fumonisin (Ferrigo *et al.,* 2014).

Physical Methods

The most widely used and straightforward technique is grain moisture adjustment. Specifically, it is important to modify grain moisture soon after harvest in order to guarantee low microbial activity. This may cause the quantities of mycotoxin in cereals to rise. Reducing the concentration of mycotoxins in cereals and cereal-based products can involve unit processes including sorting, washing, and milling (Grenier *et al.,* 2014). Sorting can be done non-invasively using opto-electronic sorting or UV light illumination. Cleaning, husking, and residue removal are effective ways to eliminate some of the mycotoxins that have collected in grain surface tissues (Karlovsky *et al.,* 2016). Consequently, dust, fine debris, and damaged grains all contain high quantities of mycotoxins (Cheli *et al.,* 2013; Chulze, 2010). Cleaning the surface of the grains stops Fusarium fungus from colonising and their mycotoxins from building up. Fungal growth and mycotoxin levels are significantly influenced by optimum humidity and seed storage temperature (Tanaka *et al.,* 2001). Studies on the absorption of toxins in vitro and in vivo have demonstrated the effectiveness of several adsorbents, such as polymers, aluminosilicates, and activated carbon (Huwig *et al.,* 2001). We can manipulate the atmosphere of storage room by using antioxidants and essential oils. Both of these controls the fungal growth and accumulation of mycotoxin. But this process is expensive one.

Radiation

It is often referred to as either non-ionizing or ionising radiation (Shanakhat *et al.,* 2018) that has the ability to lessen or eradicate harmful bacteria. Because radiation may be used in industrial settings, it is a useful tool for bulkier, heavier commodities. Through a sequence of reactions, it modifies the molecular structure of food items (Karlovsky *et al.,* 2016). An extremely significant finding was found in fruit juices like as orange, pineapple, and tomato that were contaminated with Zen and irradiated distilled water. Specifically, ZEN toxicity was decreased. However, the fruit juices' quality was impacted by radiation at a dose greater than 10 kGy (Kalagatur, Kamasani and Mudili, 2018). In naturally infected maize, 50 kGy of electron beam radiation resulted in 71.1% and 67.9%, respectively, destruction of ZEN and OTA (Shi *et al.,* 2018). Alfatoxin decrease (>95%) at 6 kGy was observed in rice processing, and gamma irradiation is another option (Gonclaves *et al.,* 2019). Following a 5-minute radiation exposure, PAT concentrations in apple juice were lowered by 83% (Walravens *et al.,* 2014). The wider employment of radiation techniques in the food business is still dubious, though, as it can result in chemical, biological, and physical impacts that are potentially dangerous to people and animals after molecular interactions (Shi *et al.,* 2018).

Conclusion and Future prospects

Mycotoxins have been widely studied and their implications in foods are enormous. Regulatory control and fast and effective analyses and detection will go a long way in reducing the danger of mycotoxins in foods especially in developing countries where there is shortage of food. It will be necessary to continuously evaluate food safety in relation to various (regulated and so-far unmonitored) secondary metabolites that are dangerous to humans and animals due to changes in the weather combined with fungal alterations in cereals and maize. Higher environmental consciousness, however, is likely to relegate the ecological aspects of food production. A greater demand for cereals and food will also increase the need for bio-acceptable solutions for larger cropping fields and a reduction in ecological damage through the use of fungicides. It is possible to use biological control, and a plethora of innovative techniques are emerging, primarily from microbiological studies and the use of microbes that can detoxify mycotoxins and inhibit fungal development. Thus far, agrotechnical interventions have shown to be effective when implemented correctly. Applying biocontrol agents in storage facilities is also a good agricultural practice that completes the cycle from stable to table. This review presents practical approaches, the most dependable of which is moisture reduction during grain storage as of right now. In order to guarantee food safety, it is critical to look for substitute techniques and agents and investigate various strategies. Bio-acceptable practices should benefit farmers, consumers, the environment, and the crop. One of the most crucial initiatives to maintain human health and eco-friendly food production will undoubtedly be the introduction and use of biological and natural protection agents against fungal contamination.

Mycotoxins still have many unanswered questions, including why fungi make them, what regulates their development, how farmers and the food business can identify and steer clear of them, and what impact they have on health. The reduction of mycotoxin dangers is a long-term goal and a daily task for organic farmers and food manufacturers, just like it is for all other farmers and food enterprises. More information and new instruments will be very helpful in the continuous fight to keep mycotoxins out of the food supply and out of customers' minds.

Reference:

A. Kodisch, M. Oberforster, A. Raditschnig. Covariation of ergot severity and alkaloid content measured by HPLC and one Elisa method in inoculated winter rye across three isolates and three European countries. *Toxins*. 12: 676.

A. Moretti, A.F. Logrieco, A. Susca. (2017). Mycotoxins: an underhand food problem. *Humana Press*, New York.

A.K. Korn, M. Gross, E. Usleber. (2014). Dietary ergot alkaloids as a possible cause of tail necrosis in rabbits. *Mycotoxin Research*. 30: 241-250.

Avery D T. (1998). The hidden dangers of organic food, American Outlook, Fall, 19–22. Axberg K, Jansson G, Svensson G and Hult K (1997), 'Varietal differences in accumulation of ochratoxin A in barley and wheat cultivars after inoculation of *Penicillium verrucosum*. *Acta Agriculture Scandinavuca*. 47: 229–237.

Bekalu, Z.E.; Krogh Madsen, C.; Dionisio, G.; Bæksted Holme, I.; Jørgensen, L.N.S.; Fomsgaard, I.; Brinch-Pedersen, H. (2020). Overexpression of nepenthesin HvNEP-1 in barley endosperm reduces Fusarium head blight and mycotoxin accumulation. *Agronomy*. 10: 203.

Benbrook, C M. (2005). Breaking the mold – impacts of organic and conventional farming systems on mycotoxins in food and livestock feed. *An Organic Center State of Science Review*. pp 58.

Bennett, J. W. and Klich, M. (2003). Mycotoxins. *Clinlcal Microbiological Rev*. 16(3): 497-516.

Berleth M, Backes F and Krämer J (1998), 'Schimmelpilzspektrum und Mykotoxine (Deoxynivalenol und Ochatoxin A) in Getreideproben aus ökologischem und integriertem Anbau', Agribiol. Res., 51, 369–376.

Bernhoft, A.; Torp, M.; Clasen, P.-E.; Løes, A.-K. Influence of agronomic and climatic factors on Fusarium infestation and mycotoxin contamination of cereals in Norway. Food Addit. Contam. 2012, 29, 1129–1140.

Birzele B and Prange A (2003), Fusarium spp. and storage fungi in suboptimally stored wheat: mycotoxins and influence on wheat gluten proteins. Mycotoxin Res., 19, 162–170.

Boutigny, A.L.; Richard-Forget, F.; Barreau, C. Natural mechanisms for cereal resistance to the accumulation of Fusarium trichothecenes. Eur. J. Plant Pathol. 2008, 121, 411–423. [CrossRef]

Buerstmayr, H.; Lemmens, M.; Hartl, L.; Doldi, L.; Steiner, B.; Stierschneider, M.; Ruckenbauer, P. Molecular mapping of QTLs for Fusarium head blight resistance in spring wheat. I. Resistance to fungal spread (Type II resistance). Theor. Appl. Genet. 2002, 104, 84–91.

Champeil, A.; Fourbet, J.F.; Dore, T.; Rossignol, L. Influence of cropping system on Fusarium head blight and mycotoxin levels in winter wheat. Crop Prot. 2004, 23, 531–537.

Cheli, F.; Pinotti, L.; Rossi, L.; Dell'Orto, V. Effect of milling procedures on mycotoxin distribution in wheat fractions: A review. LWTFood Sci. Technol. 2013, 54, 307–314.

Chelkowski J and Perkowski J. (1992). Mycotoxins in cereal grain (Part 15). Distribution of deoxynivalenol in naturally contaminated wheat kernels. *Mycotoxin Research*. 8: 27–31.

Chulze, S.N. (2010). Strategies to reduce mycotoxin levels in maize during storage: A review. *Food Additive and Contaminants*. Part A. 27: 651–657.

Chulze, S.N. Strategies to reduce mycotoxin levels in maize during storage: A review. Food Addit. Contam. Part A 2010, 27, 651–657.

Cunfer B, Mathre D E and Hockett E A (1975). Factors influencing the susceptibility of male-sterile barley to ergot. *Crop Science*. 15: 194–196.

D.Kifer, M. Sulyok, D. Jaksi c, et al., Fungi and their metabolites in grain from individual households in Croatia. *Food Additives Contaminants*. Part B. 16: 1-12.

D'Mello J P F. (2003). Mycotoxins in cereal grains, nuts and other plant products. *Food Safety: Contaminants and Toxins*. CAB International. Wallingford. 65–90.

Doll S, Valenta H, Danicke S and F. lachowsky G. (2002., Fusarium mycotoxins in conventionally and organically grown grain from Thuringia/Germany. *Landbauforschung Volkenrode*. 52: 91–96.

Dutta, D.; Puzari, K.C.; Gogoi, R.; Dutta, P. Endophytes: Exploitation as a tool in plant protection. Braz. Arch. Biol. Technol. 2014, 57, 5.

E. Janik, M. Niemcewicz, M. Ceremuga. (2020). Molecular aspects of mycotoxins—a serious problem for human health. *International Journal of Molecular Sciences*. 21: 8187.

Edwards, S.G. Fusarium mycotoxin content of UK organic and conventional oats. Food Addit. Contam. 2009, 26, 1063–1069.

Edwards, S.G. Influence of agricultural practices on fusarium infection of cereals and subsequent contamination of grain by trichothecene mycotoxins. Toxicol. Lett. 2004, 153, 29–35. [CrossRef]

Elmholt S and Rasmussen P H (2005), 'Penicillium verrucosum occurrence and Ochratoxin A contents in organically cultivated grain with special reference to ancient wheat types and drying practice', Mycopathologia, 159, 421–432.

F. Debegnach, S. Patriarca, C. Brera. (2019). Ergot alkaloids in wheat and rye derived products in Italy. *Foods*. 8:150.

Ferrigo, D.; Raiola, A.; Causin, R. Fusarium toxins in cereals: Occurrence, legislation, factors promoting the appearance and their management. Molecules 2016, 21, 627.

Ferrigo, D.; Raiola, A.; Picollo, E.; Scopel, C.; Causin, R. Trichoderma harzianum T22 induces in maize systemic resistance against Fusarium verticillioides. J. Plant Pathol. 2014, 96, 133–142.

Fink-Gremmels J. (1999). Mycotoxins: their implications for human and animal health. Veterinary Quarterly. 21 (4): 115–120.

Fink-Gremmels, J. (1999). Mycotoxins: their implications for human and animal health. *Veterinary Quarterly*. 21(4): 115-120.

Forsberg G, Johnsson L and Lagerholm J (2005), 'Effects of aerated steam seed treatment on cereal seed-borne diseases and crop yield', J. Plant Dis. Protect., 112 (3), 247–256.

Frank M. (1999). Mycotoxin prevention and decontamination. HACCP and its mycotoxin control potential: an evaluation of ochratoxin A in coffee production. *Third Joint FAO/WHO/UNEP International Conference on Mycotoxins*. Tunis. Tunisia. 3-3.

Frisvad J C and Thrane U (2000), 'Mycotoxin production by common filamentous fungi', in Samson R A, Hoekstra E S, Frisvad J C and Filtenborg O, Introduction to Food and Airborne Fungi, Centraalbureau Voor Schimmelcultures, Utrecht, 321–332.

Fung F and Clark R F. (2004). Health effects of mycotoxins: a toxicological overview. Journal of Toxicology. Clin. Toxicol., 42 (2), 217–234

Fung, F. and Clark, R. F. (2004). Health effects of mycotoxins: a toxicological overview. *Journal of Toxicology: Clinical Toxicology*. 42(2): 217-234.

Gajecki, M.; Jakimiuk, E.; Gajecka, M.; Motyka, J.; Obremski, K. Praktyczne metody zmniejszania aktywnosci mikotoksyn w paszach. Mag. Weter. Monogr. 2010, 6, 605–610.

Gautam, P.; Dill-Macky, R. Impact of moisture, host genetics and Fusarium graminearum isolates on Fusarium head blight development and trichothecene accumulation in spring wheat. Mycotoxin Res. 2012, 28, 45–58. [CrossRef]

Gonçalves, A.; Gkrillas, A.; Dorne, J.L.; Dall'Asta, C.; Palumbo, R.; Lima, N.; Battilani, P.; Venâncio, A.; Giorni, P. Pre- and Postharvest Strategies to Minimize Mycotoxin Contamination in the Rice Food Chain. Compr. Rev. Food Sci. Food Saf. 2019, 18, 441–454

Grenier, B.; Bracarense, A.P.; Leslie, J.F.; Oswald, I.P. Physical and chemical methods for mycotoxin decontamination in maize. In Mycotoxin Reduction in Grain Chains; Leslie, J.F., Logrieco, A.F., Eds.; Wiley Blackwell: New Delhi, IA, USA, 2014; pp. 116–129.

H. Hulvova, P. Galuszka, J. Frebortova. (2013). Parasitic fungus Claviceps as a source for biotechnological production of ergot alkaloids. *Biotechnology Advancement.* 31: 79-89.

H. Oguz, Mikotoksinler ve Onemi, Turkiye Klinikleri. (2017). Veterinary Sciences-Pharma. *Toxicology* 3(2) :113-119.

H. Shi, W. Schwab, N. Liu. (2019). Major ergot alkaloids in naturally contaminated cool-season barley grain grown under a cold climate condition in western Canada, explored with near-infrared (NIR) and fourier transform mid-infrared (ATR-FT/MIR) spectroscopy. *Food Control.* 102: 221-230.

H.J. Lee, D. Ryu, Worldwide occurrence of mycotoxins in cereals and cereal derived food products: public health perspectives of their co-occurrence. *Journal of Agriculture and Food Chemistry*. 65: 7034-7051.

Hökby E, Hult K, Gatenbeck S and Rutqvist L (1979), 'Ochratoxin A and citrinin in 1976 crop of barley stored on farms in Sweden', Acta Agricu. Scand., 29, 174–178.

Huwig, A.; Freimund, S.; Käppeli, O.; Dutler, H. Mycotoxin detoxication of animal feed by different adsorbents. Toxicol. Lett. 2001, 122, 179–188.

I. Holderied, M. Rychlik, P.W. Elsinghorst. (2019). Optimized analysis of ergot alkaloids in rye products by liquid chromatography fluorescence detection applying lysergic acid diethylamide as an internal standard. *Toxins* 11: 184.

I.D. Storm, P.H. Rasmussen, B.W. Strobel. (2008). Ergot alkaloids in rye flour determined by solid-phase cation-exchange and high-pressure liquid chromatography with fluorescence detection. *Food Additive and Contaminants.* 25: 338-346.

J.D. Di Mavungu, S.V. Malysheva, M. Sanders. (20170. Development and validation of a new LC-MS/MS method for the simultaneous determination of six major ergot alkaloids and their corresponding epimers. Application to some food and feed commodities. *Food Chemistry.* 135: 292-303.

J.I.Pitt, J.D. Miller. (2017). A concise history of mycotoxin research, *Journal of Agriculture and Food Chemistry*. 65: 7021-7033.

Jahn M (2002), 'Saatgutbehandlung im Ökologischen Landbau', Forschungsreport, 1/2002, 12–15.

Jouany, J.P. Methods for preventing, decontaminating and minimizing the toxicity of mycotoxins in feeds. Anim. Feed Sci. Technol. 2007, 137, 342–362.

Kalagatur, N.K.; Kamasani, J.R.; Mudili, V. Assessment of detoxification effcacy of irradiation on zearalenone mycotoxin in various fruit juices by response surface methodology and elucidation of its in-vitro toxicity. Front. Microbiol. 2018, 9, 1–13.

Karlovsky, P.; Suman, M.; Berthiller, F.; Meester, J.; Eisenbrand, G.; Perrin, I.; Oswald, I.P.; Speijers, G.; Chiodini, A.; Recker, T.; et al. Impact of food processing and detoxification treatments on mycotoxin contamination. Mycotoxin Res. 2016, 32, 179–205.

Kluger, B.; Bueschl, C.; Lemmens, M.; Michlmayr, H.; Malachova, A.; Koutnik, A.; Maloku, I.; Berthiller, F.; Adam, G.; Krska, R.; et al. Biotransformation of the mycotoxin deoxynivalenol in Fusarium resistant and susceptible near isogenic wheat lines. PLoS ONE 2015, 10, e0119656.

Krebs H, Dubois D and Külling C (2000), 'Effects of preceding crop and tillage on the incidence of Fusarium spp. and mycotoxin deoxynivalenol content in winter wheat grain', AGRARForschung, 7 (6), 264–268.

Krogh P. (1978). Causual association of mycotoxic nephropathy. *Acta Pathologica et Microbiologica Scandinavica*. Section A. Supplement 269. Pp. 28

Kuzniar, A.; Włodarczyk, K.; Wolinska, A. Agricultural and Other Biotechnological Applications Resulting from Trophic Plant-Endophyte Interactions. Agronomy 2019, 9, 779.

Larsen T O, Svendsen A and Smedsgaard J. (2001). Biochemical characterization of ochratoxin A producing strains of the genus Penicillium. *Applied and Environmental Microbiology*. 67: 3630–3635.

Lepschy J (1992), 'Fusarientoxine in Getreide – ihre Entstehung und Vorbeugungs maßnahmen', Gesunde Pflanze 44, 35–39.

Lepschy J, Dietrich R, Märtlbauer E, Schuster M, Süß A and Terplan G (1989), 'A survey on the occurrence of Fusarium mycotoxins in Bavarian cereals from the 1987 harvest', Lebensm. Unters. Forsch. 188, 521–526.

Lindner K, Burth U and Roeder O (1996), 'Einführung der Saatgutbehandlung von Winterweizen mit niederenergetischen Elektronen in die landwirtschaftliche Praxis', Mitt. aus der Biol. Bundesanstalt H., 321, 50.

M.G. Shahid, M. Nadeem, A.Gulzar. (2020). Novel ergot alkaloids production from *Penicillium citrinum* employing response surface methodology technique. *Toxins* 12: 427.

Machado, A.K.; Brown, N.A.; Urban, M.; Kanyuka, K.; Hammond-Kosack, K.E. RNAi as an emerging approach to control Fusarium head blight disease and mycotoxin contamination in cereals. Pest Manag. Sci. 2018, 74, 790–799.

Magan, N.; Aldred, D. Post-harvest control strategies: Minimizing mycotoxins in the food chain. Int. J. Food Microbiol. 2007, 119, 131–139.

Majumdar, R.; Rajasekaran, K.; Cary, J.W. RNA interference (RNAi) as a potential tool for control of mycotoxin contamination in crop plants: Concepts and considerations. Front. Plant Sci. 2017, 8, 200.

Marquardt, R. R. and Frohlich, A. A. (1992). A review of recent advances in understanding ochratoxicosis. Journal of Animal Science. 70(12): 3968-3988.

Marx H, Gedek B and Kollarczik B (1995), 'Vergleichende Untersuchungen zum mykotoxikologischen Status von ökologisch und konventionell angebautem Getreide', Lebensm. Unters. Forsch., 201, 83–86.

McMullen, M.; Bergstrom, G.C.; DeWolf, E.; Dill-Macky, R.; Hershman, D.; Shaner, G.; Van Sanford, D. A unified effort to fight an enemy of wheat and barley: Fusarium head blight. Plant. Dis. 2012, 96, 1712–1728.

Meier A, Birzele B, Oerke E C and Dehne H W (1999), 'Auftreten von Fusarium spp. und Mykotoxingehalte von Winterweizen und Möglichkeiten zur Bekämpfung', Proccedings 21. Mykotoxinworkshop, Jena, pp 19–25.

Mesterhazy, A. Types and components of resistance to Fusarium head blight of wheat. Plant Breed. 1995, 114, 377–386. [CrossRef]

Mielniczuk, E.; Skwaryło-Bednarz, B. Fusarium Head Blight, Mycotoxins and Strategies for Their Reduction. Agronomy 2020, 10, 509.

Miraglia, M., de Dominicis, A., Brera, C., Corneli, S., Cava, E., Menghetti, E., and Miraglia, E. (1995). Ochratoxin A levels in human milk and related food samples: an exposure assessment. *Natural Toxins in Food*. 3(6): 436-444.

Moss, M. O. (2002). Mycotoxin Review - 2. Fusarium. 16(4).

N. Arroyo-Manzanares, L. Gámiz-Gracia, A.M. García-Campaña. (2017). Ergot alkaloids: Chemistry, biosynthesis, bioactivity, and methods of analysis. In Fungal Metabolites, Reference Series in Phytochemistry. *Springer: Berlin/ Heidelberg*. Germany. pp. 887-929.

Nicholson P, Simpson D R, Wilson A H, Chandler E and Thomsett M. (2004). Detection and differentiation of trichothecene enniatin-producing Fusarium species on small grain cereals. *European Journal of Plant Pathology*. 110: 503–514.

Nonaka Y, Saito K, Hanioka N, Narimatsu S, Kataoka H. (2009). Determination of aflatoxins in food samples by automated on-line in-tube solid-phase microextraction coupled with liquid chromatography–mass. *Journal of Chromatography A*. 1216: 4416–4422.

Pageau D, Collin J and Wauthy J M (1994), 'Evaluation of barley cultivars for resistance to ergot fungus, Claviceps purpurea (Fr.) Tul.', Can. J. Plant Sc., 74, 663–665.

Perincherry, L.; Lalak-Kanczugowska, J.; Stepien, Ł. Fusarium-Produced mycotoxins in plant-pathogen interactions. Toxins 2019, 11, 664. [CrossRef]

Perkowski J (1998), 'Distribution of deoxynivalenol in barley kernels infected by Fusarium', Nahrung, 42, 81–83.

Piorr H P (1990), 'Saatgutqualität im Organischen Landbau', in Finke L and Linscheid J, Vorträge der 42. Hochschultagung der Landwirtschaftlichen Fakultät der Universität Bonn, 221–236.

Piorr H P (1992), 'Phytopathological advantages and risks of organic farming systems: Future perspectives to improve organic cropping systems', in Altman J, Pesticide interactions in Crop Production: Beneficial and deleterious effects, CRC Press, Cleveland, 461–472.

Platford G R and Bernier C C (1976), 'Reaction of cultivated cereals to Claviceps purpurea', Can. J. Plant Sci., 56, 51–58.

Podolska, G.; Bryła, M.; Sułek, A.; Waskiewicz, A.; Szymczyk, K.; Jedrzejczak, R. Influence of the cultivar and nitrogen fertilisation level on the mycotoxin contamination in winter wheat. Qual. Assur. Saf. Crop. Foods 2017, 9, 451–461.

R. Krska, C. Crews. Significance, chemistry and determination of ergot alkaloids: a review. Food Additives and Contaminants. Part A. 25: 722-731.

R.W. Coppock, R.G. Christian, B.J. Jacobsen, In: Gupta, R.C. (Ed.). (2018). Aflatoxins in Veterinary Toxicology Basic and Clinical Principles. *Academic Press, Cambridge, United States*. pp. 983-994.

Rintelen J (1995), 'Zum Infektionszeitpunkt von Fusarien an Weizenkörnern', Gesunde Pflanze, 47 (8), 315–317.

S. Agriopoulou, E.Stamatelopoulou, T. Varzakas. (2020). Advances in occurrence, importance, and mycotoxin control strategies: prevention and detoxification in foods. *Foods* 9: 137.

Schroeder, H.W.; Christensen, J.J. Factors affecting resistance of wheat to scab by Gibberella zeae. Phytopathology 1963, 53, 831–838.

Seifert K A and Levesque C A. (2004). Phylogeny and molecular diagnosis of mycotoxigenic fungi. *European Journal of Plant Pathology*. 110: 449–471.

Shanakhat, H.; Sorrentino, A.; Raiola, A.; Romano, A.; Masi, P.; Cavella, S. Current methods for mycotoxins analysis and innovative strategies for their reduction in cereals: An overview. J. Sci. Food Agric. 2018, 98, 4003–4013.

Shi, H.; Li, S.; Bai, Y.; Prates, L.L.; Lei, Y.; Yu, P. Mycotoxin contamination of food and feed in China: Occurrence, detection techniques, toxicological effects and advances in mitigation technologies. Food Control 2018, 91, 202–215.

Sinha R C and Savard M E (1997), 'Concentration of deoxynivalenol in single kernels and various tissues of wheat heads', Can. J. Plant Pathol., 19, 8–12.

Steiner, B.; Buerstmayr, M.; Michel, S. Breeding strategies and advances in line selection for Fusarium head blight resistance in wheat. Trop. Plant Pathol. 2017, 42, 165–174. [CrossRef]

Steyn, P. S. (1995). Mycotoxins, general view, chemistry and structure. *Toxicology Letters*. 82-83, 843-851.

Tanaka T, Hasegawa A, Yamamoto S, Lee U, Sugiura Y and Ueno Y (1988), 'Worldwide contamination of cereals by Fusarium mycotoxins nivalenol, deoxynivalenol and zearalenone', 1. Survey of 19 countries, J. Agric. Food Chem., 36, 979–983.

Tanaka, K.; Sago, Y.; Zheng, Y.; Nakagawa, H.; Kushiro, M. Mycotoxins in rice. Int. J. Food Microbiol. 2007, 20, 59–66.

Tinker P B (2001), Shade of Green: A review of UK Farming System, Royal Agricultural Society of England, Stoneleigh Park, UK.

Tola M, Kebede B. (2016). Occurrence, importance and control of mycotoxins: a review. *Cogent Food Agric,* 2(1):1191103.

Torelli, E.; Firrao, G.; Bianchi, G.; Saccardo, F.; Locci, R. The influence of local factors on the prediction of fumonisin contamination in maize. J. Sci. Food Agric. 2012, 92, 1808–1814. [CrossRef]

Trewavas A (2001), 'Urban myths of organic farming', Nature, 410, 409–410

Trewavas A (2004), 'A critical assessment of organic farming-and-food assertions with particular respect to the UK and the potential environmental benefits of no-till agriculture', Crop Protect., 23, 757–781.

Turconi, G., Guarcello, M., Livieri, C., Comizzoli, S., Maccarini, L., Castellazzi, A. M., Pietri, A., Piva, G., and Roggi, C. (2004). Evaluation of xenobiotics in human milk and ingestion by the newborn--an epidemiological survey in Lombardy (Northern Italy). *European Journal of Nutrition.* 43(4): 191-197.

Walravens, J.; Mikula, H.; Rychlik, M.; Asamd, S.; Ediagea, E.N.; Di Mavungua, J.D.; Landschoote, A.V.; Vanhaeckef, L.; DeSaeger, S. (2014). Development and validation of an ultra-high-performance liquidchromatography tandem mass spectrometric method for the simultaneous determination of free andconjugated Alternaria toxins in cereal-based foodstuffs. *Journal of Chromatography.* A. 1372: 91–101.

Walravens, J.; Mikula, H.; Rychlik, M.; Asamd, S.; Ediagea, E.N.; Di Mavungua, J.D.; Landschoote, A.V.; Vanhaeckef, L.; DeSaeger, S. Development and validation of an ultra-high-performance liquidchromatography tandem mass spectrometric method for the simultaneous determination of free andconjugated Alternaria toxins in cereal-based foodstuffs. J. Chromatogr. A 2014, 1372, 91–101

Winter W, Bänziger I, Krebs H and Rüegger A (1997), 'Warm- und Heisswasserbehandlung gegen Auflaufkrankheiten', Agrarforschung 4 (11–12), 467–470

Yi, C.; Kaul, H.P.; Kübler, E.; Schwadorf, K.; Aufhammer, I. Head blight (Fusarium graminearum) and deoxynivalenol concentration in winter wheat as affected by pre-crop soil tillage and nitrogen fertilisation. Pflanzen 2001, 108, 217–230.

Chapter 17

Good Agricultural Practices and Good Manufacturing Practices for Fungal Pathogens and Mycotoxin Management

Nisha Yadav

Department of Entomology, Purdue University,
West Lafayette 47906, Indiana, USA
ny8833617@gmail.com

Abstract

In the worldwide food sector, food safety remains a paramount concern, particularly in managing fungal infections and mycotoxins. Implementing Good Agricultural Practices (GAPs) and Good Manufacturing Practices (GMPs) plays a crucial role in controlling these issues throughout the food production and supply chain. GAPs are focused on pre-harvest management, including soil, water, and hygiene practices, while GMPs regulate post-harvest handling, storage, and processing activities to ensure food safety. This chapter provides a comprehensive analysis of GAPs and GMPs related to fungal pathogens and mycotoxin control, emphasizing the importance of combining both practices for the development of a safer and more reliable food production system.

Keywords Food Safety, Fungal Infections, Mycotoxins, Good Manufacturing Practices (GMPs), Good Agricultural Practices (GAPs)

Introduction

The safety of agricultural goods is critical in today's global food systems. The formation of mycotoxins due to fungal pathogen contamination of crops is one of the biggest issues. Controlling mycotoxin and fungal infections during food production is a crucial problem that still faces the world's food industry. When ingested by humans or animals, these toxic substances—which are generated by specific fungi—can cause serious health problems such as cancer and immunological disorders

(Bennett & Klich, 2003). Because fungal diseases can produce mycotoxins, which are poisonous secondary metabolites, they lower the quality of agricultural products. Food safety has consequently taken precedence, requiring strict controls to prevent fungal contamination at every point in the food supply chain.

Good Agricultural Practices (GAPs) and Good Manufacturing Practices (GMPs) are the cornerstones for handling fungal infections and mycotoxins in food production. To minimize fungus infections in crops before harvest, Integrated Agricultural Practices (GAPs) concentrate on pre-harvest tasks such as soil management, irrigation, and sanitation. On the other hand, GMPs deal with post-harvest tasks and emphasize handling, storing, processing, and packing in a way that reduces mycotoxin growth and fungal contamination (Van Egmond et al., 2007). The foundation of an all-encompassing plan to guarantee food safety, lower financial losses, and protect public health is made up of GAPs and GMPs. The food sector needs to use both Good Manufacturing Practices (GMPs) and Good Agricultural Practices (GAPs) due to the intricacy of the issue to lower the likelihood of fungal contamination and the subsequent creation of mycotoxin. When these procedures are properly implemented, food products are kept secure throughout the pre-and post-harvest phases (Pitt & Hocking, 2009).

The issue of fungal pathogens in agriculture is a massive one. Every year, mycotoxins harm about 25% of the world's food crops, causing significant financial losses as well as general health hazards (Kumar et al., 2008). These losses might be much more severe in underdeveloped nations since it is frequently difficult to implement GAPs and GMPs effectively due to a lack of resources. Effective management techniques are essential to guaranteeing global food security as well as lowering the amount of mycotoxins in the food supply (Battilani et al., 2016). The application of GAPs and GMPs in the management of fungal infections and mycotoxins will be examined in this chapter.

Good Agricultural Practices (GAPs) in Fungal Pathogen and Mycotoxin Management

Good Agricultural Practices (GAPs) are farming practices that are intended to reduce food contamination from physical, chemical, and biological dangers. This improves food safety and quality. Given that pre-harvest contamination frequently initiates, GAPs are especially crucial for the management of fungal infections and mycotoxins. The incidence of fungal infections in crops and the chance of mycotoxin contamination can both be decreased by implementing GAPs. In order to properly manage fungal infections and mycotoxins, this section covers the numerous GAPs techniques.

Soil Management and Crop Rotation

GAPs' core component is effective soil management. Crop resistance to fungi is mostly dependent on having healthy soil. Crop rotation is one technique that can greatly minimize the accumulation of fungal infections in the soil. For instance, the synthesis of fumonisins and zearalenone is caused by Fusarium, which can be reduced by alternating maize with non-host crops such as legumes (Blandino et al., 2009). The use of monoculture techniques raises the possibility of disease buildup because fungal spores can linger in the soil and infect new crops. Crop rotation contributes to a healthier crop environment by interfering with these pathogens' biological cycle.

Apart from the practice of rotation, the incorporation of cover crops and organic amendments can enhance soil health and reduce its susceptibility to fungal infections. According to Finckh and Wolfe (2006), cover crops like rye and clover increase soil biodiversity, which can inhibit the growth of fungi. By enhancing microbial opponents, organic matter can naturally slow down the growth of dangerous fungi in the soil.

Water Management and Irrigation Practices

Another important component in controlling fungal diseases in crops is handling water. Fungal pathogens that flourish in high moisture content include *Fusarium* and *Aspergillus*. Thus, minimizing the amount of water on crop surfaces using effective irrigation techniques, like drip irrigation, can help lower the chance of fungal growth (Stevens et al., 2003). Because they moisten plant leaves and fruits, overhead irrigation systems foster the growth of fungus and illnesses such as *Botrytis cinerea*-caused gray mold.

Furthermore, since wet soils serve as a breeding ground for a variety of soil-borne fungal infections, appropriate drainage systems in fields guarantee that surplus water does not build up. Since *Aspergillus* and *Fusarium* are known to flourish in high humidity conditions, Miller (2008) assert that maintaining proper moisture levels in the soil is essential for minimizing contamination, especially during the grain-filling period.

Pest and Disease Control

Insect pests in particular are recognized to be fungal infection vectors. For instance, *Aspergillus flavus* can enter maize kernels more easily due to the attack of corn earworm (*Helicoverpa zea*), which can result in aflatoxin contamination (Dowd, 2003). Integrated pest management (IPM) techniques are used in GAPs to regulate insect populations and stop fungal contamination. To efficiently control pest populations, integrated pest management (IPM) combines chemical controls, such as the prudent use of insecticides, with biological controls, such as natural predators and parasites (Flint & Dreistadt, 1998).

To inhibit the spread of harmful fungus in the field, biological control agents are also used, including bacteria and beneficial fungi. By rivalling them for nutrients and space, *Trichoderma* species, for instance, serve as biocontrol agents versus a variety of fungal diseases, such as *Fusarium* and *Rhizoctonia* (Harman et al., 2004).

Using these biocontrol agents fits nicely with GAPs' emphasis on ecologically friendly and sustainable farming methods.

Selection of Resistant Varieties

Another important part of GAPs is planting crop varieties that are resistant to fungus infections. Through breeding initiatives, resistant strains of peanuts, wheat, and maize have been created that are less vulnerable to mycotoxin contamination and fungus diseases. To reduce fumonisin contamination, for example, maize cultivars resistant to *Fusarium* ear rot have been created (Munkvold, 2003). Aflatoxin contamination is also less common when peanut cultivars possess resistance against *Aspergillus* species (Mehan et al., 1991).

Along with conventional breeding techniques, genetic engineering advancements have made it possible to create crops that are more resistant to fungus-related diseases. To effectively reduce fungal infections, for instance, transgenic crops expressing antifungal proteins or enzymes that break down fungal cell walls are available (Lorito et al., 2010). Nonetheless, market and regulatory factors must be taken into account before GMO crops are widely used. Cultivating resistant cultivars can significantly minimize fungal infections and lower the risk of mycotoxin contamination in areas where mycotoxins are a persistent problem.

Harvesting Practices

Reducing the likelihood of fungal infection and the subsequent formation of mycotoxin is largely dependent on harvesting procedures. Harvesting must be done as soon as possible because waiting longer raises the risk of fungal infection. For instance, harvesting maize later than expected raises the possibility of contamination by Fusarium species because fungal colonization increases with crop length (Logrieco et al., 2002). Inadequate management of mechanized harvesting can physically harm crops and open doors for fungal diseases. In order to reduce damage, crops must be handled carefully during harvest. Hell and Mutegi (2011) state that in order to prevent excessive moisture buildup, which could encourage the growth of fungi, harvesting should take place under dry circumstances.

Furthermore, the timely drying of crops and other post-harvest handling measures are essential in halting the proliferation of fungi. Aflatoxin contamination risk can be considerably decreased by drying maize and peanuts to moisture content values below 14% (Magan & Aldred, 2007). In humid environments, mechanical drying techniques are frequently utilized to attain these low humidity levels.

Worker Hygiene and Field Sanitation

Because fungal diseases can be introduced into agricultural ecosystems by human activities, worker hygiene is a crucial component of GAPs. In order to stop the spread of fungus and other infections, workers should receive training on good sanitation techniques. To lower the possibility of contamination, this entails using clean tools and equipment and using the right handwashing practices (FAO, 2003). Additionally, workers ought to be made aware of the dangers posed by fungal diseases and the necessity of acting quickly when infection symptoms are noticed in crops.

Eliminating crop leftovers and plant waste through field cleaning is another crucial step in the fight against fungal infections. Fungal spores can reside in crop wastes left in the field, giving them the opportunity to infect subsequent crops. Reducing the spread of fungal diseases during seasons of growth can be achieved by employing techniques such eliminating or plowing under contaminated plant material (Gachomo & Kotchoni, 2008).

Monitoring and Surveillance

GAPs must include ongoing crop surveillance and monitoring for indications of fungal infestation. Rainfall, temperature, humidity, and other environmental factors all have a big impact on how common fungal diseases are. Farmers are better able to predict fungal outbreaks and take preventative measures when they use early warning systems that keep an eye on these factors. For example, *Aspergillus* infection forecasting models based on environmental factors might notify farmers to treat with fungicides before fungal populations get entrenched.

Cutting-edge technologies allow for real-time evaluation of environmental conditions and crop health, such as remote sensing and precision agricultural instruments. By facilitating prompt interventions and lowering the demand for extensive chemical treatments, these technologies support sustainable agriculture methods.

Timely interventions to stop the spread of illness are made possible by the early diagnosis of fungal pathogens. Before fungal contamination spreads widely, it can be detected through field inspections and the application of diagnostic instruments such DNA-based assays (Munkvold et al., 2009). Furthermore, as temperature and humidity are major determinants of fungal growth, meteorological data can also be utilized to forecast fungal outbreaks (Magan & Medina, 2016).

It is essential to do routine mycotoxin testing in the field and throughout storage to make sure that contamination levels stay below legal limits. To identify mycotoxins in crops, sampling strategies and analytical techniques including enzyme-linked immunosorbent assays (ELISA) and high-performance liquid chromatography (HPLC) are frequently employed (Trucksess & Pohland, 2001).

Sustainable Practices and Environmental Considerations

GAPs also highlight environmentally friendly farming methods that reduce agriculture's negative effects on the environment. In addition to helping manage fungal diseases, these strategies also support soil health and biodiversity preservation by reducing the use of chemical fungicides and encouraging the use of biological controls and cultural practices (Lamichhane et al., 2017). Organic farming, agroforestry, and conservation tillage are examples of sustainable farming techniques that can improve agricultural systems' resistance to fungal infections while preserving ecological balance.

Managing fungal infections and mycotoxin contamination in crops requires the application of good agricultural practices. GAPs lower the risk of fungal infections and guarantee food safety by combining pest control, resistant crop types, sustainable harvesting, post-harvest care, and soil and water administration. Farmers can safeguard the agricultural economy and the public's wellness by using these strategies to produce crops of greater quality with lower levels of mycotoxin contamination.

Good Manufacturing Practices (GMPs) in Post-Harvest Fungal and Mycotoxin Management

Importance of GMPs

Ensuring that food items are regularly produced and controlled in accordance with quality requirements is the main goal of GMPs. Significant risks to food safety, such as pathogen or chemical leftover contamination, can arise from non-compliance with GMPs and cause foodborne illnesses or even fatalities. The World Health Organization (WHO) estimates that 420,000 people die globally each year from foodborne illnesses, which impact 600 million people annually. GMP implementation is a fundamental component of any food safety management strategy and aids in reducing these hazards.

Furthermore, a lot of countries demand regulatory approval contingent on compliance with GMPs. To guarantee food safety, for example, the Food and Drug Administration (FDA) in the US requires food makers to follow GMP guidelines (FDA, 2021). Prioritizing GMPs helps businesses not only meet legal requirements but also improve their marketability, reputation, and customer happiness.

Key Principles of GMPs for Controlling Fungal Pathogens

To manage fungal infections efficiently, food producers need to concentrate on multiple essential GMP principles:

- **Quality Control of Raw Materials**

  To avoid fungal contamination in the finished product, raw material monitoring is crucial. Strict quality control procedures should be implemented by food producers for all incoming raw materials. This entails obtaining raw materials from reliable vendors and carrying out in-depth examinations as soon as they are received (Duarte et al., 2023). The Food and Drug Administration (FDA) states that in order to guarantee safety, raw materials must be examined for microbiological contamination, including fungal spores (FDA, 2021). Because grains and nuts are particularly vulnerable to fungal contamination, it is imperative to conduct routine tests for mycotoxins (Pitt & Hocking, 2009).

- **Facility Design and Maintenance**

  Managing fungal infections is greatly aided by the architecture of food processing plants. In order to prevent the danger of contamination, facilities should be designed with elements like, sufficient ventilation systems to lower humidity levels that encourage the growth of fungi (Alvarez et al., 2021). Appropriate drainage systems to avoid standing water, which can result in the growth of fungi (Niemann et al., 2021). Easy-to-clean, non-porous surfaces that minimize fungal harborage sites through easy cleaning and sanitization (Moussa et al., 2022). To eliminate any sources of contamination, facilities must undergo routine maintenance and cleaning. This entails regular examinations and timely fixes for any structural problems that might be home to fungi that cause infections (Pérez et al., 2021).

☆ **Hygiene Practices**

Enforcing strict hygienic measures is essential to managing fungal infections in the food sector. Workers need to receive training on basic personal hygiene, which includes how to wash their hands properly, how to wear clean clothes, and how to use the right personal protection equipment (PPE) (FAO, 2020). Furthermore, all tools and surfaces that come into touch with food need to be cleaned and sterilized on a regular basis. To guarantee conformity, a methodical cleaning schedule needs to be created and recorded (ISO, 2018). Adherence to hygiene standards can greatly lower the risk of microbiological contamination, including fungal infection, according to the International Organization for Standardization (ISO, 2021).

☆ **Monitoring and Control of Environmental Conditions**

According to Duarte et al. (2023), fungal pathogens prefer certain environmental conditions to flourish in, including excessive humidity and poor temperature regulation. Food makers need to keep an eye on and manage the environmental conditions in production regions in order to reduce these dangers. Frequent monitoring of humidity and temperature can assist in determining the ideal circumstances for the growth of fungi. Automated monitoring systems can improve precision and give real-time data for prompt treatments (Gupta et al., 2021). Maintaining ideal conditions for food safety can also be aided by putting devices like temperature control systems and dehumidifiers into use.

☆ **Effective Pest Control**

Rodents and insects are examples of pests that can introduce fungus spores into food processing plants, leading to contamination (Hassan et al., 2019). To reduce this danger, effective pest control strategies should be included in GMPs. According to Schmidt and Meyer (2021) this entails: Conducting routine inspections to detect indications of pest infestations; and Creating pest management plans that specify preventive actions and reaction tactics. Ensuring that any pest activity is immediately addressed using the proper control techniques. Integrated pest management (IPM) techniques can offer a comprehensive method of pest control with minimal chemical use (Pérez et al., 2021).

☆ **Training and Education**

GMPs must include ongoing education and training for staff members. Employees need to understand the dangers of fungal infection as well as the need of following safety and hygiene procedures (Ryu & Beuchat, 2005). Frequent training sessions ought to include things like correct sanitation practices. The significance of personal hygiene; and methods for recognizing and managing fungal contamination (Alvarez et al., 2021). Fostering a culture of food safety inside the company will increase staff members' knowledge of and dedication to GMPs.

☆ **Documentation and Record Keeping**

Maintaining records and documentation is essential for monitoring GMP compliance and pinpointing areas in need of improvement. Food producers need to keep thorough records of sourcing raw materials and the outcomes of testing. Hygiene procedures and times for cleaning. Data from environmental surveillance. Records of employee training (Moussa et al., 2022). These documents offer important information for audits and inspections as well as proof of adherence to food safety regulations. Finding patterns and opportunities for improvement in the management of fungal diseases can be facilitated by routinely examining and upgrading documentation.

☆ **Hazard Analysis and Critical Control Points (HACCP)**

The principles of Hazard Analysis and Critical Control Points (HACCP) when combined with GMPs can greatly improve the management of food safety. To successfully control dangers HACCP emphasizes identifying essential control points in the production process (Codex Alimentarius, 2021). Food makers can pinpoint precise locations where fungal contamination is likely to happen and set crucial thresholds to reduce these risks by carrying out in-depth hazard evaluations. For instance, keeping an eye on the moisture content of storage places might be a crucial preventative measure against the formation of fungi in grains (Pitt & Hocking, 2009).

☆ **Control of Cross-Contamination**

According to Duarte et al. (2023), cross-contamination can happen at many phases of the food manufacturing process and has a major role in the spread of fungal diseases. Food manufacturers should follow these procedures to avoid cross-contamination: During processing, raw and finished materials are kept apart. Developing explicit protocols for cleaning and sanitizing equipment in between batches; using specialized tools and equipment for certain product kinds (ISO, 2018). To make sure that these procedures are followed and to find any possible cross-contamination hazards, regular audits must be carried out (Gupta et al., 2021).

☆ **Packaging and Storage**

According to Hassan et al. (2019), packaging is essential for preventing contamination and managing fungal infections. Packaging materials resistant to moisture and fungal development should be used by food manufacturers. To avoid fungal infestation, proper storage conditions are equally crucial. Perez et al. (2021) recommended that food goods be kept in cold, dry surroundings and that inventory be rotated on a regular basis to guarantee that the oldest products be used first. First In, First Out (FIFO) inventory management techniques can reduce the possibility of fungus contaminating products that are being stored.

Challenges in Implementing GMPs for Fungal Control

Despite the importance of GMPs, several challenges can hinder their effective implementation:

1. **Cost Implications**

   Especially for small and medium-sized businesses (SMEs), implementing GMPs may necessitate large investments in facilities, training, and equipment (Gupta et al., 2021). Some manufacturers may not be able to completely comply with GMP rules due to these cost constraints.

2. **Complexity of Regulations**

   For manufacturers, navigating the intricate web of food safety rules can be difficult. Continued education and training are necessary to stay up to date with changing standards and compliance needs (FAO, 2020).

3. **Resistance to Change**

   Workers who are used to existing procedures may be resistant to the adoption of new practices and norms. To overcome this reluctance, effective change management techniques are required, such as training and clear communication (Schmidt & Meyer, 2021).

4. **Supply Chain Challenges**

   The application of GMPs may be made more difficult by the complexity of contemporary supply chains. Extensive monitoring and auditing are necessary to guarantee that all suppliers follow the same safety standards (Niemann et al., 2021).

Case Studies of GMPs in Action

1. **Grain Storage Facilities**

   Grain storage facilities are especially vulnerable to fungal infestation because stored grains have a high moisture content. The risk of fungal development can be considerably decreased by putting GMPs into practice, such as frequent cleaning, adequate ventilation, and regular monitoring of moisture levels (Liu et al., 2020). For instance, a research by Ryu & Beuchat (2005) discovered that regular inspections and moisture control techniques resulted in a 50% reduction in the levels of fungal contamination in stored grains.

2. **Dairy Processing Plants**

   Strict cleanliness regulations are necessary in dairy processing facilities to prevent fungal contamination. A dairy plant reported a considerable drop in fungal contamination in its products, which improved product quality and safety. The facility also implemented extensive GMPs, including employee training and sanitation practices (Moussa et al., 2022). In the end the food sector needs to use Good Manufacturing Practices to control fungal infections. Food manufacturers can greatly reduce the hazards associated with fungal contamination by putting into practice essential GMP concepts such quality control of raw materials, facility design, hygiene standards, and appropriate training. It is essential to follow GMPs closely and adjust to new difficulties in food safety as the food business develops. The food sector can guarantee the safety and quality of food products for consumers by cultivating a culture of food safety and putting comprehensive GMP initiatives into practice.

Facility Hygiene and Sanitation Protocols

A sterile processing facility is necessary to stop fungus spores from entering and spreading. Inadequate hygiene practices in food processing establishments may result in cross-contamination between distinct food batches, which could spread fungal infections and raise mycotoxin concentrations.

- **Cleaning and Disinfection**

  It is imperative to do routine cleaning and sanitization on surfaces, equipment, and storage facilities. It is necessary to use efficient disinfection techniques to get rid of fungal spores that could contaminate food while it is being processed. Various disinfectants (such as compounds made from quaternary ammonium or chlorine-based solutions) can be employed to control microbiological contamination, depending on the type of facility and food products handled.

- **Worker Hygiene**

  Strict cleanliness guidelines must be followed by those who process food. Wearing protective gear, often washing your hands, and avoiding contact with raw materials when you're sick are all part of this. To stop infections, such as fungal spores, from getting into food products, worker cleanliness is very crucial. Staff members should get training on the significance of cleaning and sanitization protocols, and these protocols should be documented. To guarantee adherence to set criteria, the efficacy of sanitation methods should be routinely assessed.

Documentation and Record Keeping

GMPs must include accurate record-keeping and documentation. All facets of food production, such as the procurement of raw materials, manufacturing procedures, quality assurance, and hygienic procedures, should be documented in detail (Smith et al., 2020). This documentation fulfils several functions: Traceability: It makes it possible to track products all the way through the supply chain and trace the origin of materials. Compliance: It offers proof of following GMP guidelines and legal criteria. Continuous Improvement: It makes it easier to analyze processes and pinpoint areas that need improvement.

Temperature and Humidity Control During Storage

Fungal pathogens prefer warm, humid conditions to flourish. Controlling temperature and humidity in storage facilities is, therefore, one of the most crucial parts of GMPs for minimizing fungal contamination. Mycotoxins can be produced because of fungal growth that has proliferated due to improper storage conditions.

- **Grain Storage**

  Grains are frequently contaminated by fungi; they need to be kept in a way that keeps moisture from building up. Grain growth can be inhibited by lowering the moisture content to less than 13% and keeping the grains in silos with adequate ventilation. To maintain safe storage conditions, it is also crucial to keep an eye on the temperature and humidity levels in storage spaces (Magan & Aldred, 2007).

- **Perishable Products**

  Fruits, vegetables, and dairy products are examples of perishable goods that need to be refrigerated to avoid fungal deterioration. In order to stop fungal growth, cold chain logistics make sure that these products are maintained at low temperatures during storage and transportation. Aerobic fungal development can be suppressed through the use of modified atmosphere packaging (MAP), which lowers the oxygen content surrounding the product (Miller, 2008).

Packaging Techniques to Prevent Fungal Contamination

When it comes to shielding food from environmental elements that encourage fungal infestation, packaging is essential. Food goods can have their shelf lives extended, and the likelihood of mycotoxin generation decreased with the use of appropriate packaging materials and methods.

- **Barrier Packaging**: To stop fungus from growing, packaging that blocks light, moisture, and oxygen is crucial. Grain and other dry goods, for instance, can be shielded from moisture absorption during storage and transit by polyethylene packing. Perishable items are frequently packaged using vacuum-sealed or MAP systems, which lowers the amount of oxygen available and thereby inhibits fungal growth.

- **Antifungal Packaging**: Advances in packaging technology have given rise to antifungal packaging materials that effectively impede the growth of fungal growth. These materials frequently contain chemicals or natural antifungal agents incorporated in them that can stop fungal spores from growing on food items.

Processing Techniques to Reduce Mycotoxin Levels

Fungal contamination can sometimes still happen even with the finest GAPs and GMPs, which might result in the creation of mycotoxin. The possibility that tainted goods will reach customers can be decreased or eliminated by processing methods.

- **Physical Removal**: Produce or kernels that are contaminated can be effectively separated from the remainder of the batch by sorting and washing. The total mycotoxin burden can be decreased by removing moldy or damaged grains from the line of processing using technologies like optical screening systems (Miller, 2008).
- **Thermal Processing**: Heat has the ability to destroy several mycotoxins. For instance, aflatoxin levels can be lowered by roasting or baking maize and peanuts. However, not all mycotoxins are heat-sensitive, and some may remain stable even after thermal processing, therefore combining heat treatment with additional approaches is important (Wild & Gong, 2010).
- **Chemical Detoxification**: Food products containing mycotoxins may occasionally be detoxified using substances like ozone or ammonia. In contaminated grains, Ammoniation has been demonstrated to effectively lower aflatoxin levels; however, depending on the nation and the food product in question, regulatory approval may be needed for such treatments.

Integrated Approaches to Managing Fungal Pathogens and Mycotoxins

Considering the intricacy of managing mycotoxin and fungal pathogens, food safety requires an integrated strategy that incorporates GAPs and GMPs. To reduce the negative effects of mycotoxins and fungal infections on human health and food security, the food industry must address contamination hazards at every stage of the food production process, from cultivation to consumption.

- **Combining GAPs and GMPs**

  By combining GAPs and GMPs, a thorough management plan covering the whole food supply chain is produced. The early risk of fungal infections can be reduced by using resistant types, controlling pests, and monitoring crops before harvest. Post-harvest procedures, such as facility cleanliness, temperature regulation, and appropriate packaging, help to further prevent the growth of fungal pathogens and their ability to produce mycotoxins. Real-time modifications to production and processing conditions as well as ongoing monitoring are also part of this comprehensive approach. Early warning signs of fungal dangers can be provided by sensors and automation technologies that monitor environmental factors including temperature, humidity, and moisture levels. This enables prompt actions.

Conclusion

Controlling mycotoxins and fungal infections in the food manufacturing process is a complex task that necessitates collaboration throughout the whole supply chain. A strong framework for avoiding fungal contamination and reducing mycotoxin levels in food items is provided by the integration of GAPs and GMPs. Through the application of pre-harvest and post-harvest methodologies, the utilization of sophisticated molecular technologies, and strict adherence to global rules, the food sector can guarantee the production of food that is safe, fulfills international safety standards, and is of superior quality. Collaboration between farmers, food processors, scientists, and regulatory authorities will be vital to staying ahead of the evolving difficulties posed by fungal diseases and maintaining the safety and security of global food security.

References

Alvarez, A. M., Guillen, E., Sanchez, E., & Rodriguez, J. (2021). Ventilation systems and their role in food safety. *Food Safety Journal*, 17(4), 234-245.

Bennett, J.W., & Klich, M. (2003). Mycotoxins. *Clinical Microbiology Reviews*, **16**(3), 497-516. https://doi.org/10.1128/CMR.16.3.497-516.2003

Blandino, M., Reyneri, A., & Vanara, F. (2009). Effect of plant density on toxigenic fungal infection and mycotoxin contamination of maize kernels. *Field Crops Research*, **113**(2), 330-337. https://doi.org/10.1016/j.fcr.2009.06.015

Codex Alimentarius (2021). *Hazard analysis and critical control points (HACCP) guidelines*. Codex Alimentarius Commission.

Dowd, P.F. (2003). Insect management to facilitate preharvest mycotoxin management. *Journal of Toxicology: Toxin Reviews*, **22**(2-3), 327-350. https://doi.org/10.1081/TXR-120024099

Duarte, J., Silva, L., & Torres, M. (2023). Good Manufacturing Practices (GMP) in food safety: Raw materials and cross-contamination control. *Food Quality and Safety Journal*, 45(3), 177-189.

FAO (2020). Food safety hygiene practices. *Food and Agriculture Organization of the United Nations*, Rome.

FAO. (2003). Good Agricultural Practices (GAP) for Family Agriculture. *Food and Agriculture Organization of the United Nations*. Retrieved from http://www.fao.org/

FDA (2021). Good Manufacturing Practices (GMP) regulations for the food industry. *Food and Drug Administration*, U.S. Department of Health and Human Services.

Finckh, M.R., & Wolfe, M.S. (2006). Diversification strategies. In *The Epidemiology of Plant Diseases* (pp. 269-308). Springer.

Flint, M.L., & Dreistadt, S.H. (1998). *Natural Enemies Handbook: The Illustrated Guide to Biological Pest Control*. University of California Press.

Gachomo, E.W., & Kotchoni, S.O. (2008). The use of plant pathology in rehabilitation of sub-Saharan African agriculture: Challenges and strategies. *African Journal of Biotechnology*, **7**(13), 1896-1901.

Gupta, S., Singh, R., & Prasad, N. (2021). Environmental monitoring systems in food production: A review. *Food Safety Journal*, 22(2), 114-123.

Harman, G.E., Howell, C.R., Viterbo, A., Chet, I., & Lorito, M. (2004). Trichoderma species—opportunistic, avirulent plant symbionts. *Nature Reviews Microbiology*, **2**(1), 43-56. https://doi.org/10.1038/nrmicro797

Hassan, R., Moussa, A., & Khaled, M. (2019). Pest control in food processing environments: Preventing fungal contamination. *International Journal of Food Microbiology*, 130(6), 112-118.

Hell, K., & Mutegi, C. (2011). Aflatoxin control and prevention strategies in maize for sub-Saharan Africa. *African Journal of Food, Agriculture, Nutrition and Development*, **11**(5), 1-12.

ISO (2021). *ISO 9001: Quality management systems — Requirements*. International Organization for Standardization.

ISO. (2018). *ISO 22000:2018 Food Safety Management Systems - Requirements for Any Organization in the Food Chain*. International Organization for Standardization. Retrieved from ISO Standards Website.

Kumar, P., Mahato, D.K., Kamle, M., Mohanta, T.K., & Kang, S.G. (2008). Aflatoxins: A Global Concern for Food Safety, Human Health and Their Management. *Frontiers in Microbiology*, **8**, 1505. https://doi.org/10.3389/fmicb.2017.01505

Lamichhane, J.R., Dachbrodt-Saaydeh, S., Kudsk, P., & Messéan, A. (2017). Toward a reduced reliance on conventional pesticides in European agriculture. *Plant Disease*, **101**(4), 581-598. https://doi.org/10.1094/PDIS-05-16-0693-FE

Logrieco, A., Mule, G., Moretti, A., & Bottalico, A. (2002). Toxigenic Fusarium species and mycotoxins associated with maize ear rot in Europe. *European Journal of Plant Pathology*, **108**(7), 597-609. https://doi.org/10.1023/A:1020679029993

Lorito, M., Woo, S.L., Harman, G.E., & Monte, E. (2010). Translational research on Trichoderma: From 'omics' to the field. *Annual Review of Phytopathology*, **48**, 395-417. https://doi.org/10.1146/annurev-phyto-073009-114314

Magan, N., & Aldred, D. (2007). Post-harvest control strategies: Minimizing mycotoxins in the food chain. *International Journal of Food Microbiology*, **119**(1-2), 131-139. https://doi.org/10.1016/j.ijfoodmicro.2007.07.034

Magan, N., & Medina, A. (2016). Integrated strategies to control seedborne fungi and ochratoxin A in crops. In *Seedborne diseases: ecology and control* (pp. 115-136). Springer.

Mehan, V.K., McDonald, D., Ramakrishna, N., & Subrahmanyam, P. (1991). Resistance in groundnuts to Aspergillus flavus infection and aflatoxin production. *Asp. Flavus in Grdnts (ICRISAT)*, **24**, 265-276.

Miller, J. D. (2008). Mycotoxins in small grains and maize: Old problems, new challenges. *Food Additives & Contaminants*, 25(2), 219-230.

Moussa, A., Perez, S., & Hassan, R. (2022). Facility design and its role in preventing fungal contamination. *Journal of Food Safety Management*, 29(1), 89-102.

Munkvold, G.P. (2003). Cultural and genetic approaches to managing mycotoxins in maize. *Annual Review of Phytopathology*, **41**, 99-116. https://doi.org/10.1146/annurev.phyto.41.052002.095510

Munkvold, G.P., Hellmich, R.L., & Showers, W.B. (2009). Reduced Fusarium ear rot and symptomless infection in transgenic maize with insect resistance derived from *Bacillus thuringiensis*. *Plant Disease*, **93**(5), 488-495. https://doi.org/10.1094/PDIS-93-5-0488

Pérez, S., Hassan, R., & Moussa, A. (2021). Supply chain complexities in GMP implementation. *Journal of Food Safety and Logistics*, 33(1), 123-135.

Pérez, S., Moussa, A., & Alvarez, A. (2021). Integrated pest management strategies in food processing facilities: A focus on fungal contamination. *Food Control Journal*, 48(3), 221-232.

Pitt, J.I., & Hocking, A.D. (2009). *Fungi and Food Spoilage* (3rd ed.). Springer.

Ryu, J. H., & Beuchat, L. R. (2005). Biofilm formation by foodborne pathogens and their control in food-processing environments. *Journal of Food Protection*, 68(11), 2511-2522.

Schmidt, C., & Meyer, R. (2021). Effective pest control methods for food safety: A focus on fungal pathogens. *Journal of Pest Control Science*, 41(2), 98-110.

Smith, K., Harris, D., & Anderson, P. (2020). Record-keeping and traceability in food manufacturing: Best practices for GMP compliance. *Food Quality Assurance Journal*, 28(3), 213-230.

Stevens, R.G., Coates, B.S., & Hellmich, R.L. (2003). Irrigation systems, drainage, and water use. *Crop Management*. https://doi.org/10.1094/CM-2003-1024-01-RS

Trucksess, M.W., & Pohland, A.E. (2001). Mycotoxin protocols. *Methods in Molecular Biology*, **157**, 1-16.

Van Egmond, H. P., Schothorst, R. C., & Jonker, M. A. (2007). Regulations relating to mycotoxins in food: Perspectives in a global and European context. *Analytical and Bioanalytical Chemistry*, 389(1), 147-157.

Wild, C.P., & Gong, Y.Y. (2010). Mycotoxins and human disease: a largely ignored global health issue. *Carcinogenesis,* **31**(1), 71-82. https://doi.org/10.1093/carcin/b

Chapter 18

Economic Impact of Fungal Pathogen and Mycotoxin in Agriculture and Food Processing

Dipayan Hazra, Waghamare Minal Bhujangrao

Ph.D. Research Scholar, Uttar Banga Krishi Vishwavidyalaya, Pundibari, Cooch Behar, West Bengal -736165, India

**Corresponding author email - dipayanhazra78@gmail.com*

Abstract

Plant diseases, particularly caused by fungi are a serious issue because they significantly reduce agricultural yields. Significant outbreaks of plant disease epidemics lead to serious economic consequences such as decreased market production and more unemployment in the food and agriculture industries. Among these pathogens some pathogens produce mycotoxins which accelerates destruction of crop yield and end product from the crop yield. In several crops, particularly wheat, maize, peanuts and other nut crops, cottonseed, and coffee, mycotoxins have a major negative economic impact. According to estimates from the Food and Agriculture Organisation, mycotoxins cause losses of about 1 billion metric tonnes of food and food products annually, affecting 25% of the world's crops in each year. Mycotoxins are gaining global attention due to their strong toxicity, and in recent times, governments, producers, processors, marketers, and consumers have been more concerned about the potential health and financial implications of these substances. Mycotoxins' diverse incidence, structure, and typical chemistry make diagnosing their effects more difficult. When mycotoxins are present in food at levels that can be harmful to health, losses connected to chronic diseases. These are heavy toxins that are carcinogenic, teratogenic, and mutagenic, all of which have detrimental effects on human health. Mycotoxin contamination of agricultural items is more common. During harvest or post-harvest, a number of fungi can create mycotoxins on agricultural products, which can have serious negative effects on humans & animals and also processed food products.

Keywords: Mycotoxin, fungal pathogen, economic losses, post-harvest diseases and processed food products

Introduction:

Certain strains of various filamentous fungi create low molecular weight, poisonous substances called mycotoxins. Under the right conditions, such as moisture and temperature, they cause enormous damage to the global grain trade as well as the marketing of foods and feeds (Windels, 2000). Mycotoxins have gained global interest owing to the substantial losses linked to their effects on animal and human health, as well as the ensuing national economic ramifications (Bhat and Vashanti, 1999; Makun *et al.,* 2009). Mycotoxins have affected humans since the start of organised crop agriculture, but until the last 40 years, their consequences have received little attention. Abiotic and biotic variables may be the cause of food commodity losses both before and after harvest. In tropical hot and humid regions, fungal bio-deterioration of stored food commodities is a persistent issue in storage systems. Several fungi including Aspergillus, *Alternaria, Fusarium, Cladosporium, Penicillium, Mucor,* and *Rhizopus,* have the potential to contaminate harvested food crops (Mateus *et al.,* 2021). Certain fungi have the ability to produce harmful byproducts known as mycotoxins, which can contaminate food under certain circumstances. The genera of *Aspergillus, Fusarium, Penicillium,* and *Alternaria* are the main producers of toxins in moulds (Khodaei *et al.,* 2021; Pandey *et al.,* 2022). Consumption of mycotoxin-contaminated food items has a negative impact on human and animal health, which lowers the marketability of food commodities and raises questions about food safety (Mateus *et al.,* 2021). Over five billion individuals are thought to eat contaminated food on a daily basis and be exposed to mycotoxins through unidentified mechanisms (Khodaei *et al.,* 2021).

The economic impact of fungal pathogens and mycotoxins in agriculture and food processing is substantial and multifaceted, affecting crop yields, food safety, and market dynamics. Here's an overview with references:

Economic Impact on Agriculture:

1. Crop Yields and Quality:

Fungal Pathogens: Fungal diseases like *Fusarium, Aspergillus,* and *Alternaria* can significantly reduce crop yields. For instance, Fusarium head blight in wheat can reduce yields by up to 30% and lower grain quality (McMullen *et al.,* 2012).

Mycotoxins: Mycotoxins produced by these fungi can damage crops, leading to reduced marketability and the need for costly remediation measures. For example, aflatoxins in maize can lead to substantial economic losses (Zain, 2011).

2. Control Costs:

Pesticides and Fungicides: Farmers often spend considerable amounts on chemical treatments to control fungal diseases. The cost of fungicides can be a significant portion of the production budget. Integrated pest management strategies, though potentially reducing long-term costs, also require initial investments (Khan *et al.,* 2013).

3. Crop Insurance and Losses:

Insurance Claims: Farmers may face higher insurance premiums or claims related to crop losses due to fungal infections. This can strain agricultural budgets and affect financial stability (Cole et al., 2020).

Mycotoxin prevalence in food products:

Numerous mycotoxigenic fungi can contaminate agricultural products, which are crucial sources of food and feed. The swift advancement of agricultural technology has resulted in the direct contamination of food items such as fruits, vegetables, grains, and pulses with mycotoxins. No location in the world can be regarded as mycotoxin-free, although tropical places appear to have a bigger problem than temperate ones. This could have to do with the movement of different foodstuffs from one nation to another. Specifically, the following foods have been found to naturally contain mycotoxins.

Rice:

One of the most significant food crops in the world is Paddy. The environment and storage conditions have a significant impact on mycotoxin contaminations in rice cultivation. Aspergillus, Fusarium, and Penicillium species are thought to be the mycotoxigenic fungus found in rice. These fungal infections cause grains and glumes to become polluted with toxins, lose their quality and viability, and discolour. Numerous researchers have documented the existence of AFs, OTA, cyclopiazonic acid, FBs, TCT, ZEA, DON, CT, gliotoxin, and STC in rice (Siruguri *et al.,* 2012). According to Jayaraman and Kalyanasundaram (1990), parboiled rice bran and rice bran samples had an AFB1 level of up to 35%. A 1990 study found that samples of rice had higher levels of AFs (184 to 2,830 μg/kg) than samples of wheat and maize (Pande et al., 1990). AFB1 levels in the PAU 201 variety of preserved rice were found to be less than 15 μg/kg, reported by Siruguri *et al.,* (2012).

Maize:

Corn is frequently infected by spoilage fungus in pre- and post-harvest settings. The relative quantity of these fungi is influenced by biotic and abiotic stressors that impact the pace at which mycotoxin is produced. While AFs were abundant in tropical and subtropical locations (Muriuki and Siboe, 1995), ZEA, FBs, and TCT were the most common mycotoxins contaminating maize in Mediterranean countries (Jestoi, 2008).

Wheat:

Numerous fungal species also harm wheat during storage and transportation. For the first time, Fusarium species-induced wheat grains were found to contain the Fusarium mycotoxin Fusarenon X-glucoside (FUXGlc) (Nakagawa *et al.,* 2011). Additionally, they stated that the amounts of OTA and DON in wheat were 12 and 53 μg/kg, respectively. Badiale-Furlong *et al.* also found the T-2 and AFB1 toxins in wheat flour (2003). According to a different study, FB1 levels varied between 0.6 and 2.3 μg/kg of flour and between 0.5 and 3.9 μg/kg of wheat (Birck *et al.,* 2003). Toxin levels ranged from 603.2 to 850.4 μg/kg, and 74% (24.91%) of wheat samples from the southern region were infected with DON. There have also been reports of several newly discovered mycotoxins in wheat samples. For example, it was discovered that *Fusarium tricinctum* produced a variety of ENNs, including ENN A, A1, B, and B1 in wheat (n = 470; at levels 47 to 142 μg/kg) and durum wheat (n = 260; at levels 55 to 596 μg/kg (Orlando *et al.,* 2019). More recently, 23% of the wheat samples (n = 71) were discovered to be often contaminated with *Fusarium* toxin, also known as DON, according to Topi et al. (2021). One wheat sample in their analysis had a DON concentration of 1916 μg/kg, exceeding the EU limit allowed level of 1,250 μg/kg.

Small Cereal food crops:

Other cereals, such as barley (Hordeum vulgare L.), sorghum (*Sorghum vulgare L.*), oats (*Avena sativa L.*), and millets (*Pennisetum glaucum L.*), have also been linked to mycotoxigenic fungus under specific circumstances, including during storage, transit, and even in the field. These grains have been shown to be an appropriate substrate for the growth of AFs. Of the 116 oat and barley samples that were gathered from eastern Canada, 84 (72%) had DON, with values ranging from 8 to 9 mg/kg (Campbell *et al.,* 2000). 34% of the 73 oat samples, 34% of the 53 barley samples, and 15% of the 26 oat samples contained DON and nivalenol, respectively.

Pulses and Oilseeds:

Along with other vital minerals, pulses offer nutritional protein. However, mycotoxin contamination may result from inappropriate handling and storage. According to Begum and Samajpati (2000), Aspergillus species have been shown to produce between 333 to 10,416 μg/kg of AFB1 in oilseed and pulse crops. Tseng and Tu (1997) examined whether FB1 was present in samples of green grammes (*Vigna radiata*) and adzuki beans (*Vigna angularis*) in Ontario, Canada, but not in the healthy samples. The adzuki and mungbean samples were found to have 261 and 230 μg/g of FB1, respectively, by HPLC. Therefore, further research on mycotoxin contamination of several foods, including legumes, is necessary. Not all fungi that infect food crops produce mycotoxins. Accordingly, Ahmad and Sinha (2002) discovered that only 13 of the 34 isolates of *Fusarium moniliforme* could create ZEA (range 1.2 to 4.0 μg/ml), while only four of the 12 isolates of *Penicillium citrinum* could produce CT (range 1.0 to 3.0 μg/ml). According to Kirchinger *et al.* (2002), cowpea cultivars from South Africa had 0.12 and 0.61 μg/g FB1, whereas cultivars from Benin had no FBs. The maximum amount of FB1 was found at 16.86 μg/g, while the total FB concentration in the pulses ranged from 0.8 to 25.30 μg/g (Kritzinger *et al.,* 2002).

Economic Impact on Food Processing:

1. Food Safety Regulations:

Compliance Costs: The need to comply with food safety regulations regarding mycotoxins involves testing and ensuring that food products meet safety standards. These regulations often require investment in sophisticated testing equipment and procedures (Williams et al., 2004).

2. Economic Losses from Contamination:

Rejection and Recall: Contaminated food products may face rejection by buyers or require costly recalls. For instance, aflatoxin contamination in peanuts and tree nuts has led to significant economic losses due to recalls and export restrictions (Bhat et al., 2010).

3. Market Access and Trade:

Export Restrictions: Countries with high levels of mycotoxin contamination may face trade barriers, affecting their export markets. This can lead to reduced income from international sales (Menkovska et al., 2016).

Market and Trade impacts of Mycotoxin:

The effects of mycotoxins on the market and trade The main mechanism via which these impact markets is to reduce the worth of the commodity that is being exchanged. The the cost of a specific quantity of food product is cut back or the lot is turned down. completely, or the lot needs to be handled before being sold, at an additional cost at a premium cost. This might happen. at several distinct trading levels, from regional to global in scope. Based on the requirements of the purchaser, the set of stakeholders that carries the financial burden of mycotoxin may be lone farmers, labourers, distributors, customers, and processors or authority.

Costs of mycotoxin in regional and local markets:

There are differences in the potential market costs of mycotoxins according on the complexity of local and regional commodity markets, which differs significantly throughout nations. Crop value chains, for instance, are rather straightforward in the USA. Grain elevators (for example, those that handle wheat or maize) or shellers (for example, those that handle groundnuts or tree nuts) may purchase crops from farmers. These handlers may then process the crop further and sell it to animal owners or food processors for human consumption. Following processing, the food is distributed to distributors, who subsequently sell it to retail establishments where customers can buy it. In the United States, producers grain commodities typically carry the responsibility for expenses associated with whereas shellers and mycotoxins Growers are not the ones who carry the greatest mycotoxin-related burden The price of groundnuts and (Wu et al., 2008) tree nuts.

Conclusion

Evaluating the cost of mycotoxins to human society and determining the technical and financial viability of countermeasures is very simple for modern nations. This is due to the fact that mycotoxin costs and intervention advantages are primarily limited to the market, with little to no impact on human health. Furthermore, in industrialised nations, the viability of mycotoxin control only becomes problematic in the event that the intervention is exceedingly costly; in such instances, a less costly remedy is typically accessible. For LICs, the inclusion of mycotoxin effects on human health leads to far more intricate economic evaluations. The ability to assign monetary values to human health consequences has improved with the development of health economic instruments throughout the last 20 years. This in turn helps with cost-effectiveness analysis since it allows for a comparison of the financial costs of treatments and the improvements in human health that result from their implementation. Assessing the technical viability of interventions remains intricate due to the numerous obstacles and external risks and advantages that must be taken into account.

References:

Bhat, R., V. S. Rai, and S. R. Reddy. (2010). Mycotoxins in Food and Feed: An Overview. Mycotoxin Research, 26(3), 187-192.

Cole, D. C., M. M. Cox, and K. F. Hunt. (2020). "Economic Impacts of Mycotoxins on U.S. Agriculture and Food Processing." *Agricultural Economics*, 51(3), 341-356.

Khan, R. A., P. C. Lai, and S. J. Kim. (2013). "Economic Implications of Fungicide Use in Agriculture." *Journal of Agricultural and Food Chemistry*, 61(17), 4147-4154.

McMullen, M., A. L. Bergstrom, and S. D. DeWolf. (2012). "Fusarium Head Blight Management in Wheat: An Overview." *Plant Disease*, 96(6), 649-658. doi:10.1094/PDIS-01-12-0022

Menkovska, M., T. Nakajima, and K. Nishikawa. (2016). "Trade Implications of Mycotoxin Contamination: A Global Overview." *International Journal of Food Science*, 2016, 1-10.

Williams, J. H., J. A. Phillips, T. E. J. J. J. J. J. J. J. S. A. P. M. S. J. G. J. T. M. E. G. A. (2004). "Mycotoxins: Risks in Plant, Animal, and Human Systems." *American Chemical Society*.

Cardwell KF, Desjardins A, Henry SH et al. (2001). Mycotoxins: The Cost of Achieving Food Security and Food Quality. St Paul, MN: American Phytopathological Society.

CAST (2003). Mycotoxins: Risks in Plant, Animal, and Human Systems. Ames, IA: Council for Agricultural Science and Technology (Task Force Report No. 139).

FAO (2004). Worldwide Regulations for Mycotoxins in Food and Feed in 2003. Rome: Food and Agriculture Organization of the United Nations (FAO Food and Nutrition Paper No. 81).

Gericke CA, Kurowski C, Ranson MK, Mills A (2005). Intervention complexity–a conceptual framework to inform priority-setting in health. Bull World Health Organ, 83:285–293.

Gong YY, Cardwell K, Hounsa A et al. (2002). Dietary aflatoxin exposure and impaired growth in young children from Benin and Togo: cross sectional study. BMJ, 325:20–21.

Havelaar A (2007). Methodological Choices for Calculating the Disease Burden and Cost of-illness of Foodborne Zoonoses in European Countries. Bilthoven, Netherlands: Med-Vet Net, Network for the Prevention and Control of Zoonoses (EU Report No. 07–002).

Henry SH, Bosch FX, Troxell TC, Bolger PM (1999). Policy forum: public health. Reducing liver cancer–global control of aflatoxin. Science, 286:2453–2454.

Liu Y, Wu F (2010). Global burden of aflatoxin induced hepatocellular carcinoma: a risk assessment. Environ Health Perspect, 118:818 824.

Liu Y, Chang CC, Marsh GM, Wu F (2012). Population attributable risk of aflatoxin related liver cancer: Systematic review and meta-analysis. European Journal of Cancer, 48:2125–2136.

Lubulwa ASG, Davis JS (1994). Estimating the social costs of the impacts of fungi and aflatoxins in maize and peanuts. In: Highley E, Wright EJ, Banks HJ, Champ BR, eds. Stored Product Protection, Proceedings of the 6th International Working Conference on Stored Product Protection, 17–23 April 1994,

Canberra, Australia. Wallingford, UK: CAB International, pp. 1017–1042. Meenakshi JV, Johnson N, Manyong VM et al. (2007). How Cost-Effective is Biofortification in Combating Micronutrient Malnutrition? An Ex ante Assessment. Washington, DC: HarvestPlus (HarvestPlus Working Paper No. 2).

Stein AJ, Sachdev HPS, Qaim M (2008). Genetic engineering for the poor: golden rice in public health in India. World Dev, 36:144–158.

Strosnider H, Azziz-Baumgartner E, Banziger M et al. (2006). Workgroup report: public health strategies for reducing aflatoxin exposure in developing countries. Environ Health Perspective, 114:1898–1903.

Vardon P, McLaughlin C, Nardinelli C (2003). Potential economic costs of mycotoxins in the United States. In: Mycotoxins: Risks in Plant, Animal, and Human Systems. Ames, IA: Council for Agricultural Science and Technology (Task Force Report No. 139), pp. 136–142.

WHO (1998). Aflatoxins. In: Safety Evaluation of Certain Food Additives and Contaminants: Prepared by the Forty-ninth Meeting of the Joint FAO/

WHO Expert Committee on Food Additives (JECFA). Geneva: World Health Organization (WHO Food Additives Series, No. 40).

WHO (2001). Macroeconomics and Health: Investing in Health for Economic Development. Report of the Commission on Macroeconomics and Health. Geneva: World Health Organization.

Wild CP, Gong YY (2010). Mycotoxins and human disease: a largely ignored global health issue. Carcinogenesis, 31:71–82.

Wu F (2004). Mycotoxin risk assessment for the purpose of setting international regulatory standards. Environ Sci Technol, 38:4049–4055.

Wu F (2008). A tale of two commodities: how EU mycotoxin regulations have affected food industries. World Mycotox J, 1:95–102.

Wu F, Khlangwiset P (2010a). Health economic impacts and cost-effectiveness of aflatoxin reduc tion strategies in Africa: case studies in biocontrol and postharvest interventions. Food Addit Contam Part A Chem Anal Control Expo Risk Assess, 27:496–509.

Wu F, Khlangwiset P (2010b). Evaluating the technical feasibility of aflatoxin risk reduc tion strategies in Africa. Food Addit Contam Part A Chem Anal Control Expo Risk Assess, 27:658–676.

Wu F, Liu Y, Bhatnagar D (2008). Cost effectiveness of aflatoxin control methods: economic incentives. Toxin Reviews, 27:203–225.

Ahmad, M. S., and Sinha, K. K. (2002). Mycotoxin contamination of mustard seeds during storage. *Indian Phytopath.* 55, 299–302.

Badiale-Furlong, E., Dors, G. C., Oliveira, M., Dos, S., de Souza, M. M., and Kuhn, R. C. (2003). "Avaliacaodaqualidade de farinha de trigoeprodutos de panificaçãocomercializadas no Rio Grande do Sul" in *Simpósio de Ciências de Alimentos e Saude* (Anais: Florianópolis-SC/UFSC), 1–4.

Begum, H., and Samajpati, N. (2000). Mycotoxin production on rice, pulses and oilseeds. *Naturwissenschaften* 87, 275–277.

Birck, N. M. M., Lorini, I., and Scussel, V. M. (2003). "Sanitary conditions and mycotoxins in wheat grains (*Triticum aestivum*) and flour (common and special) through milling processing" in *IV Congreso Latino americano De Mycotoxicologia* (LaHabana, Cuba: Anais)

Campbell, H., Choo, T. M., Vigier, B., and Underhill, L. (2000). Mycotoxins in barley and oat samples from Eastern Canada. *Canadian Journal of Plant Science.* 80, 977–980.

Embaby, E. M., Reda, M., Abdel-Wahhab, M. A., Omara, H., and Mokabel, A. M. (2013). Occurrence of toxigenic fungi and mycotoxins in some legume seeds. *J. Agric. Technol.* 9, 151–164.

Jayaraman, P., and Kalyanasundaram, I. (1990). Natural occurrence of toxigenic fungi and mycotoxins in rice bran. *Mycopathologia* 110, 81–85.

Jestoi, M. (2008). Emerging fusarium-mycotoxins fusaproliferin, beauvericin, enniatins, and Moniliformin-A review. *Crit. Rev. Food Sci. Nutr.* 48, 21–49.

Muthomi, and Mutitu, (2003). Occurrence of mycotoxin producing *Fusarium* species and other fungi on wheat kernels harvested in selected districts of Kenya. *Afri. Crop Sci. Conf. Proc.* 6, 335–339.

Muriuki, G. K., and Siboe, G. M. (1995). Maize flour contaminated with toxigenic fungi and mycotoxins in Kenya. *Afr. J. Health Sci.* 2, 236–241.

Nakagawa, H., Ohmichi, K., Sakamoto, S., Sago, Y., Kushiro, M., Nagashima, H., et al. (2011). Detection of a new *Fusarium* masked mycotoxin in wheat grain by high-resolution LC–Orbitrap™ MS. *Food Addit. Contam: Part A* 28, 1447–1456.

Orlando, B., Grignon, G., Vitry, C., Kashefifard, K., and Valade, R. (2019). *Fusarium* species and enniatin mycotoxins in wheat, durum wheat, triticale and barley harvested in France. *Mycotoxin Res.* 35, 369–380.

Pande, N., Saxena, J., and Pandey, H. (1990). Natural occurrence of mycotoxins in some cereals. *Mycoses* 33, 126–128.

Sulyok, M., Berthiller, F., Krska, R., and Schuhmacher, R. (2006). Development and validation of a liquid chromatography/tandem mass spectrometric method for the determination of 39 mycotoxins in wheat and maize. Rapid Commun. *Mass Spectrom.* 20, 2649–2659.

Topi, D., Babic, J., Pavsic-Vrtac, K., Tavcar-Kalcher, G., and Jakovac-Strajn, B. (2021). Incidence of *Fusarium* mycotoxins in wheat and maize from Albania. *Molecules* 26:172.

Chapter 19

Food Safety Regulations for Fungal Pathogens and Mycotoxins: A Global Perspective

Preeti Kaleramana, Jatin Dutt*

Department of Microbiology,
Chaudhary Charan Singh Haryana Agricultural University
Hisar, Haryana ,125004, India
Corresponding author: jatindutt2706@gmail.com

Introduction

Postharvest losses refer to the losses that occur along the food supply chain due to pathogens infection, handling, storage, transportation and processing, thereby resulting in the reduction in quality, quantity and market value of agricultural commodities (Kader *et al.*, 2005; Parfitt *et al.*, 2010). Food and Agriculture Organization reported that global average loss due to the food postharvest losses in North America, Europe and Oceania was about 29%, compared to an average of about 38% in industrialized Asia, Africa, Latin America and South East Asia (FAO, 2011). These fungi not only lead to visible spoilage but also produce mycotoxins which are harmful secondary metabolites that contaminate food and feed, resulting in serious health risks, including carcinogenic effects and organ damage. The economic impact of these losses is substantial, with billions of dollars lost annually due to decreased market value and disposal costs (Kumar *et al.*, 2020). The management of post-harvest fungal pathogens and mycotoxins requires a multifaceted approach encompassing detection, diagnosis, and effective management strategies. Early detection of fungal contaminants is vital to prevent the spread of infections and minimize the production of mycotoxins. Effective detection and diagnosis of these pathogens and toxins are essential for minimizing losses and ensuring food safety. Advances in molecular techniques, immunological assays, and chromatographic methods have enhanced our ability to identify and quantify fungal contaminants with greater precision and speed. Regular monitoring and testing for mycotoxins in food products further facilitate early detection and intervention, minimizing contamination risks (Bhat *et al.*, 2010). By integrating innovative technologies and sustainable practices, stakeholders can effectively reduce the impact of post-harvest fungal contamination, ensuring safer food supplies and improved economic outcomes for producers and consumers alike (Pitt & Hocking, 2009). However, the implementation of these technologies varies across regions, influenced by economic, infrastructural, and regulatory factors.

Importance of food safety regulations

Food safety regulations are crucial for several reasons:

1. **Public Health Protection**: They help prevent foodborne illnesses, which can result from contamination by pathogens, chemicals, or allergens. Regulations ensure that food is safe to consume, reducing health risks to the public.
2. **Consumer Confidence**: Clear regulations enhance consumer trust in food products. When people know there are standards in place, they are more likely to feel secure in their food choices.
3. **Quality Assurance**: Food safety regulations promote the production of high-quality food. They establish standards for cleanliness, handling, and storage, ensuring that food maintains its nutritional value and safety throughout the supply chain.
4. **Economic Impact**: Ensuring food safety can reduce healthcare costs associated with foodborne illnesses. It also protects businesses from costly recalls and lawsuits, ultimately contributing to a more stable economy.
5. **Global Trade**: Food safety regulations facilitate international trade by ensuring that exported products meet safety standards. This helps countries maintain their reputation and access new markets.
6. **Environmental Protection**: Some regulations address environmental impacts related to food production, promoting sustainable practices that protect ecosystems and reduce pollution.
7. **Innovation and Improvement**: Regulatory frameworks can drive innovation in food safety technologies and practices, leading to improved methods for detecting contamination and ensuring quality.

Global Regulatory Frameworks

Codex Alimentarius Commission (CAC)

- ☆ Established by the FAO and WHO, Codex provides international food safety standards, including limits for mycotoxins in food and guidelines for sampling and testing.
- ☆ This standard applies to any substance that meets the terms of the Codex definition for a contaminant, including contaminants in feed for food-producing animals, except:

- 1) Contaminants having only food and feed quality significance (e.g. copper), but no public health significance, in the food(s) given that the standards elaborated within the Codex Committee on Contaminants in Foods (CCCF) has the objective to protect public health.
- 2) Pesticide residues, as defined by the Codex definition that are within the terms of reference of the Codex Committee on Pesticide Residues (CCPR).
- 3) Residues of veterinary drugs, as defined by the Codex definition, that are within the terms of reference of the Codex Committee on Residues of Veterinary Drugs in Foods (CCRVDF). 4) Microbial toxins, such as botulinum toxin and staphylococcus enterotoxin, and microorganisms that are within the terms of reference of the Codex Committee on Food Hygiene (CCFH).
- 5) Residues of processing aids that are within the terms of reference of the Codex Committee on Food Additives (CCFA)
- Relevant Codex documents include:
    - **Codex General Standard for Contaminants and Toxins in Food** (CODEX STAN 193-1995)
    - **Code of Practice for the Prevention and Reduction of Mycotoxin Contamination in Cereals** (CAC/RCP 51-2003)

2. European Food Safety Authority (EFSA)

The European Food Safety Authority (EFSA) is the agency of the European Union (EU) that provides independent scientific advice and communicates on existing and emerging risks associated with the food chain.

EFSA evaluates risks related to mycotoxins and provides scientific advice for the European Union (EU). The work of EFSA covers all matters with a direct or indirect impact on food and feed safety, including animal health and welfare, plant protection and plant health and nutrition.

Specific regulations include:

- **Regulation (EC) No. 1881/2006** setting maximum levels for certain mycotoxins in foodstuffs.
- Regular reports on the prevalence of mycotoxins in food products.

3. U.S. Food and Drug Administration (FDA)

The FDA establishes guidelines for mycotoxin levels in food and animal feed, including:

- **Guidance for Industry: Action Levels for Aflatoxins in Animal Food**.
- **Mycotoxin Regulations** for various products, focusing on detection methods and safety levels.

4. World Health Organization (WHO)

WHO efforts include the development of comprehensive risk assessments that evaluate the health impacts of mycotoxin exposure, which help countries establish regulatory limits and safety standards (World Health Organization, 2020).

Furthermore, WHO collaborates with various stakeholders, including the Food and Agriculture Organization (FAO) and the World Organisation for Animal Health (OIE), to enhance food safety practices globally and improve surveillance systems for monitoring mycotoxin contamination (World Health Organization, 2019). By providing technical support, training, and resources, WHO aims to strengthen national capacities in managing fungal threats and ensuring the safety of food supplies, ultimately contributing to improved public health outcomes worldwide (World Health Organization, 2021).

5. International Organization for Standardization (ISO)

The International Organization for Standardization (ISO) plays a crucial role in the control of fungal pathogens and mycotoxins through the development of internationally recognized standards that ensure food safety and quality. ISO standards provide guidelines for the testing, monitoring, and management of mycotoxins in various food products, helping organizations implement effective quality control measures (ISO 21527). These standards facilitate consistent methodologies for detecting mycotoxin contamination, enabling better risk assessment and regulatory compliance across different countries (ISO 22000). Additionally, ISO promotes best practices in food production and processing, emphasizing the importance of good agricultural and manufacturing practices to minimize fungal contamination (ISO 9001). By fostering international cooperation and providing a framework for quality assurance, ISO helps enhance food safety and protect public health from the risks associated with fungal pathogens and mycotoxins.

ISO develops international standards for the detection and analysis of mycotoxins and fungal pathogens, such as:

- ☆ **ISO 16050** for sampling and analysis of mycotoxins.
- ☆ **ISO 21187** for the determination of mycotoxin levels.

Food Safety Regulations for Fungal Pathogens and Mycotoxins: A Global Perspective

Regulatory Approaches

Hazard Identification: Identification of fungi and related mycotoxins in foods, such as aflatoxins, ochratoxins, and fumonisins, that may pose health risks.

Hazard Characterization: Establishing a dose-response relationship through the exclusion of data from animal and human studies. Establishment of toxic reference points such as NOAEL, LOAEL and BMDL and estimation of health guideline values (HBGV) such as tolerable daily intake (TDI) or temporary maximum acceptable daily intake (PMTDI).

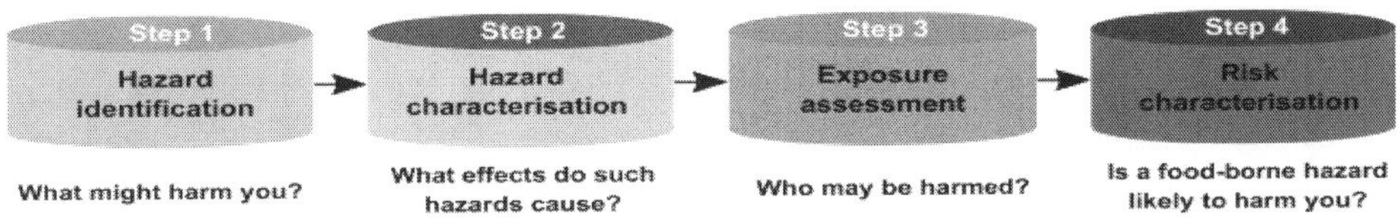

The four steps of human health risk assessment (Malir *et al.*,2023)

Exposure Assessment: Human exposure to mycotoxins is assessed through dietary consumption data and pollution levels.

Risk Characterization: A combination of hazard characterization and risk assessment to assess the likelihood and severity of adverse health effect, tools such as Margin of Exposure (MOE) and Uncertainty Factor (UF) are used to quantify overall risk.

Maximum Residue Limits (MRLs) and Tolerances: Maximum Residue Limits (MRLs) are the highest levels of pesticide residues legally allowed in or on food products. In the United States, these limits are referred to as "tolerances." MRLs are not directly about safety; instead, they are established based on the assumption that if good agricultural practices (GAP) are followed, pesticide residues will fall within these limits. MRLs are determined through a process that includes toxicological testing, field trials, and analysis of how residues behave on crops from application to harvest. These limits help ensure that food products do not exceed safe levels of pesticide exposure for consumers.

While these levels are not direct safety limits, they reflect what is considered acceptable based on the assumption that agricultural producers follow good agricultural practices (GAP). MRLs are determined through toxicological assessments and field trials to ensure that pesticide residues remain below levels that could pose a risk to human health. The process includes calculating the no observable adverse effect level (NOAEL), which represents the highest dose with no significant adverse effects in toxicological studies and multiplying it by a safety factor to establish safe exposure levels.

For instance, the U.S. Environmental Protection Agency (EPA) evaluates data from residue studies on various crops and establishes MRLs based on expected pesticide levels when used according to GAP. These limits help manage the risk of pesticide exposure to consumers by ensuring that any residues present are unlikely to cause harm based on established safety standards. The European Union and other regions have similar regulations, often with varying MRLs for different foods and pesticides.

Monitoring and Surveillance Programs - Monitoring and surveillance programs are essential for ensuring compliance with MRLs. These programs often involve testing food samples from various stages of the supply chain, including production, import, and retail levels. If residues exceed the MRL, the product is usually deemed non-compliant, which may result in recalls or restrictions on sale. This monitoring helps in identifying trends, enforcing regulations, and ensuring that agricultural practices align with safety standards. Surveillance programs not only protect consumers but also contribute to improving agricultural practices by providing feedback to farmers and producers.

Regional Regulations and Enforcement

European Union: Regulation (EC) No 1881/2006 and Updates

The EU regulates MRLs for pesticides through Regulation (EC) No 1881/2006, which sets the maximum allowable levels of contaminants, including pesticides, in foodstuffs. This regulation is updated regularly to reflect new scientific data, improve food safety, and align with international standards. EU Member States enforce these limits by monitoring food products in their markets and using the Rapid Alert System for Food and Feed (RASFF) to communicate and respond to any violations.

United States: FDA and USDA Regulations

In the United States, the FDA and USDA regulate MRLs through different mechanisms. The EPA sets MRLs (referred to as tolerances) under the Federal Food, Drug, and Cosmetic Act (FFDCA), while the FDA enforces these limits for most foods, and the USDA monitors residues in meat, poultry, and egg products. These agencies also conduct routine testing and sampling to ensure compliance, and they can issue recalls or warnings if violations are detected.

Asia-Pacific: Country-Specific Regulations and Harmonization Efforts

The Asia-Pacific region has a variety of country-specific regulations regarding MRLs. For instance, Japan follows its Food Sanitation Act, and Australia adheres to regulations set by Food Standards Australia New Zealand (FSANZ). While each country enforces its own standards, there are also ongoing efforts to harmonize MRLs through initiatives like ASEAN's guidelines and the APEC Food Safety Cooperation Forum. These efforts aim to streamline trade and improve food safety across the region.

Table-1.Regulatory Limits for Major Mycotoxins in Grains (Beia *et al.*, 2023; Porsnicu *et al.*, 2023)

Mycotoxin	*USDA (ppm)*	*European Union (ppm)*	*ISO/Global Standards (ppm)*
Aflatoxin B1	20 (Corn & peanut)	2 (cereals and nuts)	2 (aligned with Codex)
Total Aflatoxins	20 (All foods)	4 (peanuts and cereals)	2
Deoxynivalenol (DON)	1 (grain for human consumption)	0.75–1.25(grains, baby food)	0.75–1.25
Fumonisins	2-4 (corn product)	1 (processed maize, baby food)	1–4 (Codex alignment)
Ochratoxin	Not specified	0.003 (infant food)	0.003–0.010
Zearalenone	Not specified	0.05 (baby food)	0.05 (aligned with Codex)

Case Histories:

1. Aflatoxin in Nuts (EU-India Trade):

EU's stringent aflatoxin limits have led to frequent rejections of Indian nut consignments. In response, India upgraded its testing labs to ISO/IEC 17025 standards to meet EU requirements (Osaili *et al.*,2023).

2. Fumonisins in Corn (USA-Mexico Dispute):

- ✰ Mexican imports of US corn faced delays due to higher fumonisin levels in American shipments compared to Mexican standards (Li *et al.*,2024).

3. Ochratoxin in Cocoa (West Africa-EU Trade): (Leslie *et al.*, 2021)

- ✰ The EU's low ochratoxin limits affected cocoa exports from West African nations, requiring stricter pre-shipment testing and fumigation protocols.

4. Regulatory Alignment in China-EU Peanut Trade:

- ✰ EU-China collaboration on harmonizing aflatoxin levels facilitated smoother peanut trade, reducing rejection rates significantly (Omari *et al.*, 2024).

5. Climate Change and Mycotoxins:

- ✰ Changes in climate have heightened contamination risks, challenging countries to revise standards. This affects African and Asian countries trading with EU (Casu *et al.*,2024).

6. Indian Rice & Mycotoxins (EU Import Rules):

- ☆ EU regulations led to stricter testing protocols in India for aflatoxin and arsenic, promoting ISO accreditation of testing labs (Santos *et al.*, 2022)

Challenges and Future Directions

Emerging fungal pathogens and mycotoxins

Emerging fungal pathogens and their production of mycotoxins present an evolving threat to food safety, as new fungal species continue to adapt and spread due to changes in environmental conditions and agricultural practices. Mycotoxins, such as aflatoxins and ochratoxins, are known to cause severe health issues, including carcinogenic, nephrotoxic, and immunosuppressive effects in humans. The rise of these contaminants has spurred a need for enhanced detection methods and regulatory responses to address their prevalence effectively.

Several studies highlight these concerns. For example, Chakraborty *et al.* (2021) discuss how climate change and agricultural practices influence fungal adaptation, leading to increased contamination risks. In addition, Paterson and Lima (2010) elaborate on the health impacts and the need for monitoring these pathogens as they spread to new regions due to climate change. Lastly, Wu *et al.* (2014) emphasize the importance of developing robust regulatory frameworks to control mycotoxin contamination in food, noting the growing complexity of this challenge as fungal species evolve.

Climate Change and Food Safety

Climate change significantly influences the prevalence and distribution of foodborne pathogens and mycotoxins. Rising temperatures, altered precipitation patterns, and extreme weather events like floods can promote fungal growth and increase contamination risks across food supply chains. Additionally, climate-driven shifts can lead to the emergence of pathogens in regions previously unaffected, thus increasing the scope and complexity of food safety management (FAO, 2020; Cabanes *et al.*, 2010). This highlights the need for climate-resilient food safety strategies, including updated monitoring systems and adaptive management plans to handle potential new threats (El Samra *et al.*, 2017).

International Cooperation and Harmonization

Addressing the global challenges posed by fungal pathogens and climate change on food safety requires a united front. International cooperation is crucial, as no single country can effectively manage these issues alone. By harmonizing regulations and fostering collaborative efforts, countries can enhance their monitoring and control systems, enabling quicker and more effective responses to emerging threats.

The Food and Agriculture Organization (FAO) highlights the necessity of cross-border collaboration to ensure that food safety standards are consistent globally. This uniformity is vital for mitigating risks associated with imported foods, especially as climate change impacts ecosystems shared by multiple nations. When countries work together to align their food safety regulations, they create a stronger safety net for consumers worldwide.

For instance, establishing common guidelines can help prevent foodborne illnesses that arise from contaminated imports, which is particularly important as trade increases. Moreover, shared resources and expertise can lead to more robust surveillance systems, allowing for the early detection of potential food safety issues before they escalate. As the FAO states, such cooperation not only protects public health but also supports the sustainability of food systems in the face of climate-related challenges (World Health Organization (WHO).

References

Anonymous Stop neglecting fungi. *Nat. Microbiol.* 2017;**2**:17120.

ASEAN. (2015). *Guidelines for the Harmonization of Pesticide Maximum Residue Limits in ASEAN Countries.* Retrieved from https://asean.org

Bandyopadhyay, R., Ortega-Beltran, A., Akande, J., et al. (2016). Biological control of aflatoxins in Africa: Current status and potential challenges in the face of climate change. *World Mycotoxin Journal*, 9(5), 771-789. doi:10.3920/WMJ2016.2023.

Beia, I. S., Ciceoi, R., Beia, V. E., Bădulescu, L., Smedescu, D. I. & Fîntîneru, A. (2023). Mycotoxins-Incidence, Impact ON Feed, Food Safety, Food Chain and Economic Losses. Scientific Papers Series Management, Economic Engineering in Agriculture & Rural Development, 23(4).

Bhat, R., Bhat, A., & Bhat, J. A. (2010). Mycotoxins in Food: An Overview. *Food Safety and Quality*.

Casu, A., Camardo Leggieri, M., Toscano, P., & Battilani, P. (2024). Changing climate, shifting mycotoxins: A comprehensive review of climate change impact on mycotoxin contamination. Comprehensive Reviews in Food Science and Food Safety, 23(2), e13323.

Fisher M.C., Henk D.A., Briggs C.J., Brownstein J.S., Madoff L.C., McCraw S.L., Gurr S.J. Emerging fungal threats to animal, plant and ecosystem health. *Nature.* 2012;484:186–194.

Food and Agriculture Organization of the United Nations . *Global Food Losses and Food Waste: Extent, Causes and Prevention.* FAO; Rome, Italy: 2011.

Food and Agriculture Organization. (2020). *The impact of climate change on food security and nutrition*. Retrieved from FAO.

Food Standards Australia New Zealand (FSANZ). (n.d.). *Australia New Zealand Food Standards Code – Standard 1.4.2*. Retrieved from https://www.foodstandards.gov.au

Gallo, A., Giuberti, G., Frisvad, J. C., Bertuzzi, T., & Nielsen, K. F. (2015). Review on mycotoxin issues in ruminants: Occurrence in forages, effects of mycotoxin ingestion on health status and animal performance and practical strategies to counteract their negative effects. *Toxins, 7*(8), 3057-3111.

IARC Working Group on the Evaluation of Carcinogenic Risks to Humans. (2018). *Some naturally occurring substances: Food items and constituents, heterocyclic aromatic amines and mycotoxins.* IARC Monographs on the Evaluation of Carcinogenic Risks to Humans, 56, 345-368. https://www.iarc.fr/wp-content/uploads/2018/06/mono56.pdf

ISO 21527: *Microbiology of food and animal feeding stuffs – Horizontal method for the determination of the number of yeasts and/or moulds.*

ISO 22000: *Food safety management systems – Requirements for any organization in the food chain.*

ISO 9001: *Quality management systems – Requirements.*

Joffe, A. Z. 1978. *Fusarium poae* and *F. sporotrichioides* as principal causal agents of alimentary toxic aleukia, p. 21-86. *In* T. D. Wyllie and L. G. Morehouse (ed.), Mycotoxic fungi, mycotoxins, mycotoxicoses, vol. 3. Marcel Dekker, New York, N.Y.

Kader A.A. Increasing food availability by reducing postharvest losses of fresh produce. *Int. Postharvest Symp. Acta Hort.* 2005;682:2169–2178. doi: 10.17660/ActaHortic.2005.682.296.

Kumar, P., Sharma, A., & Bansal, M. (2020). Mycotoxin Contamination in Food: A Global Concern. *Journal of Food Science and Technology.*

Leslie, J. F., Moretti, A., Mesterházy, Á., Ameye, M., Audenaert, K., Singh, P. K., & Logrieco, A. F. (2021). Key global actions for mycotoxin management in wheat and other small grains. Toxins, 13(10), 725.

Li, T., Li, J., Wang, J., Xue, K. S., Su, X., Qu, H., & Jiang, Y. (2024). The occurrence and management of fumonisin contamination across the food production and supply chains. Journal of Advanced Research, 60, 13-26.

Malir, F., Barlow, K., & Mally, A. (2023). Risk assessment of mycotoxins in food: A review. *Food Control,* 135, 108826. https://doi.org/10.1016/j.foodcont.2022.108826

Marin, S., *et al.* (2013). "Current methods for the detection of mycotoxins in food and feed." *Food Control,* 34(1), 1-15.

Meyer, W., *et al.* (2013). "Quantitative PCR for the detection of fungi in environmental samples." *Environmental Microbiology Reports,* 5(4), 593-604.

Omari, R., Adams, A., Baah-Tuahene, S., Karbo, R. T., Mensah, C. A., Frempong, G. K. & Quaison, K. E. (2024). Researchers' participation in the policy process: lessons from aflatoxin control in Ghana. Policy Design and Practice, 1-16.

Osaili, T. M., Odeh, W. A. B., Al Ayoubi, M., Al Ali, A. A., Al Sallagi, M. S., Obaid, R. S., & El Darra, N. (2023). Occurrence of aflatoxins in nuts and peanut butter imported to UAE. Heliyon, 9(3).

Parfitt J., Barthel M., Macnaughton S. Food waste within food supply chains: Quantification and potential for change to 2050. *Philos. Trans. R. Soc.* 2010;365:3065–3081. doi: 10.1098/rstb.2010.0126.

Paterson, R. R. M., & Lima, N. (2010). How will climate change affect mycotoxins in food? *Food Research International*, 43(7), 1902-1914. doi:10.1016/j.foodres.2009.07.010.

Pitt, J. I., & Hocking, A. D. (2009). Fungi and Food Spoilage. *Springer*.

Poroşnicu, I., Ailincăi, L. I., Neculai-Văleanu, A. S., Ariton, M. A. & Mareş, M. (2023). Legislative Aspects Regarding the Control of Mycotoxins. SCIENTIFIC PAPERS ANIMAL SCIENCE AND BIOTECHNOLOGIES, 56(2), 153-153.

Santos, A. R., Carreiró, F., Freitas, A., Barros, S., Brites, C., Ramos, F., & Sanches Silva, A. (2022). Mycotoxins contamination in rice: Analytical methods, occurrence and detoxification strategies. Toxins, 14(9), 647.

Savary S., Ficke A., Aubertot J., Hollier C. Crop losses due to diseases and their implications for global food production losses and food security. *Food Secur.* 2012;4:519–537.

United States Environmental Protection Agency. (2024). *Regulation of pesticide residues on food*. Retrieved from https://www.epa.gov

United States Food and Drug Administration. (n.d.). *Pesticide Residue Monitoring Program*. Retrieved from https://www.fda.gov

World Health Organization. (2019). *Mycotoxin control in food and agriculture.*

World Health Organization. (2020). *Mycotoxins: A global health concern.*

World Health Organization. (2021). *Food Safety and Mycotoxins.*

Wu, F., Groopman, J. D., & Pestka, J. J. (2014). *Public health impacts of foodborne mycotoxins. Annual Review of Food Science and Technology*, 5, 351-372. doi:10.1146/annurev-food-030713-092431.

Chapter 20

Future Perspectives of Mycotoxin Detection, Diagnosis, and Management

Arushi Padiyal[1], Narender K. Bharat[2] and Manish Kumar[2]*

[1]Department of Agriculture, Maharishi Markandeshwar (Deemed to be University), Mullana, Ambala, Haryana, India
[2]Department of Seed Science & Technology, Dr. Y. S. Parmar University of Horticulture and Forestry, Nauni, Solan, Himachal Pradesh, India
**Corresponding authors Email Id: aru.padiyal@gmail.com (A.P.)*

Introduction

The escalating global population necessitates the stockpiling of substantial food supplies for future consumption. However, inadequate storage conditions can lead to the proliferation of mycotoxigenic fungi, resulting in the contamination of food commodities with various mycotoxins, posing a pressing global concern. Mycotoxins predominantly contaminate cereals and processed foods, with species like *Fusarium* affecting widely consumed grains such as wheat, maize, barley, and oats (Awuchi et al., 2021). The occurrence and concentration of mycotoxins are influenced by geographic and climatic factors, necessitating extensive studies on their co-occurrence and interactions. This chapter delves into future directions in mycotoxin research, emphasizing detection, diagnosis, and management strategies pertinent to post-harvest fungal pathogens and mycotoxins.

Advancements in Rapid Detection and quantification and AI-Driven Mycotoxin Monitoring

Accurate detection of mycotoxins in diverse food matrices remains a formidable challenge due to the complexity of certain food matrices, such as fruit juices and edible oils, and the relatively low concentrations of mycotoxins present. Traditional chromatographic techniques, such as high-performance liquid chromatography (HPLC) and gas chromatography-mass spectrometry (GC-MS), have been extensively utilized for analyzing mycotoxins in staples like rice, wheat, oilseeds, and peanuts. However, these methods are time-consuming and require sophisticated equipment. Emerging technologies, including aptamer-based assays and surface plasmon resonance, have demonstrated high affinity, specificity, and low detection limits in select commodities, such as vegetable oils (Szelenberger et al., 2024). Future research should focus on expanding the application of these novel detection methods across a broader spectrum of food products, enhancing sensitivity, and streamlining sample preparation processes.

Moreover, techniques such as hyperspectral imaging and Raman spectroscopy allow for real-time, non-invasive mycotoxin detection in stored products, preserving their integrity while ensuring food safety during post-harvest management (Chen et al., 2024). Additionally, the integration of nanomaterials in biosensors, lab-on-a-chip (LoC) devices, and point-of-care testing platforms holds great promise for on-site detection. Electrochemical biosensors offer high sensitivity and specificity for on-site detection, providing rapid and cost-effective alternatives to conventional methods. A small description has been given in various chapter about these technology across the book.

In present era, artificial intelligence (AI) and machine learning (ML) are revolutionizing mycotoxin detection, prediction, and management. AI-powered image recognition systems utilizing convolutional neural networks (CNNs) can diagnose plant diseases with high accuracy, surpassing human visual examination. Machine learning algorithms, including support vector machines (SVM) and random forests (RF), have shown success in predicting disease outbreaks based on environmental parameters such as temperature, humidity, and soil conditions (Siddique et al., 2024). The integration of AI with Internet of Things (IoT)-enabled sensors can enhance real-time monitoring and decision-making in post-harvest storage. Future research should focus on refining AI models for real-world applications and exploring predictive analytics to mitigate contamination risks proactively.

Molecular Insights into Mycotoxin Biosynthesis

Advances in genomic and transcriptomic analysis have identified key regulatory genes within mycotoxin biosynthesis pathways, offering new targets for disrupting toxin production through genetic or chemical interventions. CRISPR-Cas9 and RNA interference (RNAi) technologies are being explored for functional studies and potential mycotoxin control strategies (Godana et al., 2023). Epigenetic modifications influencing toxin biosynthesis present new opportunities for disruption of mycotoxin production. Future research should continue to explore transcriptional and signaling mechanisms that govern fungal secondary metabolism, paving the way for innovative control approaches. However, further extensive explorations are required in this field.

Innovations in Mycotoxin Management

Reducing mycotoxin levels in food commodities to safe consumption thresholds involves chemical, physical, and biological decontamination methods traditionally.

It is imperative that these interventions preserve the organoleptic properties and nutritional value of foods without generating toxic degradation byproducts. Specifically, biological control agents (BCAs), particularly plant extracts and essential oils, offer eco-friendly alternatives to synthetic chemicals, aligning with sustainable agricultural practices. However, comprehensive research is required to standardize the quality of these natural extracts, assess their safety, and determine optimal application timings and concentrations for various food commodities.

In the same sequence, beneficial microorganisms such as endophytes are being explored for their ability to suppress mycotoxin-producing fungi. These natural bio-control agents outcompete pathogens or produce antifungal metabolites, reducing mycotoxin levels in agricultural products. Advances in microbial-based detoxification methods, utilizing enzymatic and metabolic pathways to degrade mycotoxins, are also under investigation. Microbial enzymes like hydrolases, oxidases, reductases, and aminotransferases play a key role in this process. Hydrolases break down fumonisin and ZEN without co-substrates, while oxidases and reductases require cofactors. Whole microorganisms, such as *Saccharomyces cerevisiae*, naturally regenerate cofactors and transform mycotoxins during fermentation (Abraham et al., 2022). Future research should aim at optimizing these biological strategies for large-scale implementation while ensuring regulatory compliance and consumer acceptance.

Innovative physical methods, such as cold plasma technology, have garnered attention for their potential to control fungal growth and mycotoxin contamination. Cold plasma, a non-thermal technique, effectively disrupts fungal DNA and cell walls, leading to the rapid degradation of mycotoxins. CAP breaks down mycotoxins like AFB1 and DON, reducing their toxicity while preventing contamination by inhibiting toxigenic fungi and disrupting mycotoxin biosynthesis (Oliveira et al., 2025). With minimal impact on food quality, further research is warranted to optimize this technology for various food systems, ensuring efficacy and safety without compromising food quality.

Impact of Climate Change on Mycotoxin Production

Climate change is altering the distribution and severity of mycotoxin contamination in agricultural commodities. Rising temperatures, increased humidity, and erratic precipitation patterns create favorable conditions for fungal growth and toxin production. Predictive modeling that incorporates climate variables is essential for forecasting mycotoxin risks. However, current models like APSIM have limitations in capturing the complexity of environmental influences on aflatoxin production. Climate-resilient storage techniques, including hermetic storage and controlled atmosphere systems, are being developed to counteract environmental fluctuations that favor fungal growth. Strengthening surveillance, advancing detection technologies, and promoting climate-resistant crops are crucial for mitigating aflatoxin risks (Bunny et al., 2024). Research should continue to explore the impact of climate change on fungal pathogens and mycotoxins, informing policies for sustainable food production and storage.

Regulatory Enhancements and Global Standardization in Policy and Trade

The harmonization of global mycotoxin regulations is critical for ensuring food safety and facilitating international trade. Research efforts are focused on developing comprehensive risk assessment frameworks that integrate human biomonitoring data and exposure modeling. The impact of mycotoxin mixtures and their potential synergistic effects require further investigation to refine regulatory thresholds. Moreover, blockchain technology offers potential solutions for enhancing traceability and transparency in the food supply chain, enabling real-time monitoring of contamination risks. By enabling real-time traceability and accountability, blockchain minimizes food safety risks and strengthens regulatory compliance (Oriekhoe et al., 2024). Collaboration among stakeholders is essential to fully harness its potential, ensuring a safer and more sustainable global food system for future generations. Strengthening regulatory policies and consumer awareness programs will be key to mitigating mycotoxin-related health hazards.

Conclusion

Addressing mycotoxin challenges requires a comprehensive approach involving advances in detection, molecular research, decontamination strategies, and climate adaptation. Rapid detection methods, such as AI-driven monitoring and innovative biosensors, enhance food safety and streamline management. Molecular insights into biosynthesis offer new avenues for disrupting toxin production through genetic or chemical interventions. Biological, microbial, and physical decontamination techniques, including cold plasma, present sustainable solutions to reduce mycotoxin levels. As climate change impacts mycotoxin production, predictive modeling and climate-resilient storage practices are critical to mitigating risks. Strengthening global regulatory frameworks, enhancing traceability through blockchain, and fostering stakeholder collaboration will be key to ensuring a safer, sustainable food system for the future.

References

Abraham, N., Chan, E. T. S., Zhou, T., & Seah, S. Y. (2022). Microbial detoxification of mycotoxins in food. *Frontiers in Microbiology, 13*, 957148.

Awuchi, C. G., Ondari, E. N., Ogbonna, C. U., Upadhyay, A. K., Baran, K., Okpala, C. O. R., & Guiné, R. P. (2021). Mycotoxins affecting animals, foods, humans, and plants: Types, occurrence, toxicities, action mechanisms, prevention, and detoxification strategies—A revisit. *Foods, 10*(6), 1279.

Bunny, S. M., Umar, A., Bhatti, H. S., & Honey, S. F. (2024). Aflatoxin risk in the era of climatic change-a comprehensive review. *CABI Agriculture and Bioscience, 5*(1), 105.

Chen, J., Zhang, J., Wang, N., Xiao, B., Sun, X., Li, J., ... & Chen, A. (2024). Critical review and recent advances of emerging real-time and non-destructive strategies for meat spoilage monitoring. *Food Chemistry*, 138755.

Godana, E. A., Yang, Q., Zhang, X., Zhao, L., Wang, K., Dhanasekaran, S., ... & Zhang, H. (2023). Biotechnological and biocontrol approaches for mitigating postharvest diseases caused by fungal pathogens and their mycotoxins in fruits: A review. *Journal of Agricultural and Food Chemistry*, *71*(46), 17584-17596.

Oliveira, A. C. D. D., Ali, S., Corassin, C. H., Ullah, S., Pereira, K. N., Walsh, J. L., ... & Oliveira, C. A. F. D. (2025). Application of cold atmospheric plasma for decontamination of toxigenic fungi and mycotoxins: a systematic review. *Frontiers in Microbiology*, *15*, 1502915.

Oriekhoe, O. I., Ilugbusi, B. S., & Adisa, O. (2024). Ensuring global food safety: integrating blockchain technology into food supply chains. *Engineering Science & Technology Journal*, *5*(3), 811-820.

Siddique, M. M., Seraj, M. M. B., Adnan, M. N., & Galib, S. M. (2024). Artificial Intelligence for Infectious Disease Detection: Prospects and Challenges. *Surveillance, Prevention, and Control of Infectious Diseases: An AI Perspective*, 1-22.

Szelenberger, R., Cichoń, N., Zajaczkowski, W., & Bijak, M. (2024). Application of Biosensors for the Detection of Mycotoxins for the Improvement of Food Safety. *Toxins*, *16*(6), 249.

Index